FUNDAMENTALS OF TRANSDUCERS

Other TAB Books by the Authors

By Stan Gibilisco

No. 1505 *Understanding Einstein's Theories of Relativity: Man's New Perspective on the Cosmos*
No. 1525 *Black Holes, Quasars and Other Mysteries of the Universe*
No. 1605 *Basic Transistor Course—2nd Edition*
No. 1805 *Violent Weather: Hurricanes, Tornadoes and Storms*
No. 1905 *Comets, Meteors and Asteroids—How They Affect Earth*
No. 2000 *Encyclopedia of Electronics*

By R.H. Warring

No. 1142 *84 Practical IC Projects You Can Build*
No. 1493 *Electronic Components Handbook for Circuit Designers*
No. 1553 *Understanding Electronics—2nd Edition*
No. 1593 *Understanding Digital Electronics*
No. 1673 *Robots & Robotology*
No. 1793 *A Beginner's Guide to Making Electronic Gadgets—2nd Edition*

No. 1693
$21.95

FUNDAMENTALS OF TRANSDUCERS

R. H. WARRING & STAN GIBILISCO

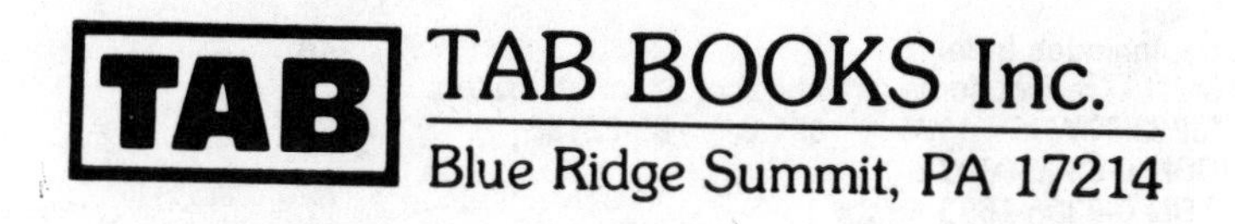

FIRST EDITION

FIRST PRINTING

Printed in the United States of America

Library of Congress Cataloging in Publication Data

Warring, R. H. (Ronald Horace), 1920-
Fundamentals of transducers.

Includes index.
1. Transducers. I. Gibilisco, Stan. II. Title.
TJ223.T7W36 1985 621.3 85-22185
ISBN 0-8306-0693-9
ISBN 0-8306-1693-4 (pbk.)

Contents

FUNDAMENTALS OF TRANSDUCERS

Introduction

Transducers are devices that change one form of energy into another. Because there are many different forms of energy—movement, acceleration, pressure, sound, radio waves, infrared, visible light, ultraviolet, X-rays, and gamma rays, to name just a few—the variety of transducers is enormous. Their applications are even more diverse.

Surprisingly, there is little collective information on the subject of transducers. That deficiency in the literature is the primary motivation for writing this book. It is divided into 30 chapters, covering virtually all types of transducers and sensors. Design, characteristics, and applications are discussed. Related circuitry and its design are also included. Some practical projects are suggested.

We begin with an overview of transducer technology and then show various examples of transducers. Subsequent chapters discuss potentiometric transducers, strain gauges, capacitive transducers, piezoelectric devices, load cells, magnetic and inductive transducers, differential transformers, digital transducers, photoelectric devices, sensors, thermocouples, bimetal strips, infrared devices, lasers, pressure transducers, accelerometers, force/torque measurements, electrical transducers, measurement of noise, transducer applications (general), control devices, proximity sensors, amplifiers for use with transducers, position encoders, vibration monitors, processors and displays, and methods of testing the behavior of structures.

This book is intended as a general reference for anyone interested in the theory and practice of transducer technology.

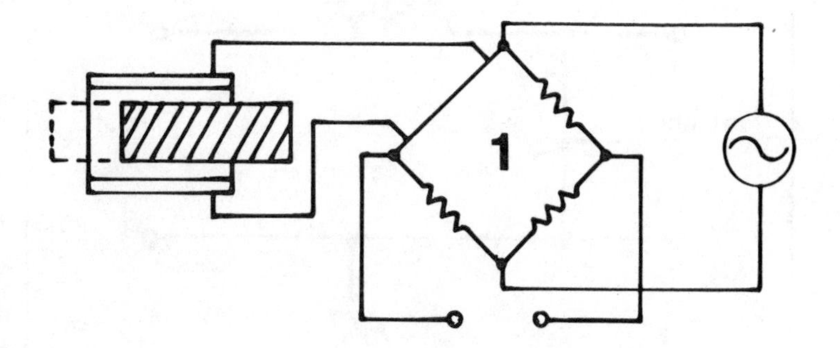

Overview of Transducer Technology

A transducer is a device that converts one form of energy into another form for some specific purpose. It might seem like such a device is quite specialized, but in fact transducers can be found everywhere. Their variety is practically unlimited.

Transducers can be categorized as passive versus active, or as input versus output, or according to the type of energy with which they are supplied or the type that they produce at the output.

PASSIVE AND ACTIVE TRANSDUCERS

Any device that operates without the need for a source of external power is called a passive device. We might, for example, build an audio low-pass filter using a coil and capacitor (Fig. 1-1A). No battery is necessary for this circuit to work; all we need to do is supply audio power at the input and connect a speaker or headset to the output. Thus, the coil-capacitor filter is a passive network. It is also a transducer, being actuated by audio power of one kind (broadband) and supplying audio power of another kind (restricted-band).

You can no doubt think of many examples of passive transducers right away: microphones, speakers, headsets, light bulbs, antennas.

An active device must draw power from an external source in order to work. This power may be derived from an electrochemical

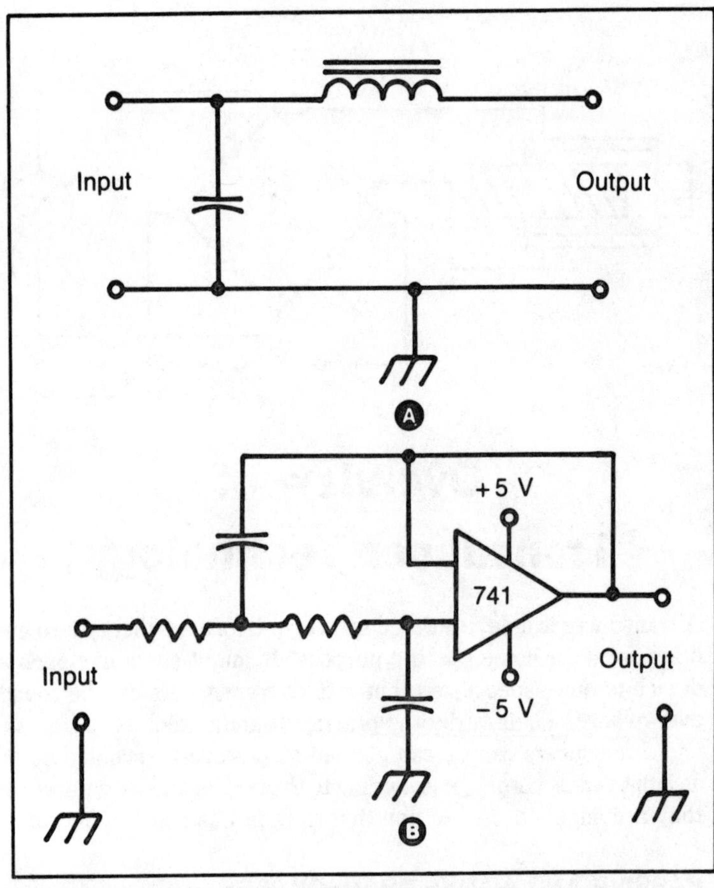

Fig. 1-1. Passive (A) and active (B) audio low-pass filters.

battery or from a power supply that operates from the commercial mains. An active transducer thus has at least three ports: input, output, and power. We can design an audio filter, having a characteristic response almost identical to that of the passive device in Fig. 1-1A, using an operational amplifier. Such a circuit is shown in Fig. 1-1B.

Other examples of active transducers include amplifiers, oscillators, tape recorders, radio receivers and transmitters, and electric eyes.

INPUT AND OUTPUT TRANSDUCERS

A transducer may be used to provide the power needed by a

device or circuit. In this application we call the transducer an input device. A common example of an input transducer is a microphone. We connect it to the input of an amplifier.

When a transducer is used to convert some quantity or variable into something we can sense directly (see or hear), we call it an output transducer. A speaker is an output transducer.

Some transducers can be used interchangeably as input or output devices. An example of this is an electric generator, which is normally used as an input transducer to provide electric current to a circuit. When certain kinds of electric generators are supplied with the power they normally produce from mechanical power, they operate as motors, which are output devices.

EXAMPLES OF TRANSDUCERS

The following are commonly used transducers:

Accelerometer: Converts acceleration into variable electric current.

Amplifier: Converts alternating electric currents or voltages into currents or voltages having greater amplitude.

Antenna: Converts electromagnetic fields into electric currents, and vice versa.

Attenuator: Reduces the amplitude of an alternating current or voltage.

Bimetallic strip: Converts temperature into physical displacement.

Capacitor (variable): Allows regulation of circuit characteristics on the basis of rotational displacement.

Electrical-output mercury switch: Converts temperature into an on-off state. (Below a given preset threshold temperature, the switch is off; above that limit it is on. Figure 1-2 diagrams the operation of this kind of thermal switch.)

Electric generator: Converts mechanical energy into electric current.

Electric motor: Converts electric current into mechanical energy.

Electrochemical cell: Converts chemical energy into electric current.

Filter: Modifies the characteristics of a certain type of energy (such as audio-frequency currents).

Laser: Converts electrical energy or magnetic energy into infrared light, visible light, ultraviolet light, or X-rays.

Light bulb: Converts electric current into visible light.

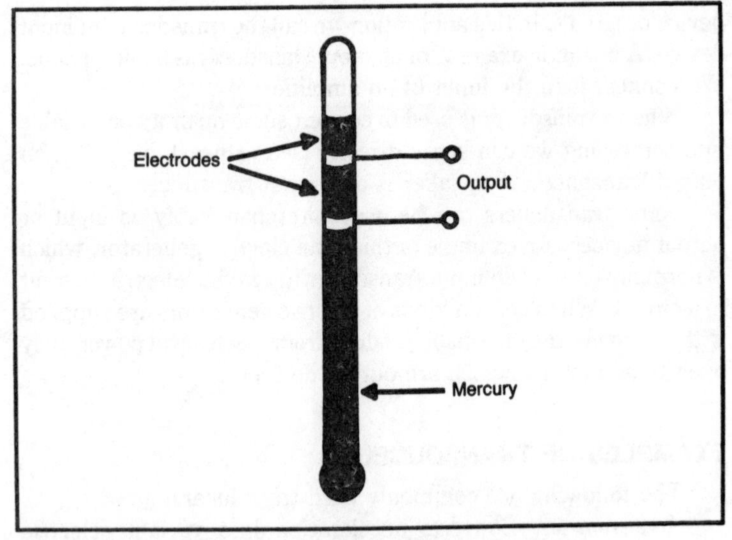

Fig. 1-2. A simple mercury-activated temperature sensor.

Light-emitting diode: Converts electric current into visible light.

Meter: Converts various quantities into mechanical displacement or a digital display.

Microphone: Converts acoustic waves into alternating electric current.

Nuclear reactor: Converts atomic energy into electricity, heat, or mechanical energy.

Photoelectric cell: Allows regulation of electric current according to the intensity of visible light.

Photovoltaic cell: Converts visible-light energy into direct-current electricity.

Piezoelectric crystal: Converts mechanical vibration into electric current, and vice versa.

Potentiometer: Allows regulation of electric current based on rotational displacement.

Pressure-sensitive device: Allows regulation of electric current based on physical pressure on a surface.

Radio receiver: Converts electromagnetic energy into acoustic waves or electric currents. Actually, a radio receiver consists of several individual transducers operating together (Fig. 1-3).

Radio transmitter: Converts electricity into electromagnetic energy. As with the radio receiver, a transmitter is made up of several transducers that perform specific signal functions (Fig. 1-4).

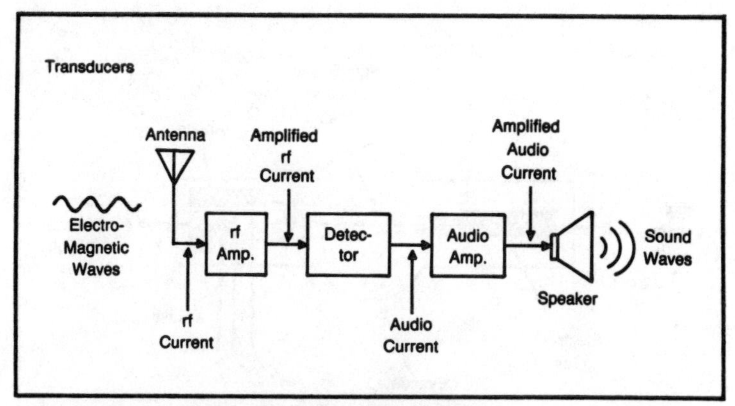

Fig. 1-3. Block diagram of a simple radio receiver.

Rectifier: Converts alternating current to direct current. A specialized form of rectifier, known as a detector, is used in certain kinds of radio transmitters and receivers.

Solenoid: Converts electric current into mechanical displacement.

Speaker: Converts alternating currents into acoustic waves.

Strain gauge: Converts mechanical stress into variable electric current.

Telephone set: Converts voices into signals suitable for transmission over wire; also converts electrical impulses into intelligible voice signals.

Television receiver: Converts electromagnetic energy into audiovisual information.

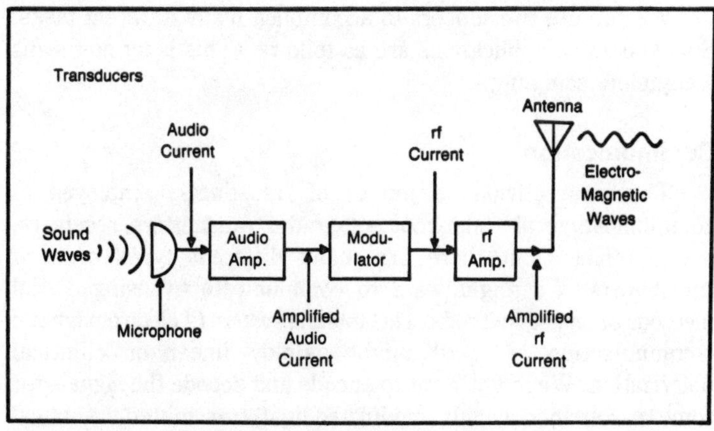

Fig. 1-4. Block diagram of a simple radio transmitter.

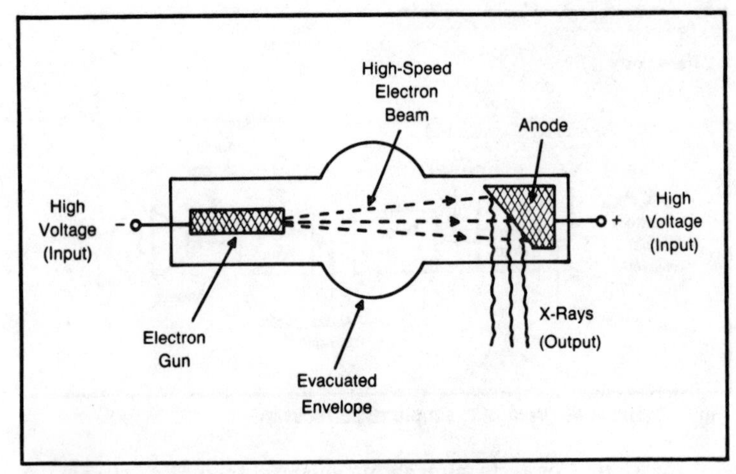

Fig. 1-5. Cross-sectional diagram of an X-ray tube.

Television transmitter: Converts audio-visual information into electromagnetic signals.

Thermistor: Allows regulation of electric current based on temperature.

Thermocouple: Converts thermal energy into electric current.

Transformer: Alters the voltage or impedance in an alternating-current circuit.

X-ray tube: Converts electrical energy into X-rays. (Figure 1-5 illustrates the operation of this kind of transducer.)

TRANSDUCER APPLICATIONS

We can use transducers to accomplish many different tasks. Some common applications are as follows. (This is by no means a complete sampling.)

Communication

The most obvious examples of transducers employed in communications are telephone sets, radio transmitters, receivers, and antennas. But there are many different types of such transducers. We might want to communicate by using digital methods or analog methods. The antenna system (if electromagnetic communication is used) might employ linear or elliptical polarization. We might want to encode and decode the signals for some reason. Increasingly, modulated light, transmitted via optical fibers, is being used for communications. Lasers and photoelectric cells thus replace antennas and carrier-current converters. An example of a complete communications system is shown in Fig. 1-6.

Control

Transducers are routinely used to regulate some variable, such as temperature, electric current or voltage, frequency, brightness, or loudness. Control may be accomplished manually, as with the volume control on a hi-fi set, or automatically, with a thermostat for example.

An automatic control system generally requires a sensor transducer that reacts to changes in the parameter under control. The sensor is connected to another transducer that operates according to the sensor output. The sensor in a thermostat is a thermal switch, usually a bimetallic strip, although a thermocouple or thermistor may also be employed. The sensor regulates the operation of a furnace or air conditioner, which is itself a form of transducer that generates thermal energy from electricity or from the potential energy in a fossil fuel.

Electrochemical

Potential energy in chemical form can be converted into electrical energy by means of a common cell or battery (Fig. 1-7A). The electrochemical cell consists of a solution or paste of some chemical compound, called the *electrolyte,* and two metal electrodes

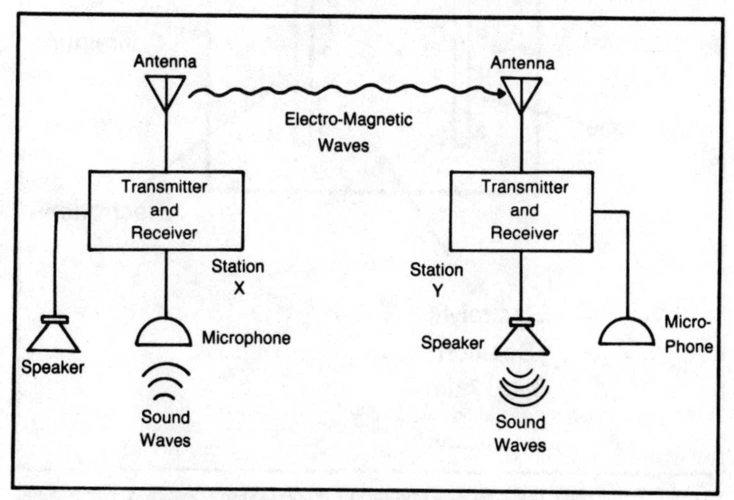

Fig. 1-6. A complete, two-way electromagnetic communications system.

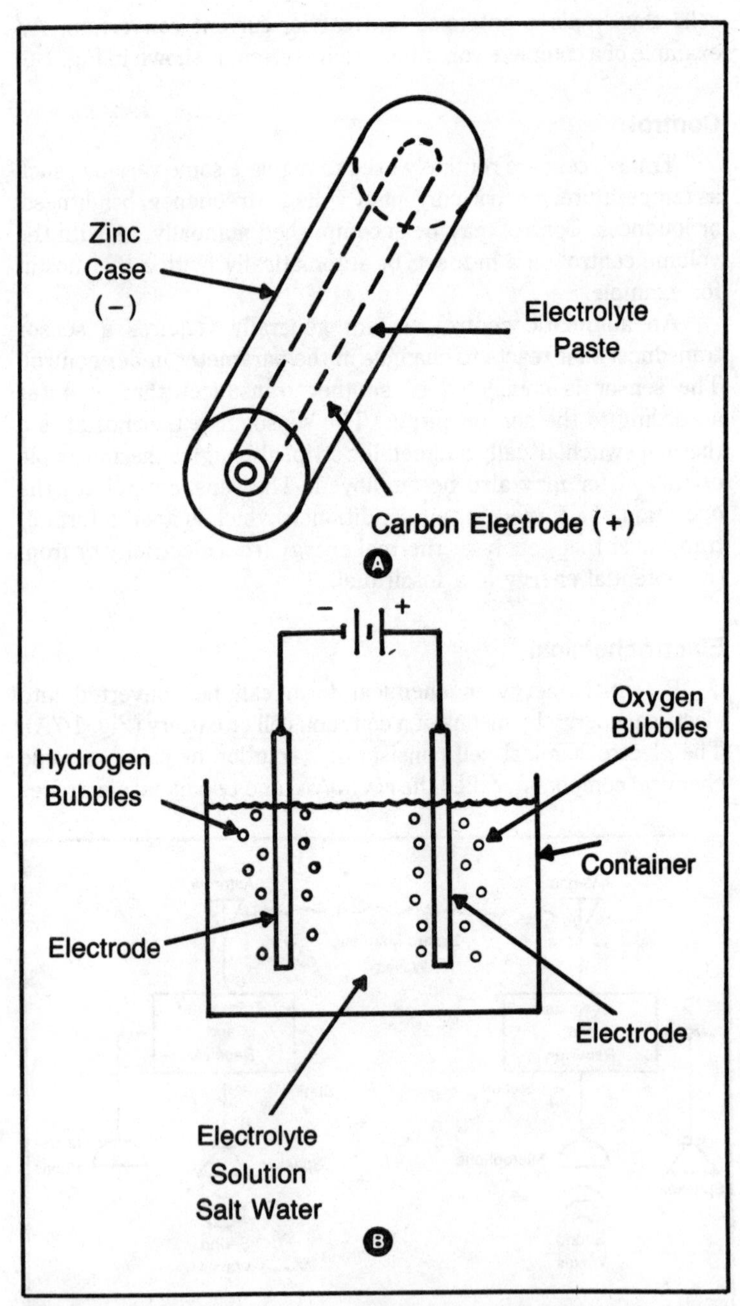

Fig. 1-7. Electrolytic cells. At A, a common zinc-carbon dry cell; at B, electrolysis apparatus generating hydrogen and oxygen from an electric current.

with which the electrolyte reacts. An excess of electrons appears at one electrode, and a deficiency of electrons exists at the other electrode. Thus, a potential difference is produced; when the two electrodes are connected to a circuit, current flows and power is generated.

Some electrochemical transducers convert electrical energy into chemical form. An example of this is the electrolysis apparatus shown in Fig. 1-7B. This device is very much like the electrochemical cell shown in Fig. 1-7A. Some electrochemical transducers can work in both modes: electrical-to-chemical and chemical-to-electrical. Your automobile battery is an example of such a device, as is any rechargeable cell or battery.

Electromechanical

The motor and generator are the most common examples of electromechanical transducers. As we mentioned before, some motors can operate as generators, and some generators can operate as motors. Motors and generators can be designed to operate from either alternating current or direct current.

Some electromechanical transducers are used to measure mechanical movement or some electrical quantity. A simple analog meter of the D'Arsonval type (Fig. 1-8A) is a form of electromechanical transducer that converts electric current into mechanical displacement. Two electromechanical transducers—a direct-current generator and a meter—can be connected to monitor linear or angular speeds (Fig. 1-8B).

Measurement and Monitoring

Any kind of transducer can, in some way, be used for measurement or monitoring purposes. We have just seen how electromechanical transducers are used to measure speed. Any measurement transducer is more commonly known as a meter, and the variety of metering devices in today's technological world is overwhelming. The control panel of any sophisticated electronic device might have a dozen or more different indicating transducers.

Testing

Testing, for quality control, troubleshooting, or empirical design, involves transducers. Of course, measurement and monitoring devices play a central role in any test procedure.

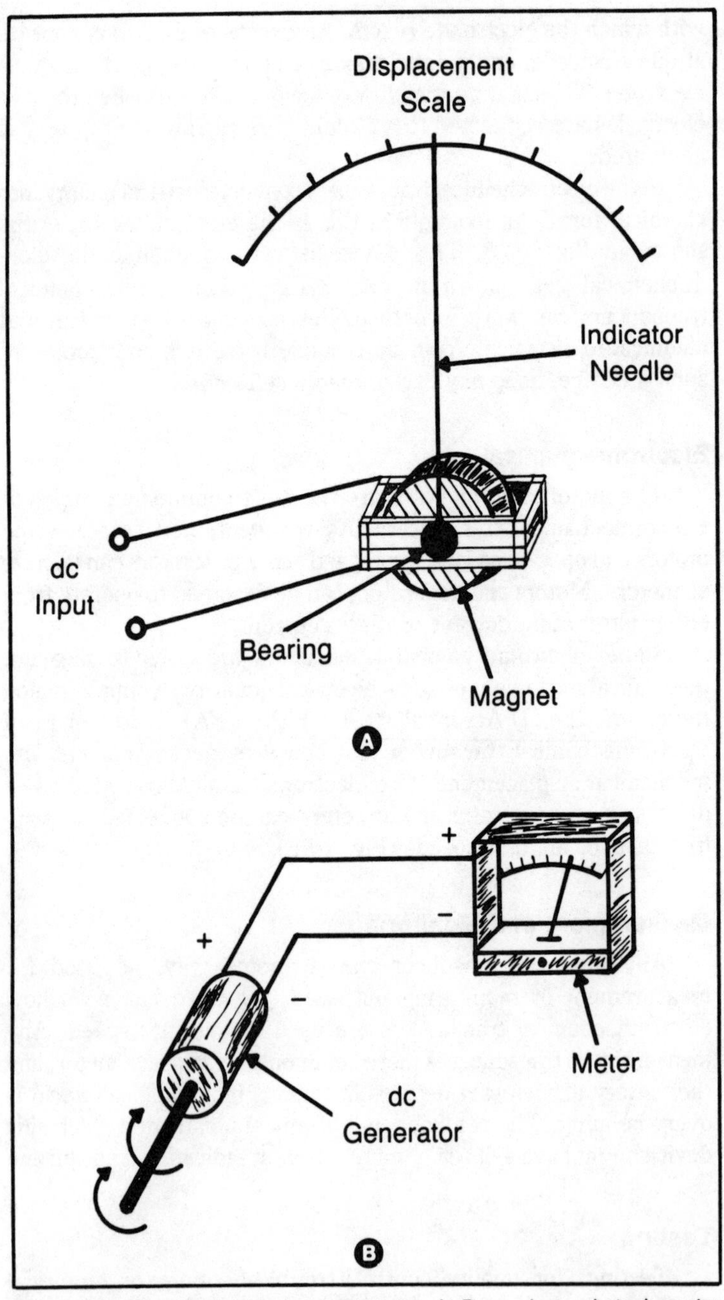

Fig. 1-8. At A, a simple direct-current meter. At B, an electronic tachometer using a generator and meter.

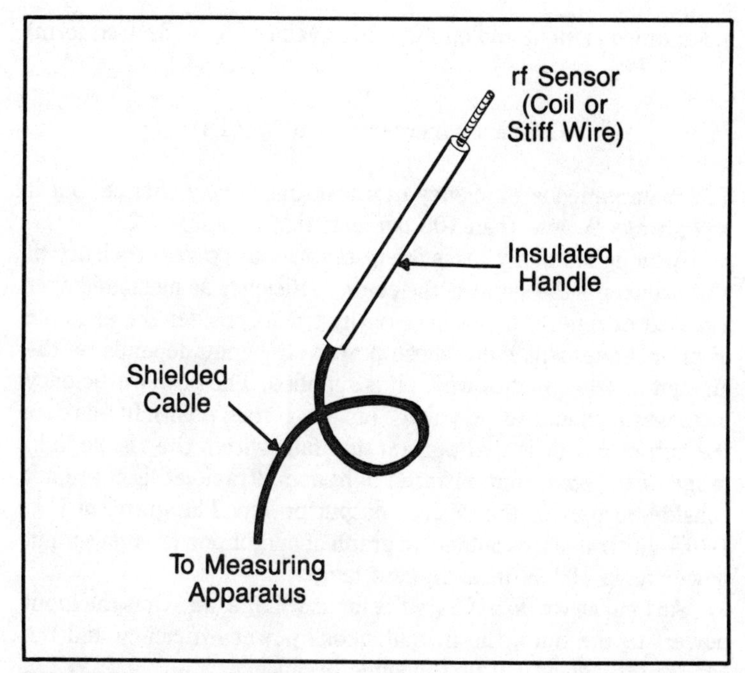

Fig. 1-9. An rf sensor probe.

Peripheral transducers, such as radio-frequency sniffers or probes (Fig. 1-9) are employed with test apparatus in many instances.

EFFICIENCY

The operation of any machine, no matter what its function, is limited by its efficiency. Efficiency is defined, in percent, as the ratio of useful energy actually realized to the theoretical ideal. An efficiency of 100 percent is never obtained in the real world, although in some cases performance is almost perfect.

The efficiency of a passive transducer is easy to define. We can simply measure the input and output energies, in joules, over a given length of time and then compare them. If the input energy to a transducer is E1 and the output energy (in the desired form) is E2, then it will always be true that $E1 > E2$. We define the efficiency according to the formula

$$\text{Efficiency (percent)} = 100(E2/E1)$$

We may, in some instances, define efficiency according to

instantaneous input and output power values (in the desired form) P1 and P2:

$$\text{Efficiency (percent)} = 100(P2/P1)$$

The instantaneous efficiency of a transducer may change, but it will always be less than 100 percent; that is, P1 > P2.

Although in many cases the instantaneous power efficiency of a transducer is the same as the energy efficiency as measured over a period of time, it is not necessarily true. Consider the example of an incandescent bulb whose power efficiency depends on the amount of voltage with which it is supplied. The power efficiency increases with increasing voltage (up to a certain point). In this case the output power is that portion that falls within the visible-light range of the spectrum; infrared light and ultraviolet light are not considered part of the desired output power. The graph of Fig. 1-10A illustrates a hypothetical graph of output power versus input power for a 100-W incandescent lamp.

As long as we do not vary the input voltage (and thus the input power) to the bulb, the instantaneous power efficiency and the energy efficiency will be the same because

$$E1 = P1t \text{ and}$$
$$E2 = P2t$$

where t is the length of time the bulb operates. But if we vary the input voltage over a period of time, the instantaneous power efficiency will change. But during the time span, the overall energy efficiency will be a constant—the average power efficiency during that time. The energy efficiency is actually the integral of the instantaneous power efficiency during the time specified. This is shown in Fig. 1-10B.

The discrepancy between instantaneous power efficiency and energy efficiency is not unique to incandescent lamps. It is, in fact, a characteristic of almost all transducers. Transducer manufacturers specify the appropriate range of input power for best operation of a device.

Efficiency is a little more complicated to define in the case of an active transducer, where the output energy/power is often larger than the input energy/power because of amplification. We must add another factor—the power-supply or battery energy/power—in order to obtain a meaningful measure of efficiency in an active transducer. If we let Eb represent the energy provided by the power supply or battery, and Pb represent the power provided by the supply at any given instant, then it will always be true that (E1 + Eb) > E2, and (P1 + Pb) > P2. The energy efficiency is then

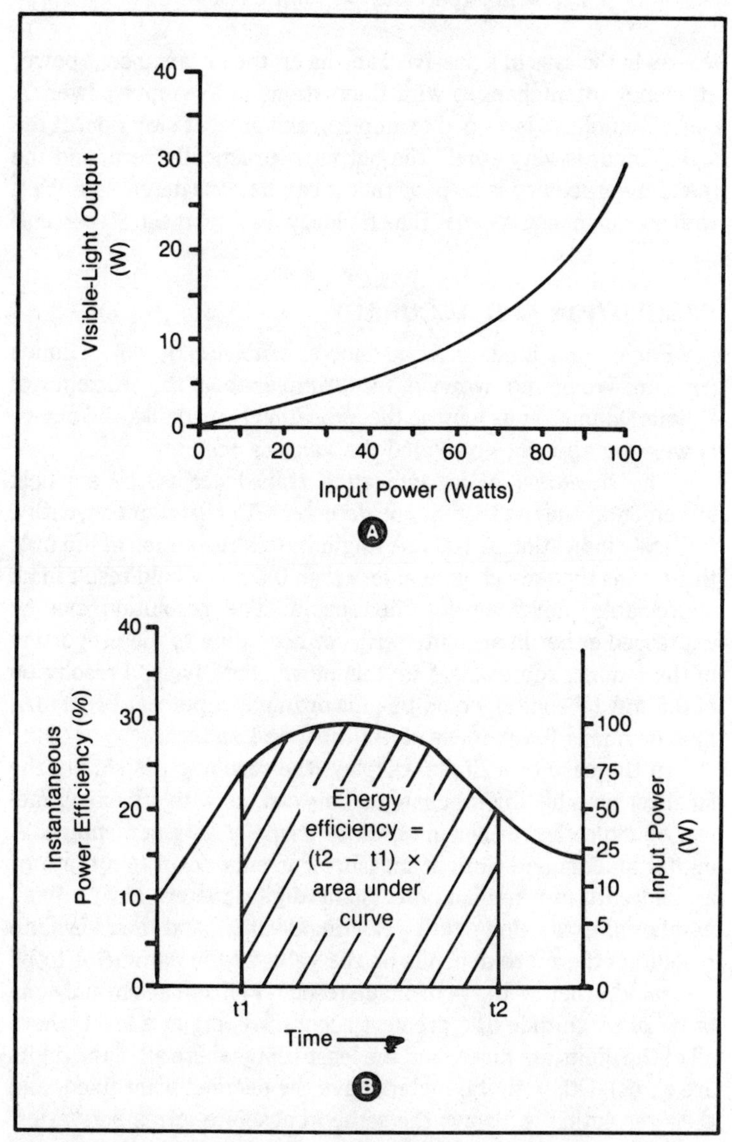

Fig. 1-10. Output versus input for a hypothetical incandescent bulb (A), and instantaneous power efficiency, integrated to obtain energy efficiency (B).

$$\text{Efficiency (percent)} = 100(E2/(E1 + Eb))$$

and the instantaneous power efficiency is

$$\text{Efficiency (percent)} = 100(P2/(P1 + Pb))$$

As in the case of a passive transducer, the instantaneous power efficiency often changes with fluctuations in the input power. A class C amplifier is a good example of this kind of situation. If the signal input is very small, the output is practically zero, and the resulting efficiency is so poor that it can be considered zero. Yet, with adequate input power, the efficiency may approach 90 percent!

RESOLUTION AND ACCURACY

For certain kinds of transducers, efficiency is not of much concern. We are not worried, for example, about the efficiency of a digital ammeter; as long as the indication is readable, the device is working all right—provided the value is accurate.

The *resolution* of an indicating transducer is the smallest incremental change that can be detected. An analog meter with a full-scale indication of 100 mA might have a resolution of 0.5 mA; this means that any change of less than 0.5 mA would result in no appreciable movement of the needle. The resolution can be expressed either in absolute terms or according to the proportion of the reading represented by this movement. Thus, a resolution of 0.5 mA (absolute) would be plus or minus 5 percent of 10 mA, plus or minus 0.5 percent of 100 mA, and so on.

In the case of a digital display, the resolution is simply the smallest possible digital change. This can, as with an analog meter, be expressed either in absolute terms or as a percentage. A digital milliammeter might measure currents down to a tenth or a hundredth of a milliampere. Some digital meters have a fixed number of digits along with a floating decimal, and their absolute resolution therefore depends on the value of the parameter to be measured, whereas the percentage resolution fluctuates by only one order of magnitude (the greatest accuracy being at a level where all of the digits are nines, and the least being where all of the digits are zeros). Other digital meters have the decimal point fixed, and their resolution is always the same in absolute terms but varies, sometimes tremendously, in terms of percentage over several orders of magnitude. (For example, a reading of 0.0003 mA in a display with four digits to the right of the decimal point has a resolution of plus 50 percent and minus 25 percent; but a reading of 100.0003 mA has a resolution of plus or minus 0.0001 percent.)

The accuracy of a metering transducer is the maximum possible difference, as a percentage, that can exist between the actual value (as determined by a standard) and the indicated value. This is usually based on the full-scale reading. The error increases (in general) as the value decreases from full scale. Accuracy depends on the resolution and also on the calibration of the metering device.

SENSITIVITY

The sensitivity of a transducer can be defined in two ways: the minimum amount of input that generates a detectable or measurable output; or the smallest amount of input change that results in a detectable or measurable change in the output. This quantity can be expressed in absolute terms, as a percentage, or as a decibel ratio.

These two sensitivity figures will often (in fact, usually) differ for a given transducer. Both factors may also change along with changes in some other parameter.

We might consider the example of a speaker that requires 15 mW of audio power in order to produce a noticeable sound at a pure sine-wave frequency of 1000 Hz. At low power levels a 5-mW change might produce a detectable change in the volume; as the power level is increased, it might require a larger and larger absolute increment to produce a noticeable change in the volume. A hypothetical response of this kind is shown in Fig. 1-11 for a frequency of 1000 Hz. But the figures may—and probably will—be different if the frequency is 500 Hz, 2500 Hz, or if the input waveform is not perfectly sinusoidal. When defining the sensitivity of a transducer, then, we must be sure that we specify the values of variables that might affect the result, or else we will end up with meaningless figures.

RESPONSE TIME

Some types of transducers do not produce an output signal until a certain time after the application of the input signal. An example of such a device is an ordinary incandescent bulb. It does not light up right away when the input power is applied, but instead takes a few hundredths or a few tenths of a second to reach full brilliance. You have probably noticed this especially with high-wattage bulbs such as movie-camera lights. Measuring devices, such as analog

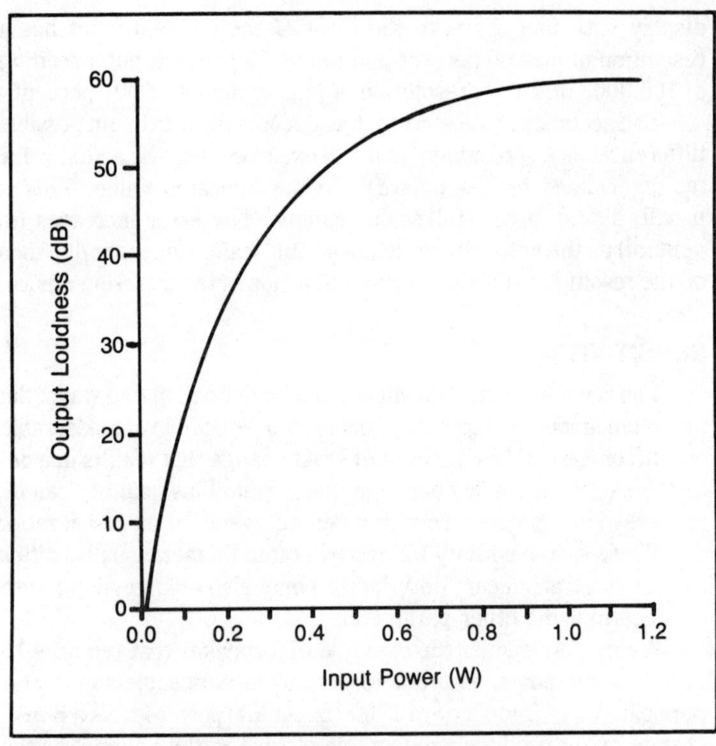

Fig. 1-11. A hypothetical speaker output-versus-input curve.

or digital meters, also require a short length of time to produce a reading. Thermostatic devices fall into this same category. There are many other examples.

The *response time* of a transducer is defined as the time t, usually measured in nanoseconds, microseconds, milliseconds, or seconds, between the time t1 at which input is applied and the time t2 at which the output stabilizes (Fig. 1-12A). Response time can also be defined as the time t′ between the power-removal time t3 and the point t4 at which the output again stabilizes (Fig. 1-12B). These two response times, t and t′, are not, in general, the same for a given transducer at a particular power level.

HYSTERESIS

Hysteresis in a transducer is closely related to response time, in that it is an indicator of the degree of sluggishness with which the device reacts to changes in the input.

A good example of hysteresis is provided by the operation of

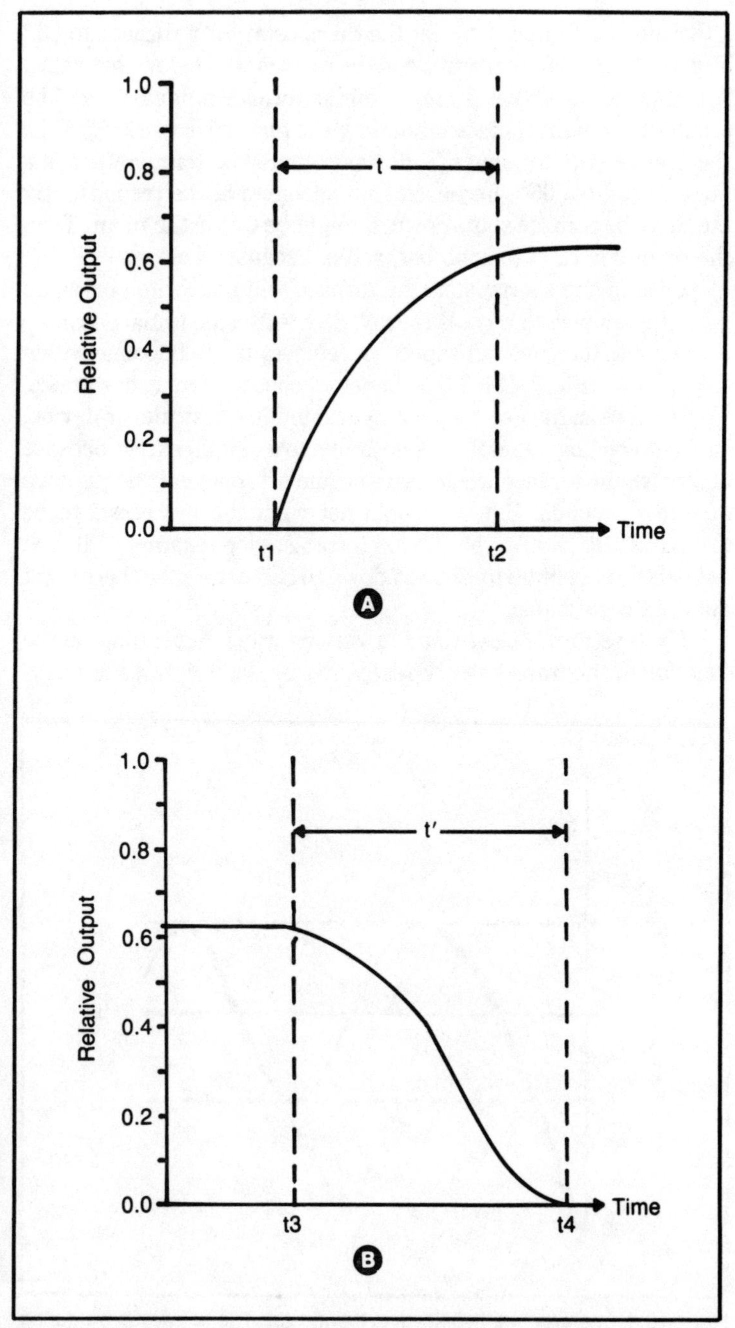

Fig. 1-12. Rise (A) and decay (B) curves. Rise time is given by t, decay time by t′.

a thermostat. Suppose we set the thermostat for a furnace to 65° F. while the initial temperature in the room is well below that value. The thermostat will at first turn on the furnace immediately. The temperature will increase (assuming the furnace is working). The thermostat will not shut off, however, until the temperature is a little more than 65°, because of sluggishness in the response. By the time the furnace shuts down, it might be 67° in the room. Then the room will start to cool, but, again, because of slowness in the response of the thermostat, the furnace will not switch on again until the temperature cools to, say, 63°. After the initial warm-up of the room, then, we can expect the temperature to fluctuate within a span of 4° (Fig. 1-13). This phenomenon is known as hysteresis.

Hysteresis may or may not be wanted in a transducer device. In the preceding example it is desirable to a certain extent because otherwise the furnace would cycle on and off constantly at intervals of a few seconds. But we would not want the hysteresis to be excessive; it would be uncomfortable, for example, if the temperature went up to 72° and down to 58° before the thermostat initiated corrections.

Hysteresis can be defined in various ways, depending on the function of the transducer. Usually, hysteresis is defined in terms

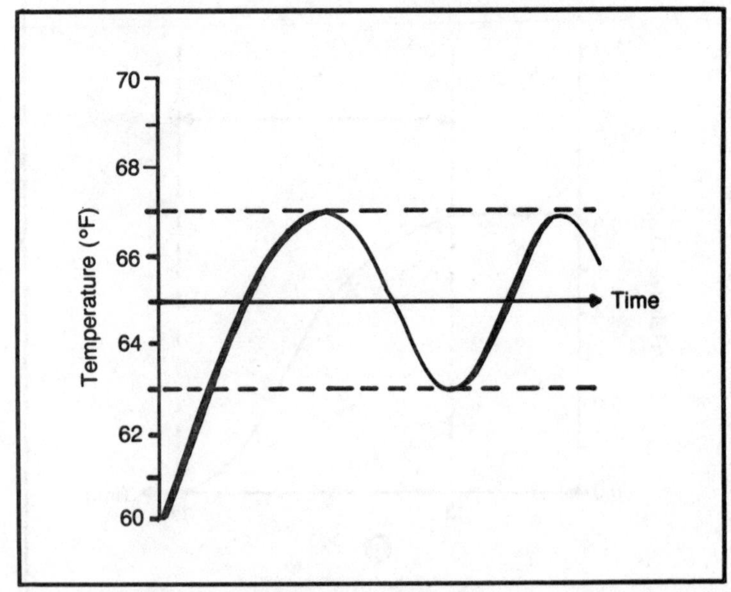

Fig. 1-13. Hysteresis in a hypothetical thermostat. The furnace is on during times indicated by the heavy line.

of the maximum fluctuation that takes place in the output of a control or metering device.

USEFUL LIFE

An important consideration in the choice of a transducer for a given application is the service life of the device. This is usually indicated in hours, but it can also be given in terms of the number of switching or control operations that the transducer can perform. Useful life is especially important for control, monitoring/measurement, and testing transducers. Expense is also a significant consideration because we do not want to have to constantly replace a component that costs a lot of money.

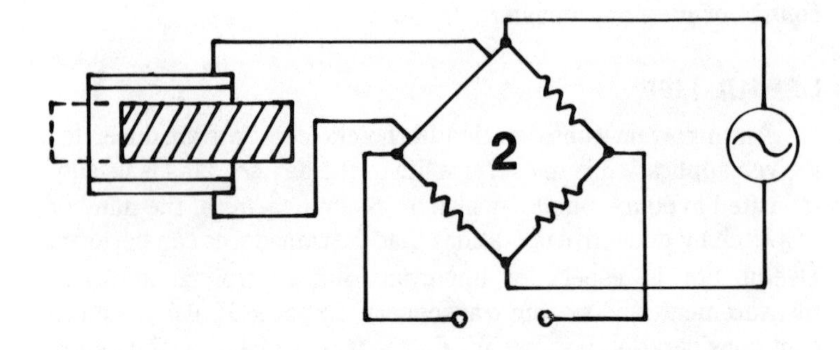

Simple Electromechanical Transducers

The most common transducer is the potentiometer, which converts mechanical movement into a proportionate electrical signal. We use them every day on radios and cassette recorders for volume, tone, and balance controls. They are also the simplest and cheapest devices for transducer projects.

A clear example of the use of a potentiometer (transducer) to transform mechanical movements into electrical signals is on the control sticks of model radio-control transmitters. Taking the typical dual-axis stick, movement up and down is connected to one potentiometer to move its wiper by the same proportionate amount. Movement side-to-side similarly adjusts the position of the wiper on a second potentiometer connected to that movement. Between them the signals from the two potentiometers correspond to any position to which the dual-axis stick is moved (Fig. 2-1). Separate potentiometers are also operated by twin levers to signal fine adjustment of control position independent of stick position (although some of the latter sets now use all-electronic trims). These are, in effect, separate overriding stick controls working in just the same way but with much smaller movement response.

Stick position (and trim position) signals devised from the transducers (potentiometers) are then encoded in digital form for transmission. At the other end of the system the signal is detected by the receiver, and the signal information is extracted to feed to the servos. The latter operates on a closed-loop circuit, where the

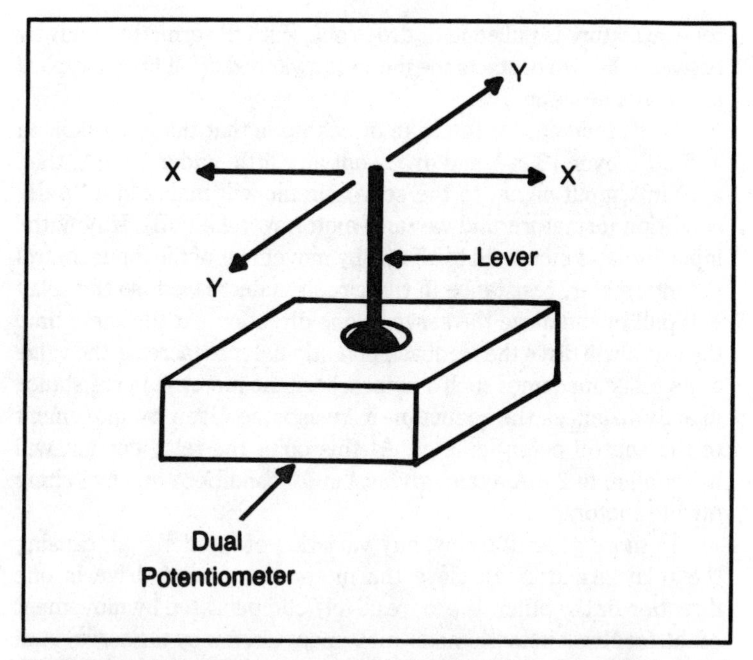

Fig. 2-1. A dual potentiometer is commonly used as a two-dimensional control or position indicator. This device is often referred to as a joystick.

actual mechanical output movement is measured, fed back into the receiver circuit, and compared with the signal information. If there is any difference, and error-signal position is present that works to correct the output movement until the error signal is reduced to zero. This servo movement faithfully duplicates the command signal. The device that detects the amount of movement and generates any necessary error signal is another potentiometer coupled to the servo control movements.

CLOSED-LOOP CIRCUIT

A closed-loop circuit that demonstrates the use of feedback, or which can equally well be used as a project requiring proportional control response, is shown in Fig. 2-2. For simplicity this circuit uses a relay (which could be replaced by an equivalent solid-state switching circuit). The command input signal is devised from one potentiometer (the control potentiometer). The servo is a dc electric motor, the output of which is mechanically linked to a second potentiometer (the feedback potentiometer). As connected, the servo motor will drive in either direction, according to whether the relay armature is pulled in or drops out. With the armature midway between the two contacts the motor is switched off. This is a typical proportional setup.

If we assume, for the sake of example, that the relay pulls in at a little over 10 mA and drops out at a little under 10 mA, then a 10-mA input signal to the servo circuit will maintain a "*null*" condition (armature midway and motor switched off). If now the input signal is increased to 15 mA by movement of the input control potentiometer, resistance in the circuit is decreased, so the relay will pull in and drive the servo in one direction. At the same time the servo will drive the feedback potentiometer to increase the value of its resistance until such a point where the increase in resistance exactly balances the reduction in resistance given by movement of the control potentiometer. At this point the relay current will have fallen to 2 mA again, giving "null" conditions and switching off the motor.

In more general terms, any variation of input signal, causing the relay armature to close the motor circuit and drive in one direction or the other, is progressively compensated by movement of the feedback potentiometer movement (driven by the servo) until null conditions are established again when the motor stops. Thus, the feedback control (potentiometer) ensures proportional movement of the servo output drive relative to the actual change in input signal. Any greater or lesser movement leaves unbalanced resistance and an "error signal" remaining in the circuit to continue driving the servo until the "null" condition is reached.

In a practical feedback circuit the input signal variation is provided by the receiver response to proportional signals rather than direct movement of a control potentiometer, with the value of the feedback potentiometer chosen accordingly. Exactly the same

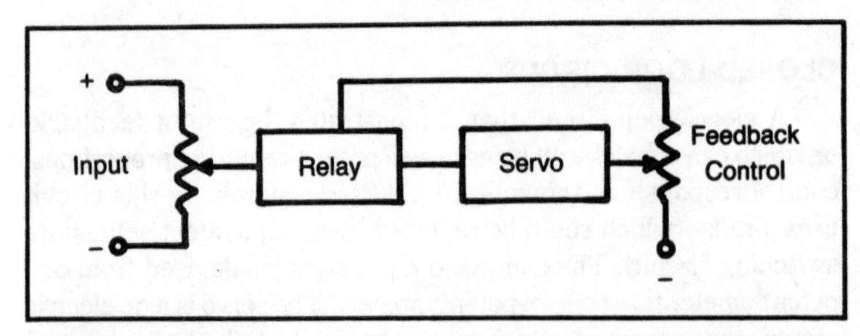

Fig. 2-2. A simple closed-loop electromechanical circuit using potentiometers as transducers.

working principle applies, however, although it may be necessary to provide some means of preventing overrun of the motor so as to avoid momentary "hunting" of the servo about the null point. This can be done by providing a damping signal (for example, by connecting the motor brush to the free end of the feedback potentiometer) or mechanical or dynamic braking across the servomotor itself.

Note that this is an *analog* system, which is the simplest to construct for general applications. For more precise proportional control, digital working is used (as on all model radio-control sets).

LINEAR OR ROTARY INPUTS

Another convenient feature of a potentiometer as a transducer is that it is available in two different geometric forms: circular with a rotary wiper, or linear with a sliding wiper. The former is easily related to rotary input movements; for example, the movement is coupled to the potentiometer spindle to give it a proportionate amount of rotation. With a slide-type potentiometer any push-pull (linear) input movement can be coupled directly to the slider.

Of course, only a simple form of crank linkage is needed to couple a linear (push-pull) movement to a rotary (circular) potentiometer or a rotary input movement to a slide-type potentiometer. However, the resulting mechanical movement of the potentiometer wiper will not then follow exactly the input movements. Only rotary-to-rotary or linear-to-linear couplings will give this.

TYPES OF POTENTIOMETERS

Besides having the basic geometric forms (circular or linear), potentiometers can also have different electrical characteristics. Thus, the electrical signal resistance may be strictly proportional to wiper/slider movement (true linear response) or nonlinear in various degrees through to a *logarithm* response where increasing movement produces more and more signal.

In addition, there are various forms of combinations used for potentiometers, some better than others for particular applications. Potentiometers are also produced specifically as *transducers*, too, rather than simple variable resistance *controls*, with particular attention given in their design to reducing wiper/slider movement friction, reducing wear, and minimizing electrical noise. These and other aspects of potentiometer transducers are described in detail in Chapter 3.

POTENTIOMETERS AS COMPUTERS

Before leaving simple applications of potentiometers as control and feedback transducers, an interesting project is to build an elementary analog computer using two potentiometers (with a third potentiometer employed as a means of calibrating the circuit). The circuit involved is very simple (Fig. 2-3).

The three potentiometers are mounted on a suitable panel and wired to a microammeter and a battery, as shown. If we select a value of 25 kΩ for potentiometers A and B and using a 0-500-μA meter, the circuit will work off a 1.5-V battery. (If you want to use a 0-1-μA or 0.5-μA meter, adjust the battery voltage upwards accordingly.)

The third potentiometer is now used to calibrate the circuit. Potentiometers A and B are turned to maximum resistance, and the "adjust" potentiometer is turned until the meter reads exactly 100 μA. A suitable value for this adjust potentiometer is 1 to 5 kΩ. Once set up it should not need further adjustment.

With the setup showing 100 μA, we now mark the pointer position for potentiometer A as 100 and for potentiometer B as 0. This corresponds to the meter reading (100) equaling A plus B (100 + 0 = 100).

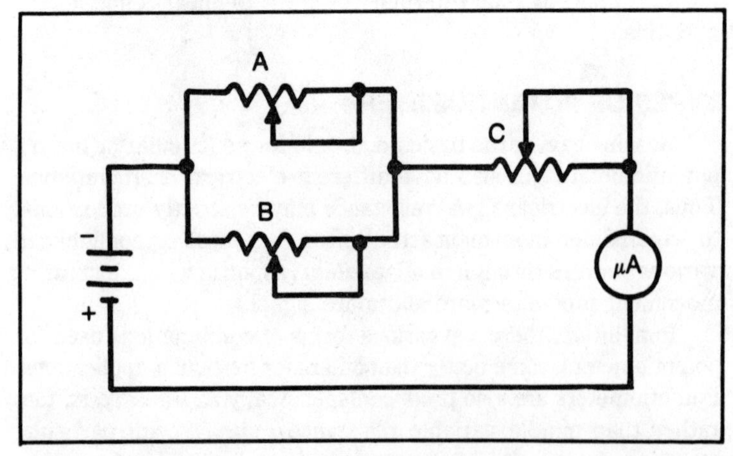

Fig. 2-3. An analog computer using two control potentiometers (A, B) and one calibration potentiometer (C).

Potentiometer A is now adjusted until the meter reading advances to 200 μA, and this position is marked as 200 on the potentiometer scale. Repeat the process to obtain calibrated positions for potentiometer A at 300, 400, and 500 μA. The potentiometer is then returned to its original (100) position.

Leaving potentiometer A in this position, we now turn potentiometer B until 200 is indicated on the meter scale. The pointer position at this setting is marked 100 for potentiometer B. This corresponds to the 100 position on A plus the 100 position on B = 200 on the meter (meter reading = A + B). Repeat the process for meter readings of 300, 400, and 500 to obtain calibration points for potentiometer B of 200, 300, and 400, respectively; that is, with potentiometer A at its 100 position the calibration value of the potentiometer B setting will always be 100 less than the meter reading.

The analog computer is now calibrated for working. Any settings of potentiometers A and B will be read on the meter as A + B. Unlike a digital computer, which can only count, the analog computer will give the sum over an infinite range of variations within the limits of the scale values.

This basic design lends itself to development in a number of ways. Different values of potentiometers and meter scales can be selected to extend the range of A + B so that they can be read or adapted to other units. The meter scale, for example, can be replaced by another pasted-on scale. Thus, in the elementary model described, a new scale could be calibrated for the meter, starting at zero for minimum reading (instead of 100). Further, if required, the *meter* scale could be calibrated to required positions (calibration values) on the two potentiometers.

The computer also need not necessarily be confined to arithmetic working. It is only necessary to render a particular problem as an analogy to the basic relationship between current flow and variable resistance that the analog computer presents to make the computer read in directly the type of information required.

To extend the principle further, a more advanced analog computer could be designed, which incorporates a number of fixed resistors rather than potentiometers, in the form of a plugboard. Each resistor value represents a different condition value or analogous data state, that is, where the variation of one factor is analogous to the effect on current (the related or analogous factor) when resistance is varied.

Further extension, in simple analog working, is also possible by varying *voltage*. Because

$$\text{current} = \frac{\text{voltage}}{\text{resistance}}$$

varying the voltage is equivalent to multiplying or dividing, as far as the current value is concerned. This considerably extends the scope both for arithmetic and analogous working. This can be done by a variable voltage supply (e.g., a tapped battery in a simple unit) for multiplication or by dropping resistors for division.

The analog computer is a very simple device, but it is surprising what it can be made to perform. It is largely a matter of individual ingenuity as to how far it can be adapted or programmed to give the solutions required. Even with very simple designs such as the one described here, its scope is enormous. The elementary analog computer is also attractive because it is neither expensive not complicated to make (although programming an elaborate pegboard or fixed resistor values can be a lengthy process), and in most cases it can be checked and calibrated directly.

Although what we have dealt with is basically an instrument circuit, it can equally well be applied to time transducer working as well, where the two potentiometers respond to mechanical input signals.

DIAPHRAGMS AND BELLOWS

Diaphragms and bellows can be called all-mechanical transducers capable of transforming one type of energy (pressure) into a mechanical movement proportional to the applied pressure. Another way of looking at it is that they can convert some form of energy or force that cannot be seen (pressure) into another form that can be seen (mechanical movement). "Seeing" in this case can be made much easier, and the system much more sensitive, by limiting to a mechanical system that amplifies the original movement of the transducer. A common example is the recording barograph with a pen on the end of an arm traversing a paper scale. A simple lever system multiplies the movement of the aneroid capsule as it expands or contracts under changing pressure so that the free movement is very much greater than the aneroid movement (Fig. 2-4).

The main limitation of diaphragms and bellows as transducers

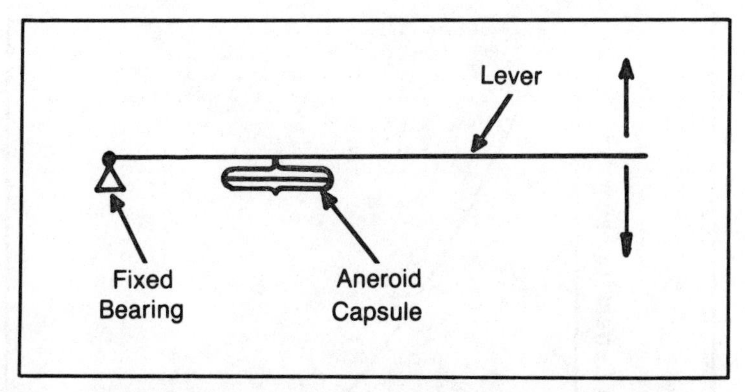

Fig. 2-4. Mechanical movement can be enhanced by the use of a simple lever.

is that they are only capable of producing mechanical movement response. There is no way they will generate an electrical signal on their own unless they are coupled to another type of transducer that directly converts mechanical movement into an electric signal. This, indeed, is the most common form of pressure transducer capable of giving a direct electrical signal output. Pressure transducers are very common and are described in detail in Chapter 18.

ELECTRIC MOTORS AS TRANSDUCERS

The electric motor is an energy converter. Supply it with electricity, and it rotates to provide a power output via its shaft. The efficiency with which it does this can vary considerably. With the very small toy-type dc electric motor, overall efficiency may be 20 percent or less. With industrial ac motors, especially large ones, the overall efficiency is usually well over 90 percent.

The energy-conversion principle of an electric motor also works the other way round. Rotate it, and it will generate electricity. Used this way, mechanical power input is converted to electric power, which is the way electric motors are used as transducers.

Normally a simple dc motor would be the choice for a transducer because this will produce a dc output signal that is simplest to process. It can be used directly to give a readout in a simple meter, for example. Also, the efficiency of energy conversion is not necessarily important. The main thing is that the output signal should be linear with respect to input variations.

This is the case over the majority of the working range of a simple low-voltage dc motor where the current consumption (with

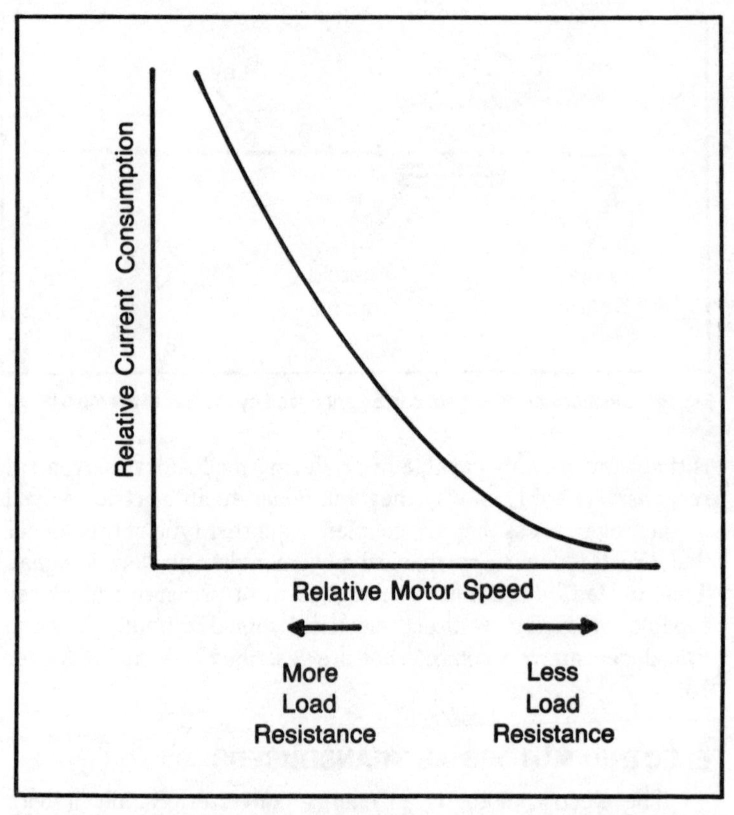

Fig. 2-5. For constant input voltage the relative current required is proportional to the load resistance and inversely proportional to the speed.

a constant supply voltage) varies linearly with speed. The faster the motor is allowed to run the lower the current it needs (Fig. 2-5). The actual speed at which it will run depends on the braking effect of any load applied to it.

It follows that if the motor is driven so that it produces rather than consumes electricity—that is, it is operated as a generator—the current it generates in a fixed (electrical) load circuit varies linearly with the speed at which it is being driven (Fig. 2-6). The actual current at any point on this characteristic graph is then a measure of speed revolutions per minute (rpm).

ELECTRIC TACHOMETERS

The electric tachometer uses a motor as a transducer. The rotor spindle is coupled to the shaft whose speed is to be measured and

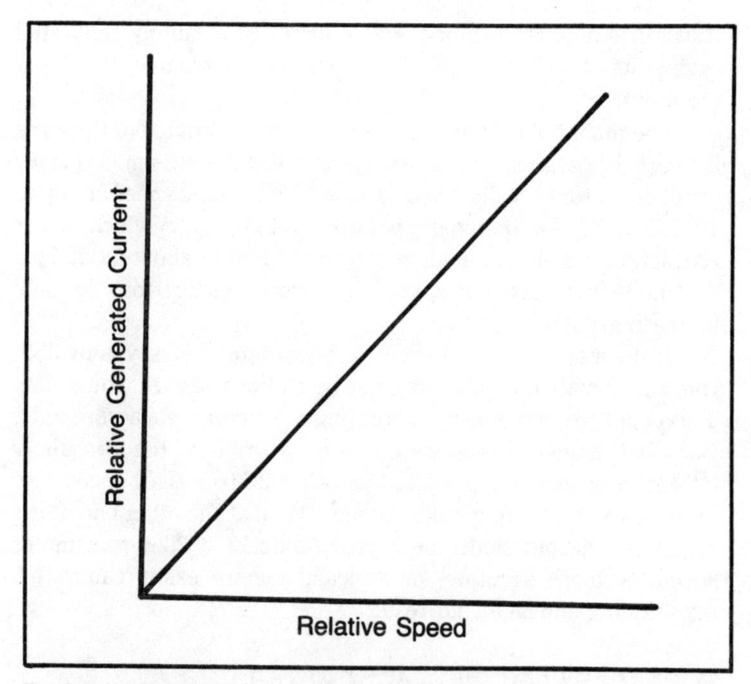

Fig. 2-6. When a motor is used as a generator, the current through a given resistance increases in proportion to the shaft speed.

connected by two wires to a milliameter and a calibrating resistor (Fig. 2-7). The latter can be a fixed value if the dynamic characteristics of the rotar are known. The resistor motor is then calculated to give maximum meter current at the maximum speed to be indicated. If the dynamic characteristics are not known, a normal resistor can be used instead to operate the circuit.

This type of transducer is very simple to set up and has maximum advantages over mechanical revolution-counting systems. A major advantage is that it can be worked remote from the point of measurement without involving mechanical cable drives.

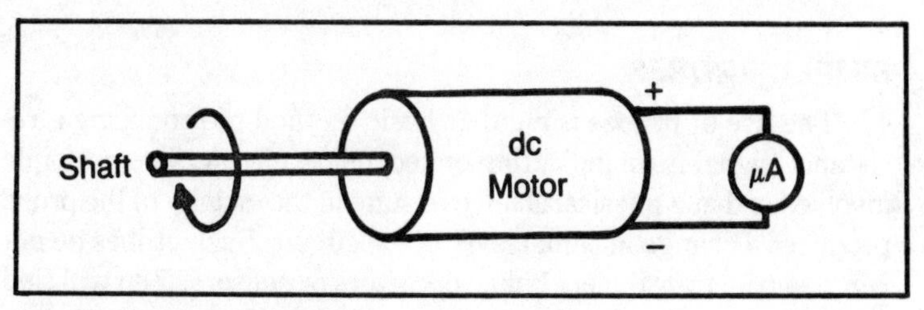

Fig. 2-7. A simple tachometer.

It also has no unknown or variable losses (it is simply calibrated against its own output), and it is generally reliable over the life of the motor brushes.

The main point in designing a revolution counter of this type is to use the optimum working range of the motor chosen. Typically small dc motors are designed to have free-running speeds of up to 10,000 rpm. An optional operating range, over which linear characteristics should hold, would then be from about 1,000 rpm to 10,000 rpm. At low speeds its characteristics may become increasingly nonlinear.

If the maximum shaft speed to be measured is, say, only 2500 rpm the signals obtained may not be as linear as we would like. This could be overcome by calibrating the meter scale accordingly. But a better answer is to use a geared-up drive for the transducer (motor). For example, a geared-up 4:1 2500-rpm shaft speed now produces a 10,000-rpm motor speed. Down at the other end of the range, a 500-rpm shaft speed rule produces a 2000-rpm motor speed. A more accurate, or at least a more easily calibrated, revolution counter should result.

PORTABLE ELECTRIC TACHOMETER

A portable electric tachometer is easily made by mounting motor, meter, and resistor together on a suitable panel, shaped to be easily held without obscuring the meter. The elaboration of a stepup drive gearbox is largely unwarranted in this case, so the motor used should be chosen to give a reasonable sign of current output over the driven revolution range it is designed for. For a less sensitive meter, a simple solid-state amplifier could be incorporated in the circuit.

This is one of those projects where cut-and-try is the best design approach. Start with the electric motor, see what sort of signal level it produces when driven at various speeds, and then finalize the circuit accordingly.

SIMPLE PROBES

The use of probes is another basic method of producing a resistance signal in an indicating or secondary circuit. The principle involved is that a physical change of state in the vicinity of the probe produces a change in signal level in the circuit. Such probes do not work as true transducers but as detectors or sensors. You will find various types described in Chapters 13 and 25.

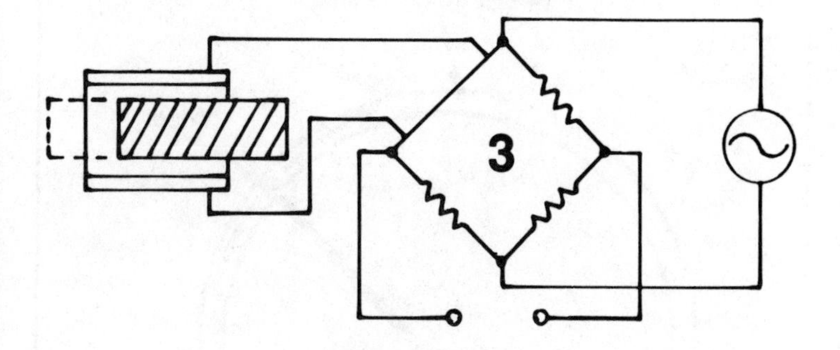

Potentiometric Devices

Potentiometers (variable resistors) are common components used in electrical circuits. They are usually circular in form and incorporate a resistance track in the form of an arc of about 270° traversed by a wiper that can be rotated over the track by turning a spindle (Fig. 3-1).

The resistance track is usually carbon, although it can also be a winding of resistance wire. Carbon-track potentiometers are cheapest, but they are generally suitable only for low-power loads (1/4-1/2 W for general circuit application). Wirewound potentiometers can have much higher power ratings, depending on the diameter size of the wire.

Another basic difference between the two is the available resistance values. Lowest values for carbon-track potentiometers are usually 0-100 Ω, extending up to 4.7 MΩ or higher. Wirewound potentiometers may be produced with full-load resistance of only a few ohms and ranging up to about 47 kΩ.

Both carbon-track and wirewound potentiometers are also produced in linear form, that is, with a straight track traversed by a sliding wiper (Fig. 3-2). Both circular and linear potentiometers are also produced with resistance elements of deposited thin film (conductive plastic).

Potentiometer characteristics can be *linear* (not to be confused with linear geometry) or *logarithmic* (Fig. 3-3). With linear response, equal movements of the wiper produce equal changes in resistance.

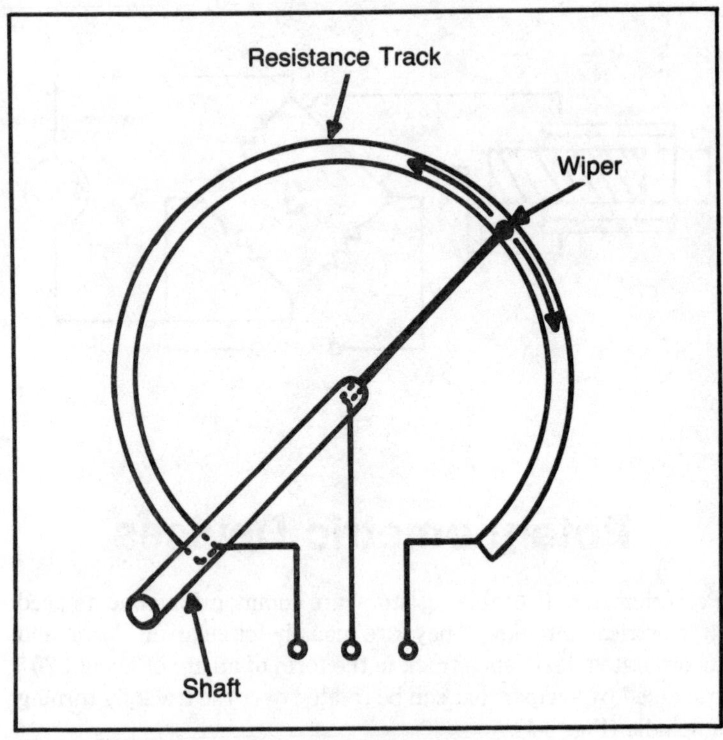

Fig. 3-1. Construction of a rotary potentiometer.

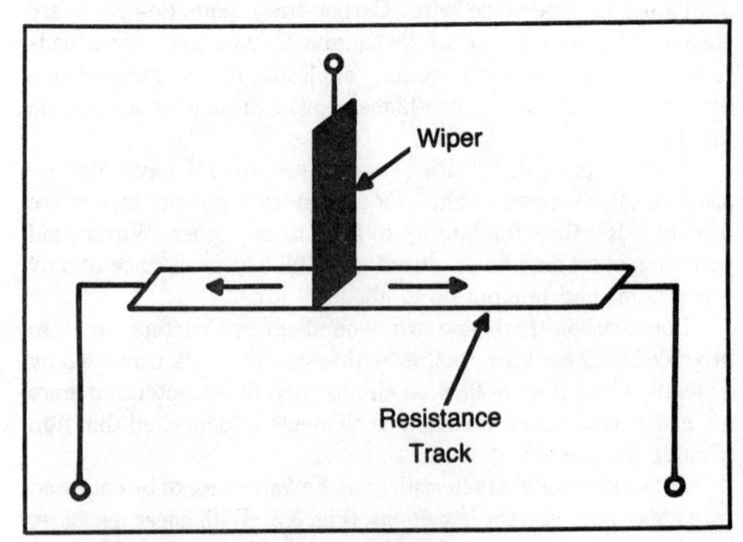

Fig. 3-2. Construction of a slide potentiometer.

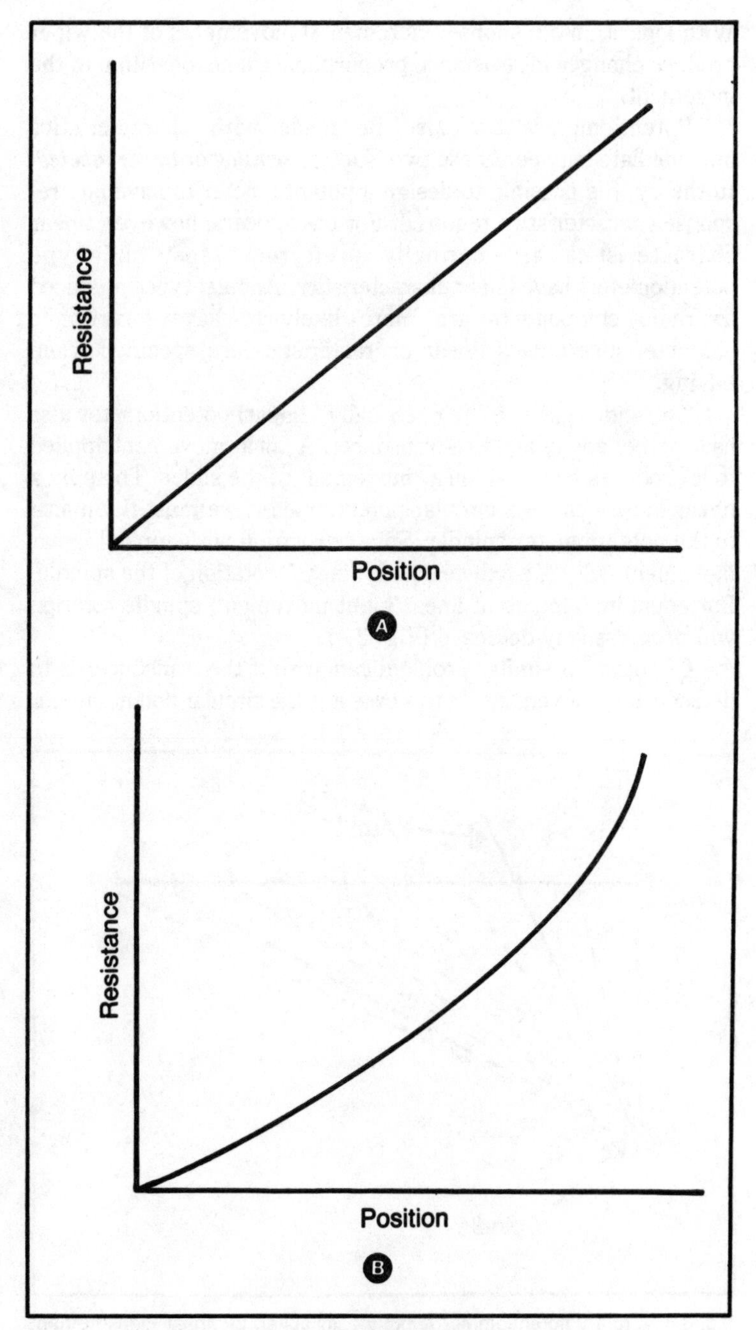

Fig. 3-3. Linear (A) and logarithmic (B) potentiometer responses.

With logarithmic response, incremental movements of the wiper produce changes in resistance proportional to the logarithm of the increment.

Potentiometers can also be made with characteristics intermediate between these two, such as *semilog* or *linear tapered*. In theory it is possible to design a potentiometer to have any response characteristics required. For transducers, however, linear characteristics are normally preferred. Most slide-type potentiometers have linear characteristics. Annular types produced as radio components are more likely to have logarithmic characteristics unless linear characteristics are specified when buying.

The slide-type (or "geometrically" linear) potentiometer also has another advantage as a transducer. A linear movement applied to it produces an equal linear movement of the slider. To apply a linear movement to a circular potentiometer, we must fit an arm to the potentiometer spindle. Spindle rotation with applied linear movement will *not* produce equal "linear" rotation of the spindle. For equal increments of linear input movement, spindle rotation will progressively decrease (Fig. 3-4).

Of course, a similar problem can arise if the transducer is to detect rotary movement. In this case it is the circular potentiometer

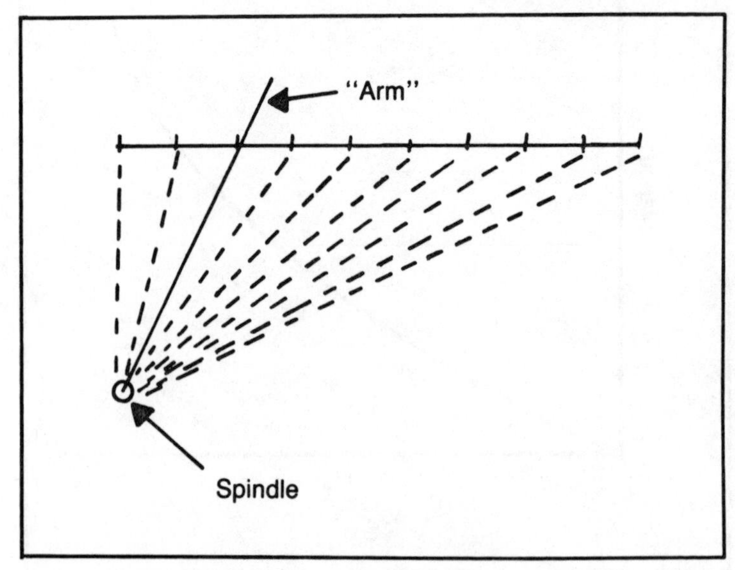

Fig. 3-4. A rotary potentiometer is not ideally suited for linear measurement of displacement.

that will produce "equal" movement response, whereas the slider type will not.

An advantage of the potentiometer as a transducer is that it is a simple device, easy to calibrate and able to provide a high signal output. It can also be used with either dc or ac excitation. On the other hand, its sensitivity can be low if it is a high resistance type, and it can be subject to wear and generation of spurious noise, being particularly susceptible to vibration. Other major disadvantages are its high friction and a tendency to wear that limits usable frequency response. In addition, it must be regarded as a limited-life device.

These disadvantages apply mainly to standard electronic circuit potentiometers. Potentiometers produced specifically as transducers overcome many of these limitations. Quite complex designs have been produced to overcome the effects of vibration on the wiper system and to exclude dirt and other abrasive materials that could aggravate wear. Cost is still relatively low, however, compared with other types of transducers.

One other point worth noting when choosing a potentiometric transducer is that carbon-track and conductive-plastic types provide infinite resolution. Wirewound types normally provide an output in small increments, determined by the wire diameter. This may or may not be significant in the application involved.

The following are further numerous potentiometer types.

CARBON-TRACK POTENTIOMETERS

Where carbon is used as the resistive element, a carbon film may be sprayed onto an insulated form or ceramic base, or a heavier coating may be used, which is subsequently compressed and molded at a high temperature into a sheet on a phenolic or ceramic base. These types are known as *sprayed carbon* and *molded carbon*, respectively. Sprayed carbon potentiometers are generally nonprecision types with limited life and stability. Molded carbon potentiometers are heavy-duty types. Both types can be produced with linear or nonlinear characteristics, the latter generally having logarithmic characteristics.

WIREWOUND POTENTIOMETERS

Wirewound potentiometers consist of bare insulating wire wound around an insulating form in either single-turn or multiturn configurations. Both precision and nonprecision types are produced. Potentiometer characteristics of this type of element are close linearity (although nonlinear windings can also be produced), high power ratings, low temperature coefficients, and close resistance tolerances. Wirewound potentiometers can also be expected to have good stability and long mechanical life and are suitable for working over a wide temperature range.

CONDUCTIVE-PLASTIC POTENTIOMETERS

The element in conductive-plastic potentiometers consists of a mixture of conductive and nonconductive plastics in a thermosetting plastic binder deposited on a ceramic substrate. Particular advantages are small size, infinite resolution, long life, and high reliability. Because the resistive element is very smooth, conformity can be nearly perfect to any required function.

The excellent life of conductive-plastic potentiometers makes them particularly suitable for servo applications where the potentiometers may be operating consistently over a few degrees of arc (which will provide high wear rates on wirewound and carbon potentiometers).

It is a general characteristic of nonwirewound potentiometers that they are less suitable for working in the voltage-divider mode because of "dc offset." This applies particularly in the case of conductive-plastic potentiometers.

CERMET POTENTIOMETERS

Cermet resistive elements contain a film of precious metal and glass fired onto a ceramic substrate. Such elements can be produced with a wide range of resistances with both linear and nonlinear characteristics, with virtually infinite resolution and high power ratings. Cermets can also be operated at higher temperatures than other types of potentiometers. A possible disadvantage with cermet potentiometers is a high temperature coefficient, although this has been substantially reduced with modern types to become almost comparable with wirewound potentiometers. Its life is limited by the life of the wiper, which may be subject to relatively high wear because of the hardness of the ceramic track.

TEMPERATURE COEFFICIENT OF RESISTANCE

The *temperature coefficient of resistance* is the unit change in resistance per degree Celsius from a stated reference temperature. It does not follow, however, that the same coefficient applies over the whole of the working temperature range. For critical ap-

applications special resistance elements may be employed (such as wirewound types with special resistance wire to provide low temperature coefficients).

NOISE

Noise can be defined as spurious variations in electrical output not representative of the input drive to transient resistors appearing between the wiper and resistive track. It can be determined quantitatively in terms of actual transient resistance values. For nonwirewound potentiometers such spurious deviations are referred to as *smoothness* and are measured in terms of voltage variations from theoretical values for specific travel movements. Smoothness values are quoted as a percentage of the input voltage.

WORKING MODES

Potentiometers may be used in the *variable-current* or *voltage-divider* modes. In the variable-current mode, shown in Fig. 3-5, the adjustability R_A is given by

$$R_A = \frac{I_A - R_T/2}{R_T} \times 100$$

where I_A is the current reading achieved
R_T is the total resistance of the potentiometer.

R_A is then the adjustability expressed as a percentage of the total resistance of the potentiometer after setting the potentiometer to

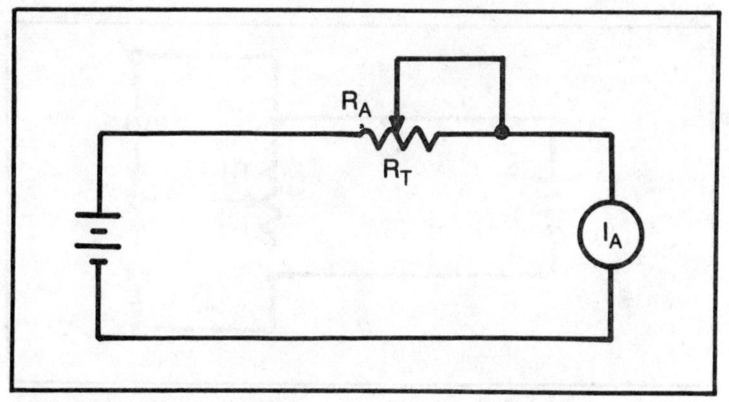

Fig. 3-5. Variable-current mode.

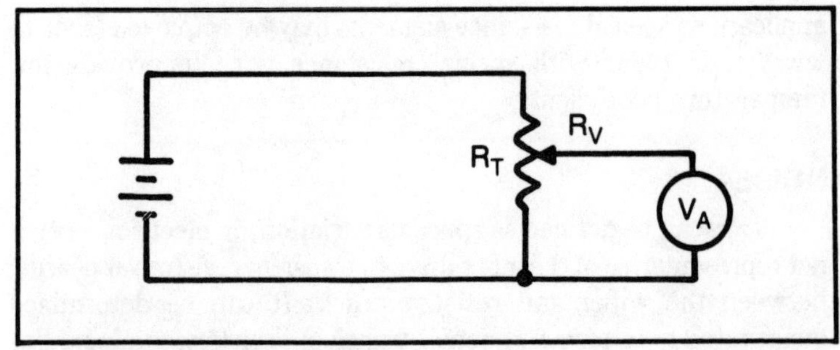

Fig. 3-6. Voltage-divider mode.

50 percent of its total resistance. Adjustability is read as signal range.

The voltage-divider mode is shown in Fig. 3-6. Here the adjustability R_V, the other adjustment to 50-percent voltage ratio, is given by

$$R_V = (V_A - 0.5) \times 100$$

where V_A is the voltage ratio achieved.

This performance may be modified by the effective resistance in the divider leads and also by contact-resistance variations. Precision-type potentiometers commonly employ multiwire wipers to minimize the latter effect.

POTENTIAL-DIVIDER CIRCUIT

The potential-divider circuit is shown in Fig. 3-7, where R1 and

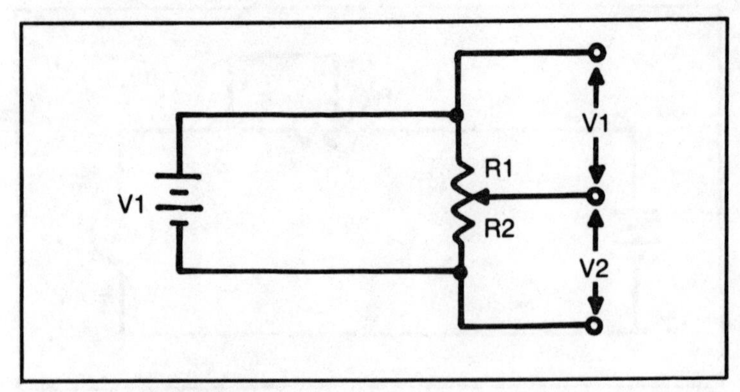

Fig. 3-7. A potential-divider circuit.

R2 represent the (variable) resistance values available on either side of the wiper of a potentiometer. Because the current flowing through the potentiometer must be the same as the current through R1 and R2, the following relationship applies:

$$V1 = \text{source voltage (such as battery voltage)}$$

$$V2 = \frac{V1}{R1 + R2} \times R1$$

$$V3 = \frac{V1}{R1 + R2} \times R2$$

(Note: $\frac{V1}{R1 + R2}$ is the current flowing through R1 and R2.)

By adjustment of the potentiometer, virtually any rate of voltage V2 or V3 can be set up, less than the supply voltage V. Also, for any initial setting of the potentiometer any change in its setting (i.e., "transducer" movement) will vary both V2 and V3 proportionately. Either V2 or V3 can, therefore, be tapped as a signal source corresponding to movement, because an output signal could be used for direct meter indication or in an alarm circuit. In the latter case the alarm circuit is designed to switch on at a particular threshold voltage level extracted from either V2 or V3.

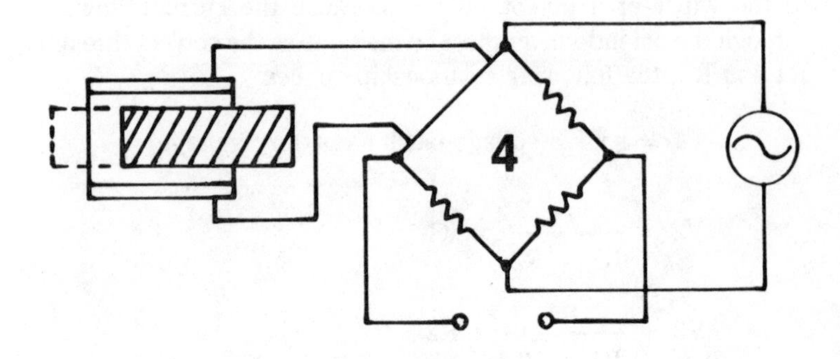

Strain Gauges

The original form of strain gauge was based on the ability of a metallic resistance wire to work as a transducer. If a length of wire is stretched, the resulting change in length is responsible for a proportionate reduction in diameter. Because the electrical resistance of a wire conductor is proportional to its cross-sectional area (and thus the square of the diameter), its effective resistance value will vary with strain. Provided the wire is not stretched beyond its yield point or elastic limit, it will recover its original length when the strain is released. Its actual electrical resistance under any strain up to its (mechanical) yield point is then a measure of the strain to which it is being subjected.

The simple, wire strain gauge thus transfers strain into another measurable quantity—electrical resistance. The usual method is to connect the strain gauge into one arm of a Wheatstone bridge. Differences in actual wire resistance are determined by the amount of imbalance produced in the bridge circuit. Ideally, to eliminate the effect of differences in temperature, the resistance wire should have a zero temperature coefficient (that is, its resistivity should not change with temperature), although this can be compensated in other ways. What the simple, wire strain gauge does suffer from, though, is hysteresis effect, or a different response under decreasing strain compared with increasing strain.

Wire strain gauges are either unbonded or bonded. Unbonded strain gauge elements are made of one or more filaments of resistance wire stretched between supporting insulators. The supports are either attached directly to an elastic member used as a sensing element or are fastened independently, with a rigid insulator coupling the elastic member to the taut filaments. The displacement (strain) of the sensing element causes a change in the filament length, with a resulting change in resistance. Although transducers using unbonded gauges are still available, they are more fragile and less frequently used than bonded gauges.

The usual form of bonded strain gauge consists of a foil of a resistance alloy, such as constantan, bonded to an epoxy backing film (see Fig. 4-1). The foil is either die cut or etched to produce a grid pattern sensitive to strain along one axis. Elongation or shortening of the gauge along the sensitive axis produces an increase or decrease in the resistance of the gauge. The backing film is then adhesive-bonded to an elastic member to sense the strain of the member due to applied stress. Foil gauges can measure tensile, compressive, or torsional stresses and so are used in transducers for practically every mechanical parameter. Foil gauges have also been constructed by thin-film techniques utilizing a ceramic film substrate on which the resistance alloy is vacuum deposited.

PIEZORESISTIVE STRAIN GAUGES

Piezoresistive strain gauges, also known as semiconductor

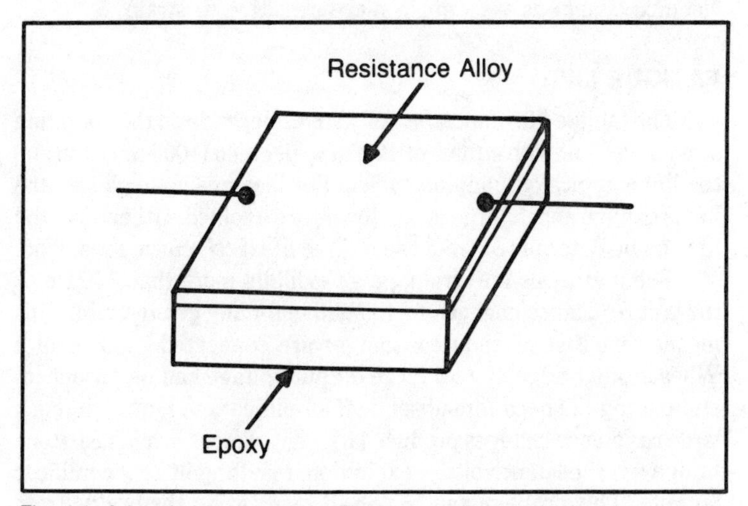

Fig. 4-1. A bonded strain gauge. Resistance alloy is attached to an epoxy backing.

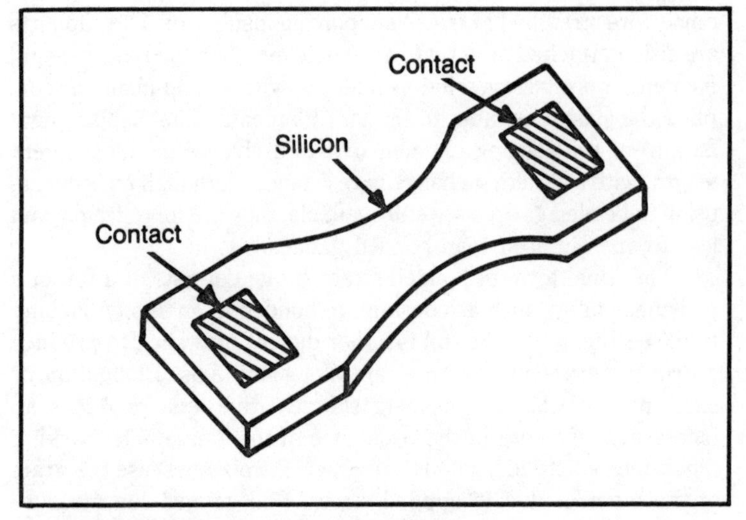

Fig. 4-2. A silicon semiconductor strain gauge.

strain gauges, are basically solid-state silicon resistors fabricated from a single piece of p- or n-doped silicon and incorporating an "active" length together with contacts at each end (Fig. 4-2).

Unlike a wire element, the resistance change in a piezoresistive strain gauge results largely from its change of resistivity with strain rather than its length and cross section. Compared with wire gauges, it is also virtually free of mechanical hysteresis, and its resistance change is very much more sensitive to strain.

FATIGUE LIFE

The fatigue life of metallic foil gauges depends on the operating strain level, but with strains of 1000 μin. per inch (1000 micro strain) the life is typically 2 million cycles. For piezoresistive gauges the life is comparable if large strain levels are avoided. Otherwise, the life can be determined from the fatigue life (S/N) curve for silicon.

The piezoresistive strain gauge exhibits more than 100 times the unit resistance change of a foil gauge for any given strain. This means that if semiconductor gauges are connected as arms of a Wheatstone bridge, a very large output voltage can be produced, eliminating the need for subsequent amplification. However, such large resistance changes produce large unbalances in a Wheatstone bridge with constant-voltage excitation, resulting in very nonlinear outputs. This problem can be solved by exciting the bridge from a constant-current supply. A constant-current supply contains more complex circuitry than a constant-voltage regulator and may not always be as readily available.

Actually, the resistance change of a semiconductor strain gauge as a function of strain is not completely linear over its total strain range. This results in nonlinearity in some transducers, even with constant-current excitation. Foil gauge transducers require amplification because of low bridge output, but the linearity of the output signal is not a problem.

TEMPERATURE EFFECTS

All resistance strain gauges are temperature sensitive to some extent. This sensitivity results from the change in the resistivity of the gauge material with temperature as well as from the differential expansion between the gauge material and the elastic member to which it is bonded. This latter effect generates a false strain input to the gauge. Self-compensated foil gauges are being made with alloys whose thermal expansion is similar to that of the sensing element and whose coefficient of resistivity is very small.

No similar choice of materials is available for semiconductor

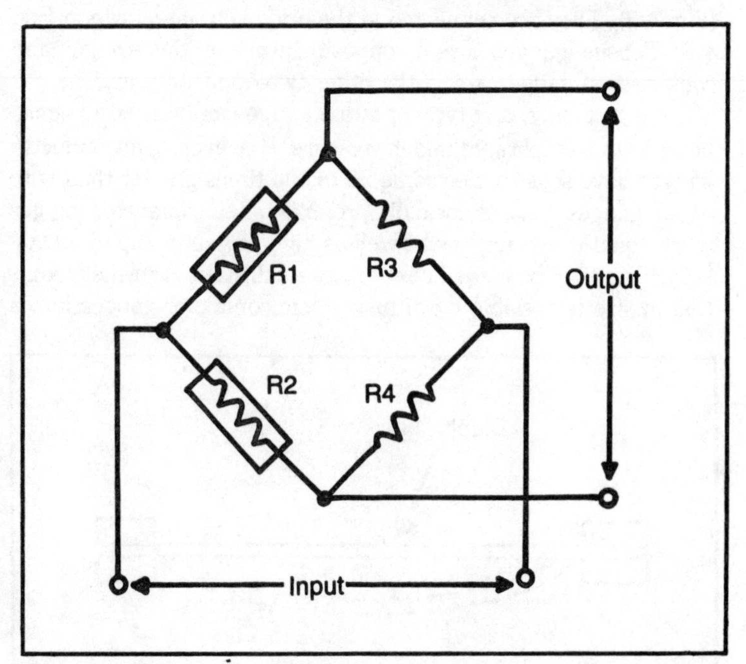

Fig. 4-3. A bridge circuit for minimizing the effects of temperature on a strain gauge.

gauges. The strain sensitivity of piezoresistive gauges is also significantly temperature sensitive, aggravating the other temperature effects.

Some of the temperature effects on strain gauges can be cancelled in the Wheatstone bridge circuit of Fig. 4-3. If R1 and R2 or R3 and R4 both experience the same increase in resistance, their ratio will still be essentially the same, cancelling much of the effect of the increased resistance. The effectiveness of bridge-circuit compensation is usually greater for self-compensated metallic strain gauges than for semiconductor strain gauges.

DIAPHRAGM BONDED STRAIN GAUGES

Strain gauges bonded to a flexible diaphragm are a common type of pressure transducer for measuring fluid pressure. With a diaphragm clamped around its edge, deflection under pressure will result in both tension and compression stresses being developed simultaneously (Fig. 4-4). Thus, two gauges placed at the center of the diaphragm will measure tensile stress, and two more gauges towards the edge of the diaphragm will measure compression stress (Fig. 4-5). They are connected in the bridge circuit shown in Fig. 4-6. Tensile gauges are in opposite arms of the bridge, and compression gauges are in the other two opposite arms.

Theoretically, any type of strain gauge can be used to sense bending of a diaphragm under pressure. However, semiconductor gauges have sensitivities some 50 to 100 times greater than wire or foil gauges and are normally preferred. Semiconductor gauges work equally well for low as well as high pressures (up to 50,000 lb/in.^{2}), and can be made in extremely small sizes. A further reduction in size is possible by diffusing semiconductor gauges into a

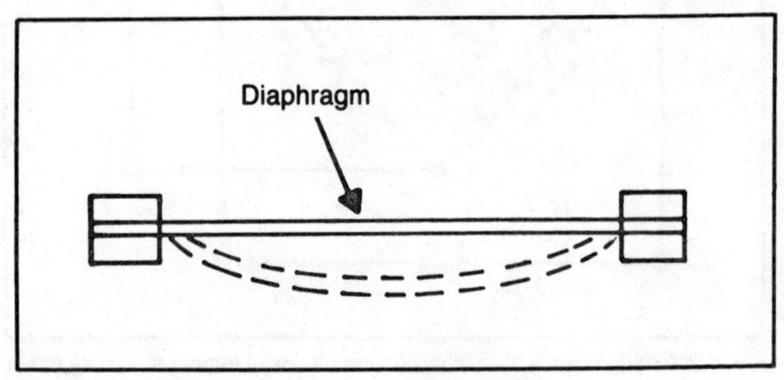

Fig. 4-4. Deflection of a diaphragm under pressure.

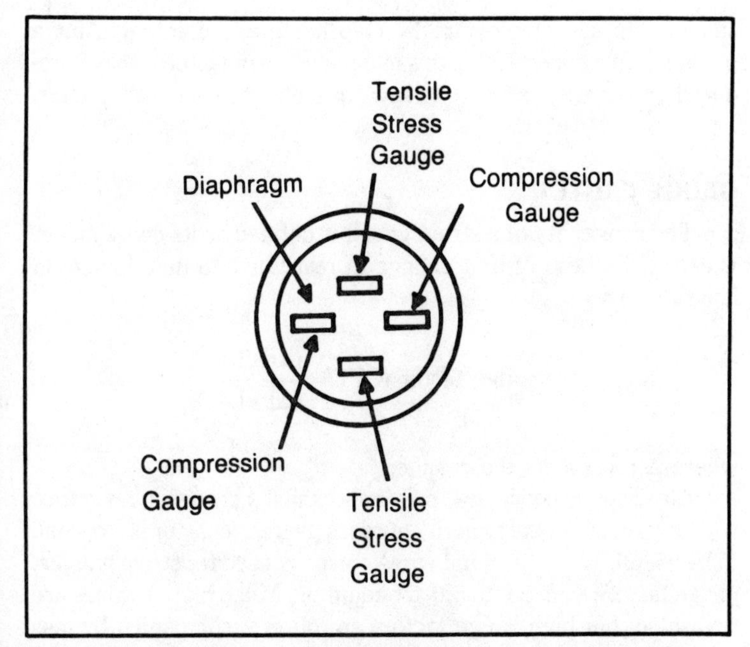

Fig. 4-5. Typical arrangement of strain gauges on a flexible diaphragm.

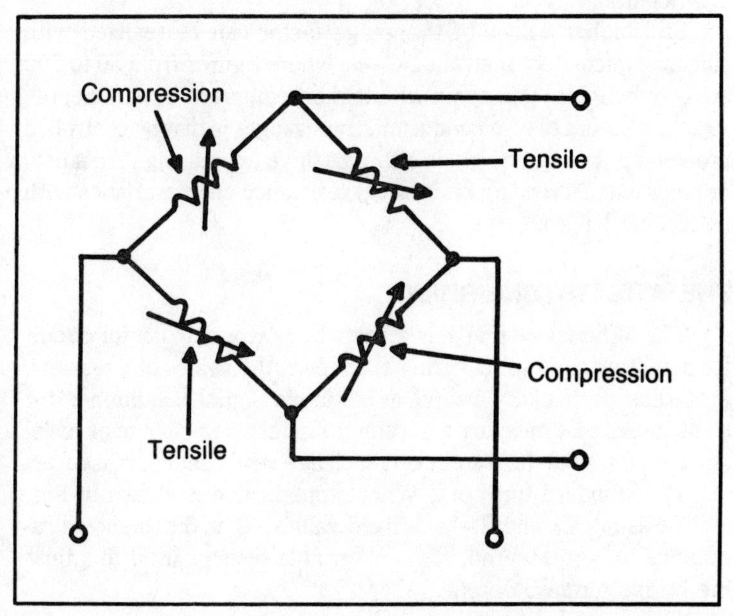

Fig. 4-6. Interconnection of strain gauges, in the configuration of Fig. 4-5, to form a bridge circuit.

silicon chip and employing the chip itself as a diaphragm for a pressure transducer. Using this technique, manufacturers have produced pressure transducers as small as 0.050 inch overall diameter.

GAUGE FACTOR

The sensitivity of a strain gauge is defined as its ***gauge factor***, which is the ratio of unit change in resistance to unit change in length:

$$\text{gauge factor (GF)} = \frac{(\Delta R)/R}{(\Delta L)/L}$$

where Δ represents the change.

All electrical conductive materials exhibit a change in resistance with length, but in the majority of cases the gauge factor is too small to be useful. With the metal wires normally used for strain gauges, gauge factors from 2.0 to 5.0 are common. Much higher values are realizable, but high gauge factors are often accompanied by loss of stability (greater sensitivity to temperature or higher temperature coefficients).

Still higher values of the gauge factor can be realized with silicon semiconductor strain gauges, where figures from 50 to 200 are commonplace (but again with higher temperature coefficients). Another feature of semiconductor strain gauges is that by controlled processing they can be given either positive (increasing resistance) or negative (decreasing resistance) resistance characteristics with increasing length.

THE WHEATSTONE BRIDGE

The *Wheatstone bridge* is a basic but extremely useful circuit for detecting and/or measuring changes in the values of a resistor. It is thus particularly useful as a simple signal conditioner for transducers that measure a parameter in terms of change of resistance produced in the transducer; such as a strain gauge transducer.

The standard form of a Wheatstone bridge is shown in Fig. 4-7. Resistors R1 and R2 have fixed values. R_X is the unknown resistance to be measured. R_A is a variable resistor used to adjust the bridge for zero output.

The relationship is then

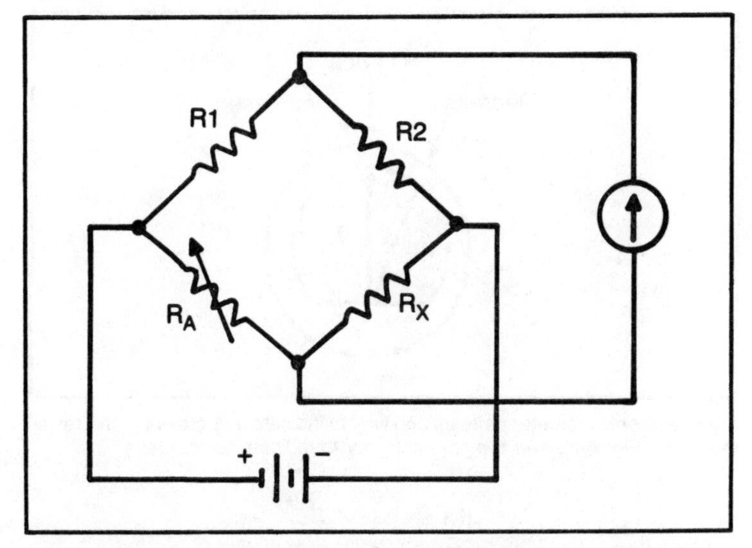

Fig. 4-7. Wheatstone bridge for detecting small changes in the value of a resistance R_X.

$$R_X = R_A \times \frac{R2}{R1}$$

If R1 and R2 are equal, then $R_X = R_A$. In other words, with the bridge balanced by adjusting R_A to give zero output, the value of R_X is then equal to the adjusted value of R_A.

Now suppose R_X is a resistance-type transducer and R_A is adjusted to the same resistance so as to balance the bridge and give zero output. Any force applied to the transducer to change its resistance will unbalance the bridge, resulting in a signal output that can be read on a meter. This reading will be a direct measure of the transducer response.

This works both ways, so that with a zero-center meter, transducer reaction providing an increase in resistance will cause proportionate needle deflection one way; a decrease in resistance, a proportionate needle deflection the other way (Fig. 4-8). The sensitivity of such a bridge is then determined by the sensitivity of the meter to small changes about the center point. If necessary the output signal can be amplified.

Looking at the bridge circuit again but reallocating resistors R1, R2, R3, and R4 (as in Fig. 4-9), we obtain the basic relationship for zero output:

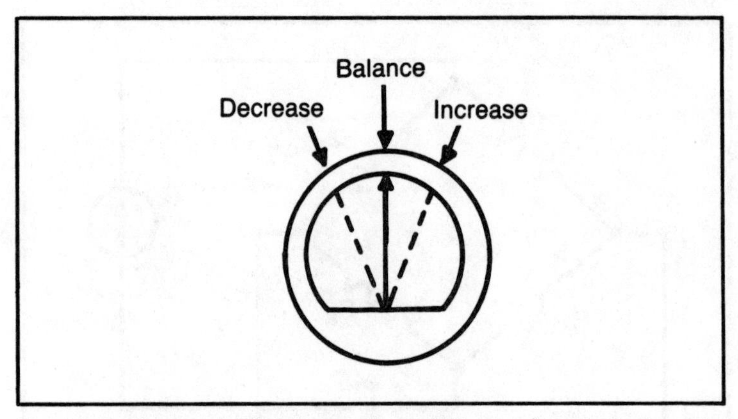

Fig. 4-8. A galvanometer deflects one way to indicate a decrease in the resistance R_X (Fig. 4-7), and the opposite way to indicate an increase.

$$R3 = R4 \times \frac{R2}{R1}$$

or

$$R1 \times R3 = R2 \times R4$$

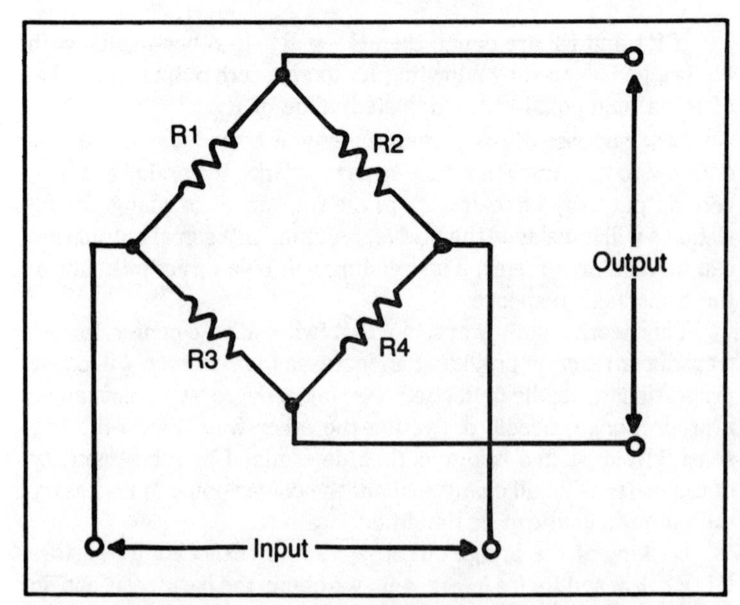

Fig. 4-9. Another arrangement of the bridge circuit for measuring strain on a flexible diaphragm (see text for discussion).

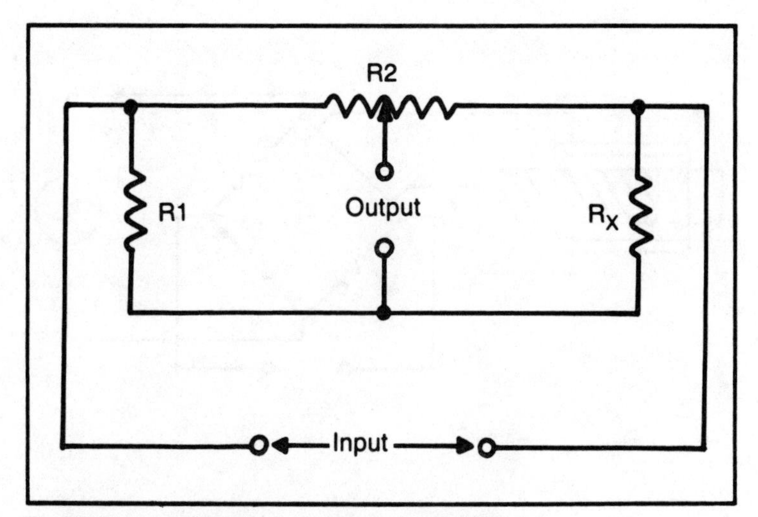

Fig. 4-10. A simplified form of strain-gauge bridge.

If R1 and R3 are strain gauges working in opposite configuration (that is, force reaction produces an increase in resistance in one and a decrease in the other), the resulting signal output from bridge unbalance is now equal to the product of the two differences. For initial balancing of the bridge, either R2 or R4 would then need to be adjustable.

For even stronger signal output, strain gauges could be used in each of the four arms. In this case an external variable resistor would be needed to set up initial bridge balance.

A simpler form of bridge capable of accommodating one strain gauge is shown in Fig. 4-10. Here R1 is a standard resistor setting the range, and R2 is variable to set up the bridge for zero output. Variation in the strain-gauge resistance R_X then generates a proportionate signal output. A particular feature of this circuit is that it can also be used with an ac input to measure the effect of changes in capacitance by replacing R1 and R2 with capacitors and R_X with a capacitive transducer.

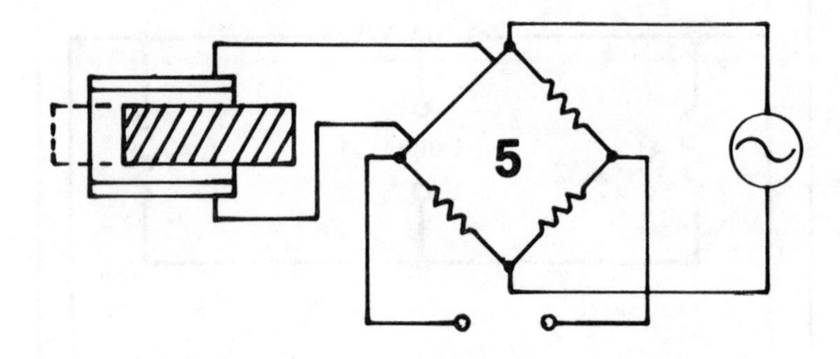

Capacitive Transducers

Capacitive displacement transducers use the electrical quantity of capacitance to convert mechanical movements into a corresponding electrical signal. They are particularly accurate and sensitive transducers and widely used for measuring position, length, or angles. They can also be used in contacting and noncontacting modes. In the former case the movement displacement to be measured is directly connected to one movable element of the capacitor. In the noncontacting mode the object under study is itself made one plate of the capacitor.

A capacitor consists of two plates or electrode areas of conductive material separated by a dielectric filling the gap between them (Fig. 5-1). The capacitance of such a device is the ratio of the current to the potential difference between them and can be calculated from the formula

$$\text{capacitance} = 0.22 \frac{Ak}{d}$$

where A is the area of each plate in square inches
d is the gap in inches
k is the dielectric constant of the gap substance
capacitance is given in picofarads (pF).

It follows that the capacitance may be varied by change of area

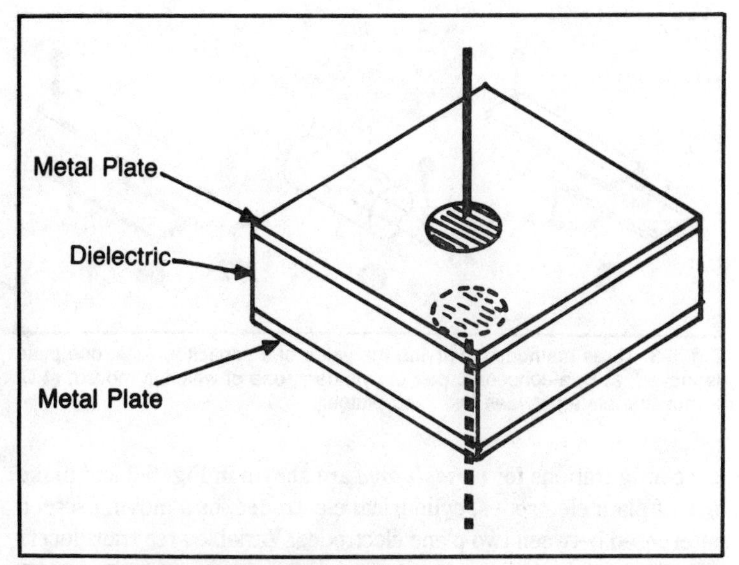

Fig. 5-1. Basic construction of a capacitor.

(A) or gap (d). Variation of capacitance with area follows a linear law, but variation with gap follows a reciprocal law (see Fig. 5-2).

Both forms of working are used in capacitance transducers. Ba-

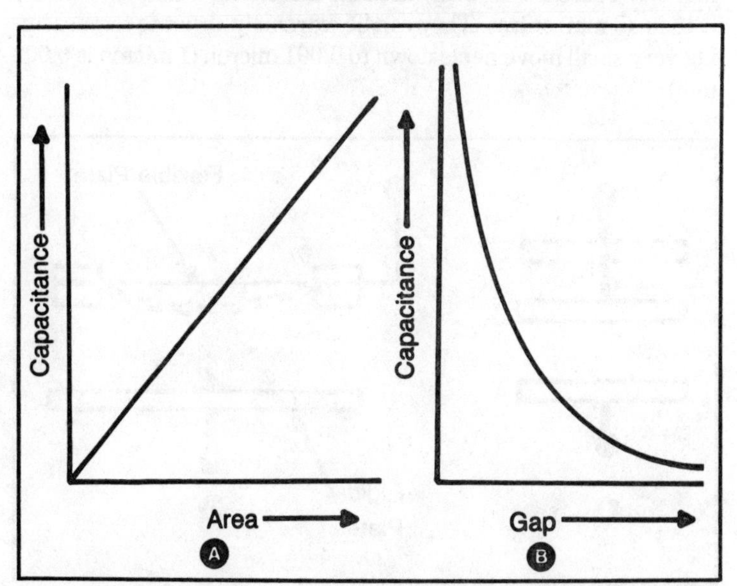

Fig. 5-2. Capacitance is a function of place surface area (A) and also of the gap between the plates (B).

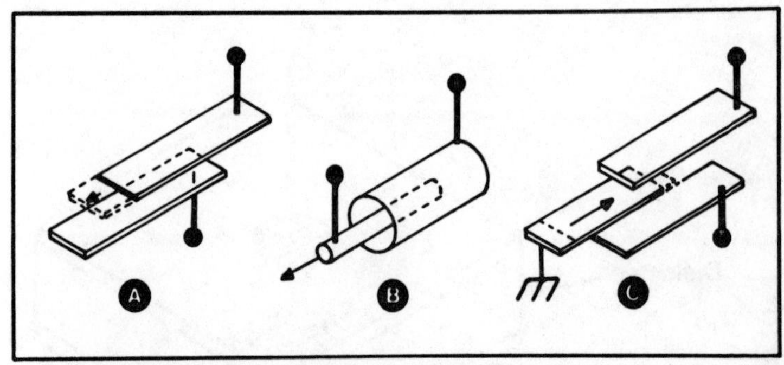

Fig. 5-3. Three methods of varying the value of a capacitor. At A, one plate is moved; at B, a concentric pair of cylinders, one of which is moved; at C, a movable plate between two fixed plates.

sic configurations for *variable area* are shown in Fig. 5-3 and make use of plain electrodes, cylindrical electrodes, or a moving screen interposed between two plane electrodes. Variable-area transducers are used for both linear and angular measurements, their most useful size being for measuring movements from about 3/16 in. (5 mm) upwards.

Basic configurations for *variable-gap* transducers are shown in Fig. 5-4. These are normally used for measuring small movements, 3/16 in. (5 mm) or less. They can be extremely accurate for measuring very small movements down to 0.001 micron (1 micron is 0.001 mm).

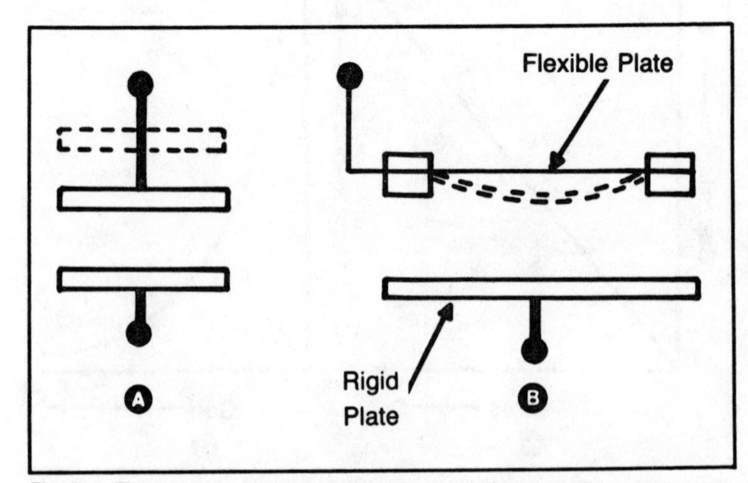

Fig. 5-4. The gap spacing may vary because of movement of a whole plate (A) or because of distortion in the shape of one of the plates (B).

Noncontacting capacitive transducers may work in either mode. With a variable-area type, introducing a screening body between two fixed electrodes (constant gap) varies the capacitance. The actual body introduced may be a conductor or a nonconductor. Also, it does not necessarily have to enter the fixed gap. Introducing it adjacent to one of the fixed electrodes will also vary the capacitance. With a variable-gap proximity transducer the body itself becomes one electrode (or carries the second electrode). These can work with much larger variable gaps than contacting type variable-gap transducers.

A basic configuration for a noncontacting proximity transducer is shown in Fig. 5-5.

LINEARITY

The linearity and repeatability of a capacitance transducer depend on the quality of machining and alignment of electrodes (they are independent of material or magnetic effects). To achieve an accuracy of 0.01 percent means that the gap in electrodes must be accurate to this degree. For example, with a 0.06 in gap, the accuracy needs to be 6 μin. Misalignments in the electrodes cause some nonlinearity.

Production transducers of this type are currently made with peak errors less than 0.01 percent for short ranges (up to 5 mm) and 0.005-0.0025 percent for longer ranges (10-100 mm).

Although the noncontacting variable-gap transducers do not

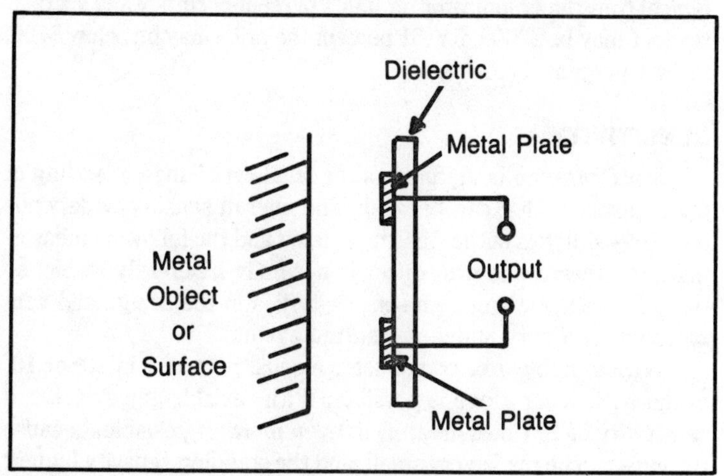

Fig. 5-5. A noncontacting proximity transducer.

have the linearity of the variable-gap type, they have other advantages that are worth exploiting in terms of stability, sensitivity, and convenience. Certain types of displacement transducers, for example, have been designed for use at high temperatures (up to 600° C.). Another advantage of capacitance measurements—namely, that the dielectric does not vary and the transducers can be made mechanically stable over this temperature range—means that very stable measurements of static or slowly changing strain or displacement can be made at high temperatures.

ACCURACY

Accuracy is defined as the maximum error in indicated change of position from zero when compared with a reference standard such as slip gauges. The errors include all sources, such as nonlinearity, slope, resolution, drift, and zero shift.

A wide range of accuracies is possible, depending on the needs and economics of the application. Typical accuracies are

0.1 percent	e.g.,	±	.001″ per inch low cost applications
0.01 percent	e.g.,	±	0.0001″ per inch in 1-in. range general measurement
	or	±	0.001 mm per mm in 10-mm range
.001 percent	e.g.,	±	0.001 mm per mm in 100-mm range high-accuracy systems

These accuracies are for the transducer. The associated instrumentation may be better or worse, depending on the cost. A typical transducer indicator for 0.001 percent accuracy and six digit readout may be $3000; for .01 percent the price may be below $700; for 0.1 percent, below $150.

SENSITIVITY

Sensitivity can be defined as the smallest change in setting of the transducer that can be read. The overall sensitivity depends on the sensitivities of the transducer itself and the following measuring instrument. The transducer is normally inherently stable, so the onus on achieving high sensitivity lies in the design and construction of a very stable measuring system.

With variable-area transducers a sensitivity of 1 in 10^6 or 10^7 of the transducer range is possible. With variable-gap systems, a sensitivity of 10 times greater, or even more, is possible, because the gaps can be made very small, and the standing capacity higher; thus, the change that must be observed is small. Hence, the sensitivity in absolute terms is very high, although the range of movement is small. For example, a pair of capacitor plates 2 in. in area with a gap of 0.01 in. has a capacity of 44 pF. A change in the gap of 10^{-10}, which is 10^{-12} in., can be observed. At such extremes of sensitivity the stability of construction is critical.

MEASURING INSTRUMENTS

Instrumentation is necessary to convert the setting of the transducer into a readable form. Typically, an ac signal is used to drive the transducer when the amplitude of this signal after passing through the transducer is a measure of the transducer setting. This is in analog form; a digital readout needs analog-to-digital conversion.

The most accurate form of transducer-to-digital converter is the self-balancing bridge (Fig. 5-6). This may be accurate to a few parts in a million and give a six or seven figure readout. An alternative is the open-loop transducer-to-dc converter (Fig. 5-7). An ac generator drives the transducer. The output is rectified and filtered and is available as a dc signal that is proportional to the transducer setting. This dc may be digitized to give a five figure readout. Either the bridge or dc converters may be used to provide dc control signals.

Other forms of measuring instruments are shown in Fig. 5-8. At A is an oscillator where the variation in capacitance causes a variation in resonant frequency. This suffers from the problems of stability of frequency, and therefore displacement readout, due

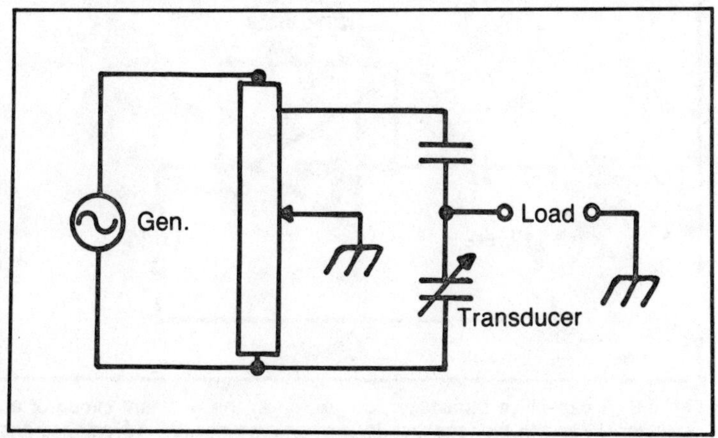

Fig. 5-6. A self-balancing bridge circuit for use with a capacitive transducer.

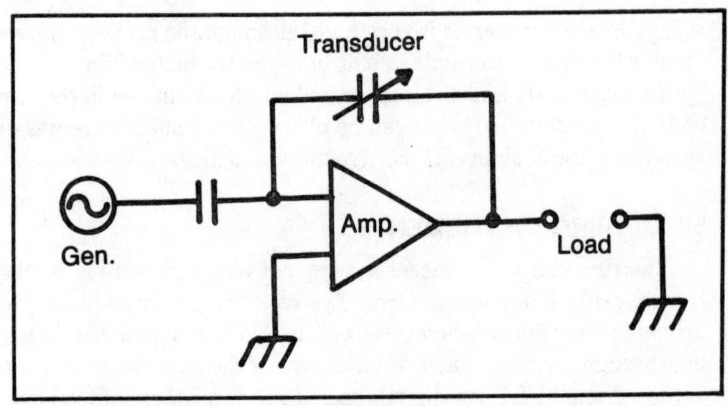

Fig. 5-7. An open-loop converter for use with a capacitive transducer.

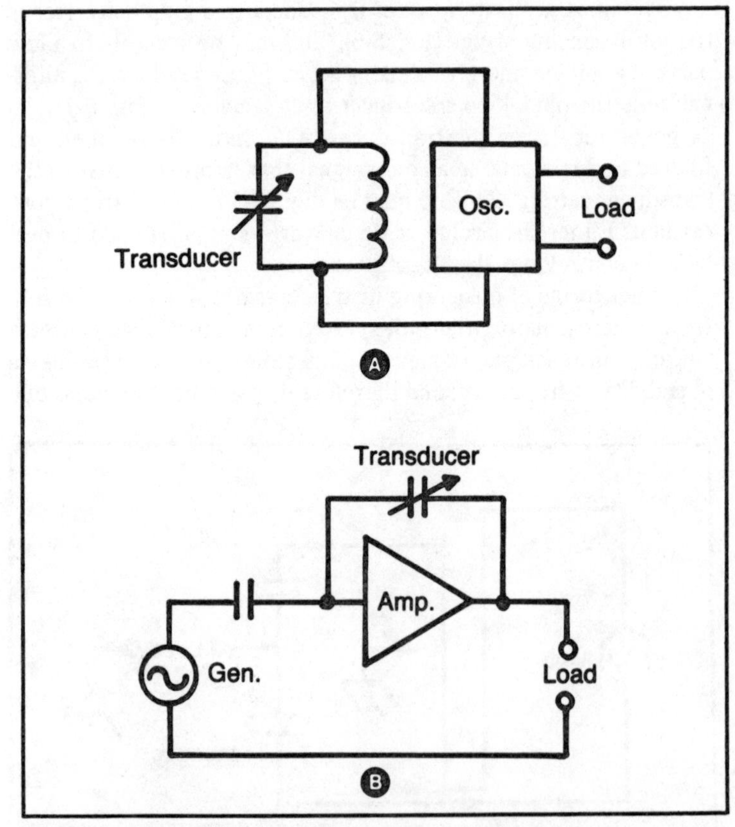

Fig. 5-8. A capacitive transducer, connected in the resonant circuit of an oscillator (A) and in the feedback loop of an amplifier (B). Advantages and disadvantages of both methods are discussed in the text.

to changes in the shunt capacity of the capacitor leads (lead length variations) and drifts in the tuned circuitry.

Shunt capacity is the term given to the capacity between core and screen of the coaxial wires used to drive the transducers.

Figure 5-8B is a useful circuit, whereby a linear output can be obtained from a variable-gap transducer using it in the feedback loop of an amplifier. Here we have the impedance of the capacitor

$$Z2 \; \alpha \; \frac{1}{C2}$$

$C \alpha \frac{1}{d}$. Therefore, Z2 α d, where d is the change in displacement between the plates. The closed-loop gain of the amplifier is $\frac{Z2}{Z1}$. The output voltage of the amplifier is

$$V = AV_{in} = \frac{Z2}{Z1} \quad V_{in}$$

Therefore

$$V \text{ out } \alpha \frac{dV_{in}}{Z1} = Kd \text{ (where K is a constant)}$$

This means that the output voltage is directly proportional to d.

A more stable circuit is shown in Fig. 5-9. This is a Blumleine bridge, which uses in its construction inductive voltage dividers as two of the ratio arms. Inductive voltage dividers are a very accurate (1 in 10^6) means of subdividing ac and exhibit very good long-term stability (1 in 10^8). They are therefore ideal components in this instance for the bridge arms. With this bridge the shunt capacities from the cables do not affect the accuracy of the measurement because they never appear in parallel with the measuring capacitances. Both the generator and detector can be made with low impedances so that the sensitivity of the instrument is not changed by large changes in shunt capacity (cable length). Another advantage of the bridge circuit is that if the bridge is self-balancing, many sources of drift, such as changes of amplifier gains and carrier level (which would cause a change in the voltage output, and thus apparent displacement, of an off-balance system), are nullified. The balance point of the bridge is independent of the carrier level or

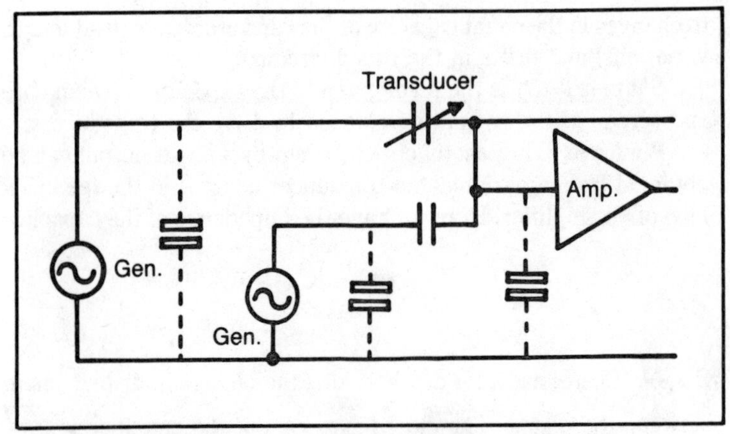

Fig. 5-9. A Blumleine bridge for use with capacitive transducers.

amplifier gains. A large dynamic range is also possible with resolution of six decades (1 in 10^6 or 1 μm in 10-mm range). Typical full-scale capacitance is around 0.5-1 pF. Typical cable capacity would be around 200-300 pF for 3 meters (m) of cable. With the correct circuit design a change of 10^{-6} pF can be measured in 1 pF, shunted by 300 pF with an accuracy (in the case cited) of around 4×10^{-6} pF.

RESOLUTION

Resolution is the least change that can be displayed by a measuring instrument. A transducer system may be capable of sensitivity of one part in 10^6 of the range. If a four decade indicator is used to digitize the transducer, then the system resolution is one part in 10^4. A seven decade instrument with a .01-percent transducer accuracy would still have useful resolution in the last three decades for small changes.

STABILITY AND DRIFT

Accuracy of measurement also depends on freedom from long- or short-term drifts, that is, the stability of the system. There are numerous potential sources of drift, such as electrical changes, temperature and other environmental changes, and mechanical changes. All these have to be considered in a measuring system requiring high accuracy. With suitable designs it is possible to achieve accuracy, stability, and resolution of the order of one part per million, but the resulting cost of equipment can be high.

ASSOCIATED CIRCUITS

Associated electronic circuits for capacitive displacement transducers can be relatively simple, due to the high output impedance of the transducer. Figure 5-10A shows a circuit for a differential capacitor with an inductive divider connected to form a bridge circuit.

The transducer draws very little current. Therefore, the impedances of the leads are not significant. Strong capacitance can be minimized by screening, as indicated. It is not necessary to

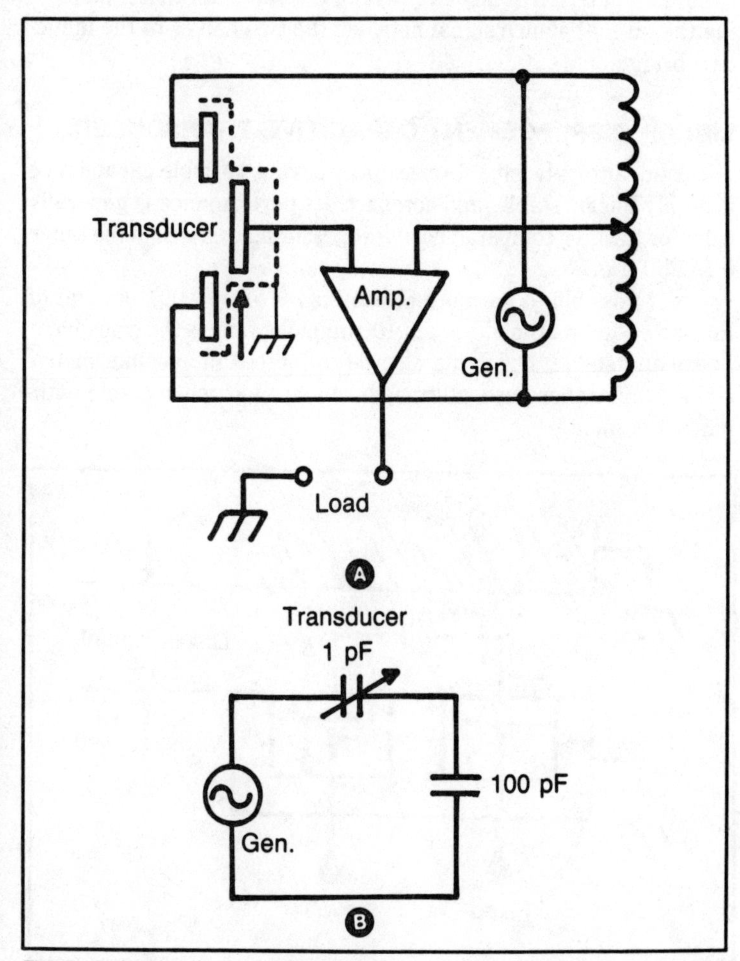

Fig. 5-10. Another form of bridge circuit for use with a differential capacitive transducer (A); the equivalent circuit (B).

screen the lead from amplifier to inductive divider because the latter has a low output impedance and is not affected by strong capacitance.

The equivalent circuit is shown in Fig. 5-10B. Typical values would be 1 pF on the transducer capacitance and 100 pF strong capacitance in shunt. This would be equivalent to a signal loss of 100 (hence the need for the amplifier).

Carrier phase and amplitude are detected by a phase-sensitive circuit, so that both the sense and amplitude of the displacement can be indicated. This readout may have either the inductive divider calibrated in terms of displacement or a zero-center meter displaying the out-of-balance signal between the two halves of the inductive bridge.

LINEAR DISPLACEMENT CAPACITIVE TRANSDUCER

A linear displacement transducer having variable capacitance is highly linear, stable, and compact. Its performance is generally superior to other comparable length-measuring systems in the range 0 to 250 mm.

It is possible to measure to accuracies of ± 0.001 percent of full scale—for example, 1 μm in 100 mm. Not only is the transducer linear and stable, but it can also be calibrated in absolute metric or inch units; therefore, all transducers are interchangeable without adjustment.

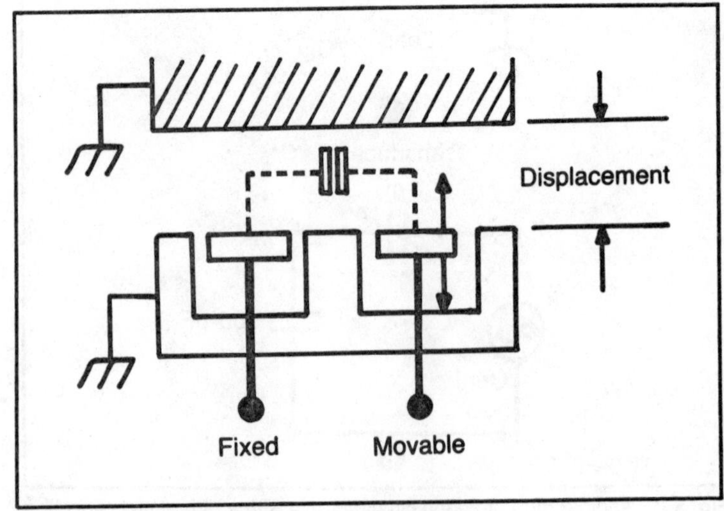

Fig. 5-11. A capacitive displacement transducer.

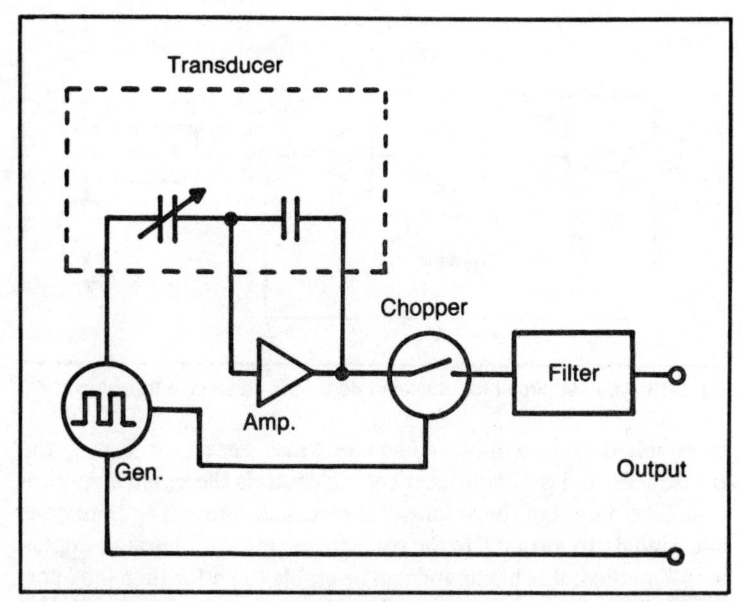

Fig. 5-12. A circuit for converting the capacitance of a transducer into a variable dc signal.

Resolution is also high. Changes in setting of less than one part per million can be reliably measured (0.01 μm in 10 mm). This is possible because of the very stable construction and the drift-free ac ratio measuring system.

An example of a capacitive type displacement transducer is shown in Fig. 5-11. It consists of one fixed capacitor and one variable capacitor. The ratio between fixed and variable capacities is measured to give the displacement in electrical form. Because both capacitors are in the same environment and close together, ambient changes such as temperature affect both in the same manner. Hence, the stability and freedom from drift are very good indeed.

The simple and robust structure of the capacitive transducer allows very high accuracy and small size. Typically, a 5-mm-range transducer accurate to 0.5 μm and repeatable to 0.05 μm can be less than 10 mm in diameter and 50 mm long. A 100-mm range repeats to 0.2 μm and is 33 mm in diameter by 300 mm long.

INSTRUMENTATION

Figure 5-12 shows a dc-dc signal conditioner that converts the variable setting of the transducer into a variable dc signal. Dc power

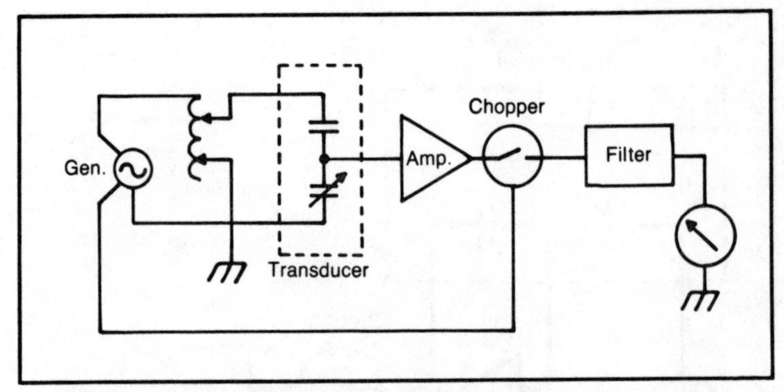

Fig. 5-13. Another circuit for obtaining dc from a capacitive transducer.

is supplied to the unit. A square-wave generator drives the transducer, and the transducer setting controls the signal amplification. The output of the amplifier is rectified, filtered, and provides a dc signal proportional to the transducer setting. This is an "open-loop" method; it is linear and can be stable to better than 0.02 percent with a dynamic range of 10,000 at a bandwidth of 250 Hz.

An even more accurate and stable device is the "closed-loop" ac bridge system (Fig. 5-13), where an inductive divider, which is accurate and stable to better than one part per million, is compared with the ratio of the capacitive transducer.

Any difference or out-of-balance between divider and transducer setting is amplified, synchronously rectified, and available as a direction-sensitive dc output. The instrument is used either with null balance (that is, a position is set on the dials and the transducer moved to balance) or with open loop, in which the dc output is displayed to show the difference between the position called for and the actual position. The great advantage of this is the very high stability and accuracy of the digital setting of the dividers. The transducer setting can be read in digital form to six figures. This system is excellent for closed-loop servo systems.

Direct digital readout is obtained by balancing the ac bridge automatically. When the transducer is moved, the out-of-balance signal operates digital switches, which restore the balance. The setting of the switches is then the exact equivalent of the transducer setting in either inch or metric digital form. This digital setting may be displayed, recorded, and generally used for digital data processing.

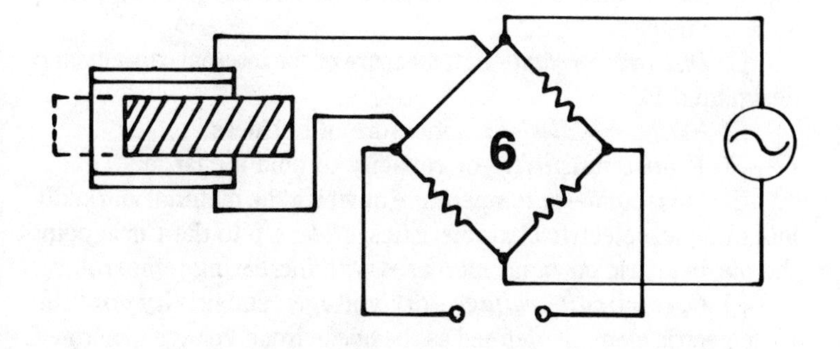

Properties of Piezoelectric Materials

Piezoelectric transducer elements are normally based on polycrystalline ceramics, such as barium titanate, lead zirconate, lead titanate, and lead metaniolbate (or what are generally known as *ferroelectric* substances), artificially polarized. This enables the piezoelectric properties to be controlled in the manufacturing process.

Manufacture starts with proportioning and mixing component ceramic powders, followed by high-temperature calcination. The now chemically combined ingredients are then repowdered, mixed with binder, and formed into pellets. These are then fired in a kiln to produce the final rough ceramic elements. These are then lapped and cleaned, and electrodes are plated on them. They are then followed by a high-voltage field under completely controlled conditions.

PIEZOELECTRIC PARAMETERS

The *piezoelectric constant*, symbol d_{ij}, is a measure of the basic sensitivity of a piezoelectric material and is defined as the change generated (g) per unit of applied force (F):

$$d_{ij} = \frac{g}{F}$$

Other important parameters are

☐ *Frequency constant*—or a monitor relating the dimension of a given piezoelectric element to its natural frequency (usually quoted as K_c in inches).

☐ *Dielectric constant*—or a measure of the internal capacity and designated E.

☐ *Elastic modules*—or a measure of stiffness.

☐ *Volume resistivity*—or capacity to hold a charge.

☐ *Curie point*—or temperature at which the material markedly loses its piezoelectric characteristics. *Note*: Up to the Curie point the piezoelectric constant increases with increasing temperature.

☐ *Open-circuit voltage*—or voltage sensitivity of the piezoelectric element, defined as the open-circuit voltage generated from unit of applied force. This is related directly to both the piezoelectric constant and dielectric constant:

$$\text{open-circuit voltage} = \frac{d_{ij}}{E}$$

EQUIVALENT CIRCUITS

The equivalent circuit for a piezoelectric transducer is shown in Fig. 6-1. The internal resistance R will be very high (normally

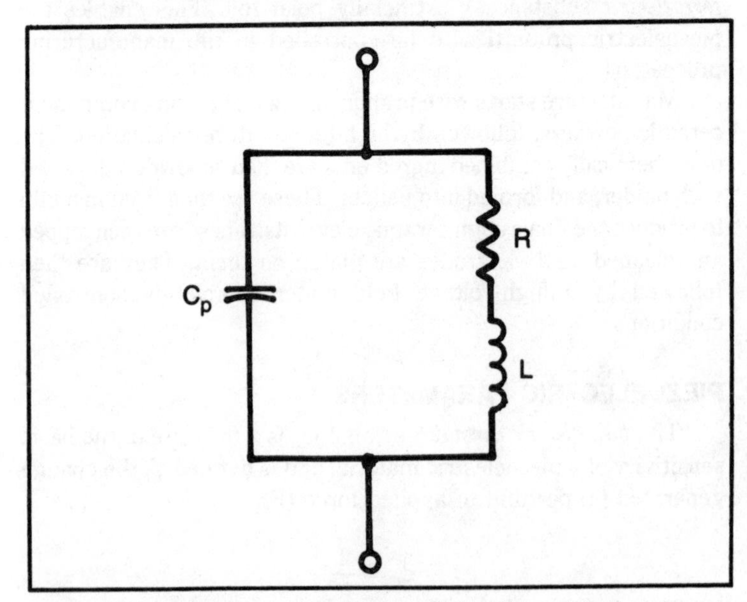

Fig. 6-1. A piezoelectric transducer is equivalent to this combination of capacitance (C_p), inductance (L), and resistance (R).

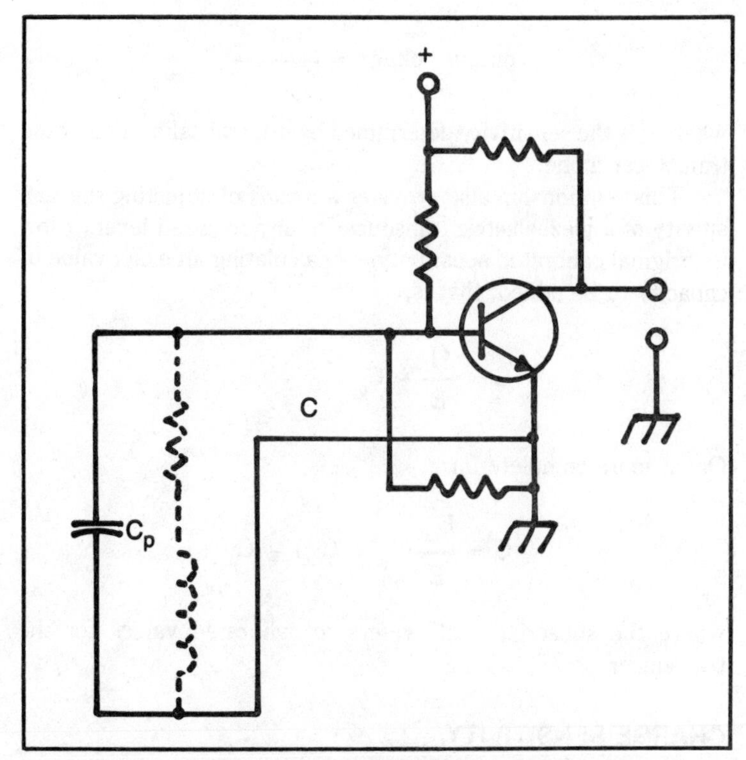

Fig. 6-2. In practice, the piezoelectric transducer behaves as a combination of internal capacitance (C_p) and parallel-wiring capacitance (C).

in excess of 20,000 MΩ); also the effect of the inductance L is insignificant at the natural frequency of the material, so both R and L can be ignored. Effectively, therefore, the piezoelectric transducer behaves as a capacitor C_p producing a charge across its plates proportional to the load force on the element.

The open-circuit voltage (E) out of the transistor is then equal to this charge divided by the transducer component:

$$E = \frac{q}{C_p}$$

In a working device the output is connected to a load (for example, amplifier circuit). The load is thus the input resistance of the amplifier. Some capacitance (C) will also be added from the wiring when the working voltage generator equivalent circuit is as shown in Fig. 6-2:

$$\text{output voltage} = \frac{q}{C_p + C}$$

where q is the sensitivity determined by original calibration of the transducer alone.

This relationship also provides a means of adjusting the sensitivity of a piezoelectric transducer to any required level (below its original calibrated sensitivity) by calculating an exact value of capacity to be added: that is,

$$C = \frac{Q}{E} - C_p$$

Or, in more complete form,

$$C = \frac{E_{cal}}{E}(C_p + C_{cal}) - C_p$$

where the subscript "cal" refers to calibrated values for the transducer.

CHARGE SENSITIVITY

Capacity added in shunt with the transducer will have no effect on its charge sensitivity. However, capacity added in series will reduce the charge output. Thus, in the circuit shown in Fig. 6-3, C_s is a series capacitor acting with a swamp effect. C1 represents any parallel capacitance ahead of C_s, and C2 any parallel capacitance beyond C_s. Neither C1 nor C2 will have any effect on charge sensitivity, but the charge appearing at the input to the amplifier will be

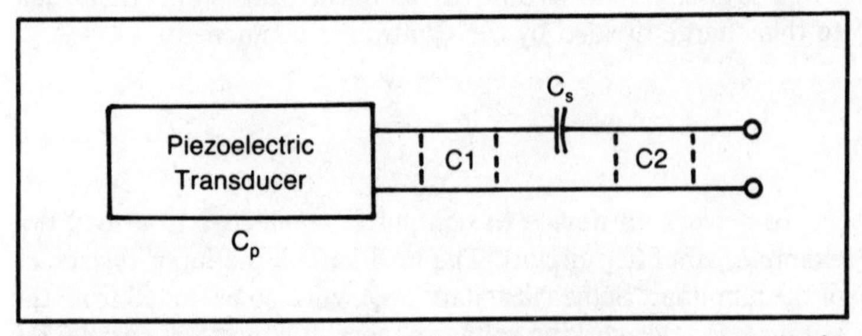

Fig. 6-3. Series capacitance affects the behavior of a transducer (see text for discussion).

$$Q^1 = Q \frac{C_s}{C_p + C1 + C_s}$$

where Q is the basic transducer charge sensitivity.

FREQUENCY RESPONSE

When a *changing* load is applied to a piezoelectric material (for example, when a piezoelectric transducer is being used as an accelerometer), it becomes a self-generating source of variable electrical signals. The significant parameter in this case is the product fRC:

where f is the frequency of (changing) load in Hz
R is the input impedance of the amplifier in ohms
C is the *total* capacitance in farads of the transistor plus additional shunt capacity (if any).

At values of fRC less than 1.0, transducer response falls off sharply (See Fig. 6-4). This can be a significant factor when using a piezoelectric transducer for low-frequency measurement.

Example. Suppose the frequency to be measured is 2 Hz. The total capacitance of the transducer and cables (C_p + C) is 500 pF, and the input impedance of the amplifier 100 M Ω. In this case

$$\begin{aligned} fRC &= 2 \times 500 \times 10^{-12} \times 100 \times 10^6 \\ &= 0.10 \end{aligned}$$

Note that a method of improving the relative response of a piezoelectric transducer at low frequencies is to increase the value of RC in the circuit by using additional shunt capacitance and/or larger cables. However, any gain in this respect will produce a loss of sensitivity.

At higher frequencies a different problem arises. With increasing frequency the relative response tends to rise *above* 100 percent. In order to maintain a linear (truu 100-percent) response it is necessary to limit the frequency to be received to about one-fifth of the resonant frequency of the transducer. For higher frequency, correction factors are involved.

LINEARITY

Linearity of response (above fRC = 1.0) can also be affected

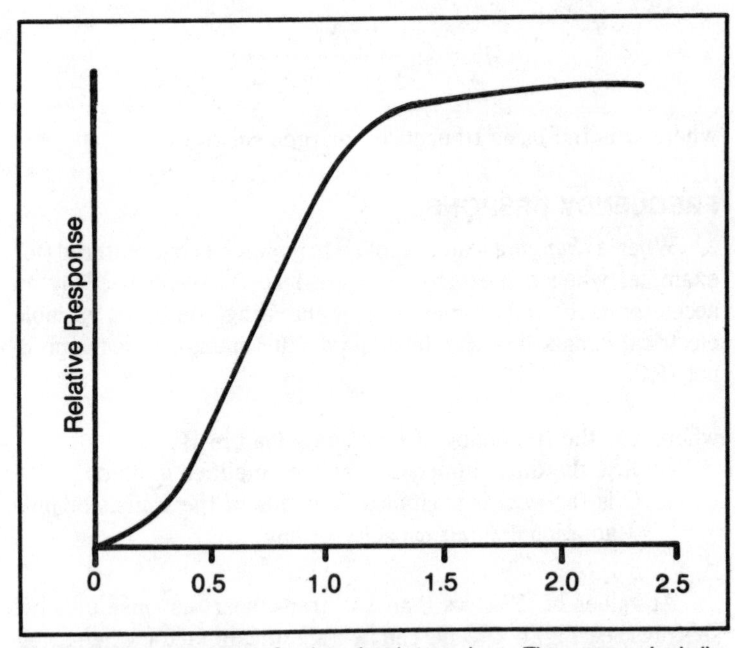

Fig. 6-4. Relative response of a piezoelectric transducer. The response is similar to that of a highpass filter with cutoff frequency f such that fRC = 1.0.

by physical effects resulting from the environment in which they are used. Such effects are normally observable well before the transducer itself suffers physical damage.

CROSS-AXIS SENSITIVITY

Because of its polarization, a piezoelectric transducer has maximum response along an input axis. Sensitivity to force exerted at an angle Θ to this axis is reduced in proportion to the cosine of that angle—that is, by $Q \cos \Theta$.

However, a more general expression for cross-axis sensitivity, related to the angle ϕ at which the load axis deviates from the polarized axis is

$$\frac{Qt}{Qxx} = (\tan \Theta \cos \phi) \times 100$$

This is given as a percentage, representing the loss of sensitivity or misalignment of applied force from the design axis, and is usually less than 5 percent.

USE IN MAGNETIC FIELDS

Magnetic fields or rf fields normally have no effect on the performance of piezoelectric transducers. At very low frequencies, however, the presence of rotating magnetic fields can cause spurious readings. In such cases magnetic shielding of the transducer, and also similar shielding of the amplifier, may be necessary.

SIGNAL TREATMENT

Piezoelectric transducers may be operational in either the voltage-sensory or charge-sensory modes. In the voltage-generating mode the following voltage amplifier may be a simple cathode follower with unity gain, or it may provide any degree of gain required. System sensitivity is reduced with added series capacitance, such as long cables.

Charge amplifiers sense the actual charge developed in the transistor and can operate at much lower input impedances than voltage amplifiers. Low-frequency response is not dependent on the RC constant and is determined only by the amplifier frequency-response characteristics. Thus, charge amplifiers eliminate problems of cable lengths affecting sensitivity as well as reducing the possibility of problems of spurious noise.

See also Chapter 19, Accelerometers.

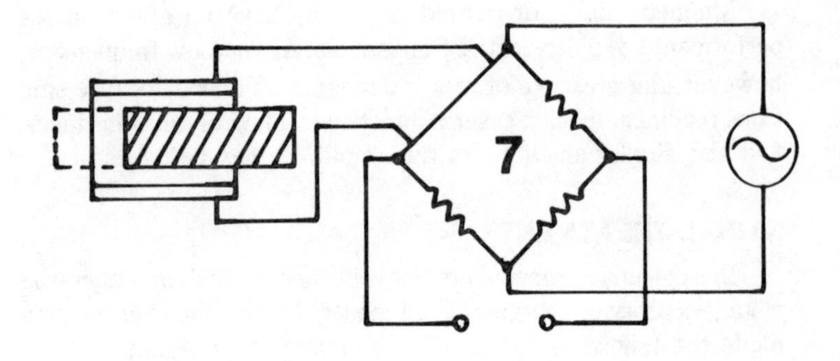

Piezoelectric Transducers

Both the nature and the magnitude of any piezoelectric effect in a crystalline substance are dependent on the direction of the applied force or electric field with respect to the crystallographic axes of the substance. Numerous crystalline materials do exhibit piezoelectric characteristics to some marked extent, but for all practical purposes choice is limited to some half dozen "natural" crystals and an even more limited range of polycrystalline ceramic materials that exhibit piezoelectric characteristics only after polarizing. The word "natural" is used in the sense that both piezoelectric substances were, and still are in certain cases, cut from the naturally occurring substances rather than a synthesized product of the same chemical formula. Quartz, for instance, is an outstanding example of a natural piezoelectric substance, and although the fabrication of piezoelectric plates and elements from natural quartz crystals still survives, synthetic quartz crystals may now be produced on a practical, controlled scale.

DISTORTION AND ELECTRICITY

Basically, the piezoelectric effect is an interrelationship between mechanical distortion and electrical effects peculiar to certain crystalline materials. If such materials are distorted by mechanical loading, they generate electricity; conversely, if charged with electricity, they undergo a dimensional change. The piezoelectric crystal, in other words, is a simple form of transducer for converting one form of energy into another, and vice versa.

PLATE ACTION

A general analysis of the performance of piezoelectric material is usually related to its behavior in plate form. With respect to the orientation of a specimen plate with respect to the axes of the crystal from which it is cut, emphasis may be given to one particular "plate" action. Plate action for various piezoelectric substances is given in Table 7-1.

SELECTION OF MATERIAL

Selection of the piezoelectric element best suited to any particular duty is only part of the overall problem. The ultimate performance depends also, to a considerable degree, on the method of mounting and driving, and to a lesser degree, on the service conditions (Table 7-2). It is particularly important that the method of anchorage and/or the driving mechanism used do not interfere with

Table 7-1. Plate Action of Typical Piezoelectric Materials.

Piezoelectrical material	Cut	Basic plate action	Notes
Rochelle salt	0° X 45° X 0° Y 45° Y	Face shear Length expansion Face shear Length expansion	Action strongest
Quartz	0° X	Thickness and length expansion	Only common cut
Ammonium dihydrogen phosphate	0° Z 45° Z	Face shear Length expansion	
Dipotassium tartrate	0° Z 45° Z	Thickness and face shear Length expansion	
Potassium dihydrogen phosphate	0° Z 45° Z	Face shear Length expansion	
Lithium sulphate	0° Y	Thickness and volume expansion	Only common cut

Table 7-2. Typical Applications of Plate Elements.

Basic plate action	Typical applications
Length expansion	Underwater sound transducers Laminated elements (bending action)— microphones, pickups
Thickness expansion	Ultrasonic transducers Underwater sound transducers
Face shear	Laminated elements (twisting action)— microphones and receivers
Volume expansion	Pressure gauges, hydrophones
Transverse expansion	Accelerometers, vibrators

the dimensional changes involved.

Limiting factors for piezoelectric elements are maximum service temperature and safe valves of relative humidity. Maximum service temperatures are given in Table 7-3. Problems with relative humidity are only present with a limited number of piezoelectric materials, notably Rochelle salt, where the safe range of use is 40-70 percent relative humidity, and dipotassium tartate, the behavior of which is affected by relative humidities around 70 percent. See Table 7-4.

CHARGE AND VOLTAGE MODES

Most modern types of piezoelectric transducers are available

Table 7-3. Maximum Service Temperatures.

Crystal	Temp°C.
Rochelle salt	45
Quartz	550
Aluminum dihydrogen phosphate	125
Dipotassium tartrate	100
Potassium dihydrogen phosphate	150
Lithium sulphate	75
Barium titanate	100

Table 7-4. Typical Piezoelectric Applications.

Piezoelectric material	Form(s) of element		Typical applications
Rochelle salt	Length expander Thickness expander Face shear	Plates	Underwater sound transducers Laminated elements for microphones, pickups, receivers, etc.
Quartz	Length expander Thickness expander	Plates	Underwater sound transducers Ultrasonic generators
Tourmaline	Thickness expander Volume expanders	Plates	Limited application pressure gauges
Barium titanate (polarized)	Thickness expander Length expander Volume expanders	Plates	Underwater sound transducers Ultrasonic generators
Ammonium dihydrogen phosphate	Length expander Thickness expander Face shear	Plates	Alternative to Rochelle salt
Dipotassium tartrate	Length expander Thickness expander Face shear	Plates	Frequency controls, radio filters, etc. (mainly)
Potassium dihydrogen phosphate	Length expander Length face shear	Plates	Fairly low piezoelectric effect
Lithium sulphate	Thickness expander Volume expanders	Plates	Largely underwater sound use Pressure gauges.

in either charge-mode or voltage-mode versions. Charge-mode transducers require external charge amplifiers and low-noise cable connections. Voltage-mode transducers may incorporate built-in microelectronic amplifiers capable of being coupled directly to the readout instrument or analyzer via ordinary coaxial cable.

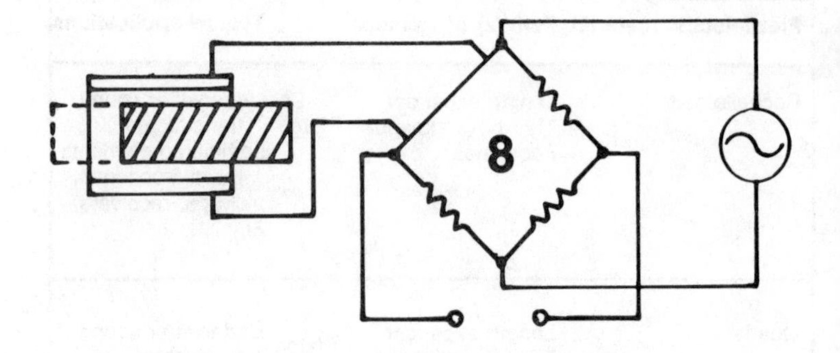

Load Cells

Load cells are transducers capable of measuring tensional or compressive loads or forces (and in some cases shear or transverse loads) in terms of some other quantity (usually an electrical signal). The main application of load cells is in weight measurement by using pressure transducers or, for measurement with slung weight systems, tension transducers.

For weight measurement the three basic types of transducers are

☐ *Hydraulic*—converts force into pressure, which can be indicated directly by a pressure gauge calibrated in units of force or weight. Such a system is relatively inexpensive, robust, flexible, and capable of accommodating high forces. It is temperature and pressure dependent because both parameters can affect the compressibility of the fluid.

☐ *Pneumatic*—converts force to air pressure either in a closed system or an open (continuous-flow) system. Pressure generated is again a measure of the applied force, but in this case the fluid (air) is highly compressible. Accuracy and consistency of a pneumatic load cell system thus tends to be inferior to a hydraulic system, although it may be eminently suitable as a control system monitoring a pneumatic controller. Generally, the working principle adopted is similar to that of a pneumatic relay.

☐ *Electrical*—normally either based on piezoelectric

transducers or on resistance strain gauges. In such systems the force is converted to a proportional electrical signal in a bridge circuit that can be fed, via an amplifier if necessary, to an indicating meter calibrated directly in terms of weight, a recorder, digital readout, or a similar device.

ELECTRONIC WEIGHING

Electronic weighing is rapidly becoming an alternative to mechanical weighing, especially in industrial systems. In small-scale applications involving low weight capacities (such as domestic weighing machines), it is inevitably more costly, but it can have specific advantages. For example, because weight is increased in terms of an electrical signal (conversion to power), a digital display is relatively simple. Also, because of the high sensitivity of piezoresistive strain gauges, load cells can be designed to measure quite small weights (under 1 lb) with excellent accuracy.

On the industrial side the cost of mechanical weighing increases substantially with increasing capacity; that of an electrical load cell, only marginally so.

There will be a crossover point at which electronic weighing becomes less costly than mechanical weighing; this generally occurs between 500 to 1000 lb, depending upon the particular application (Fig. 8-1).

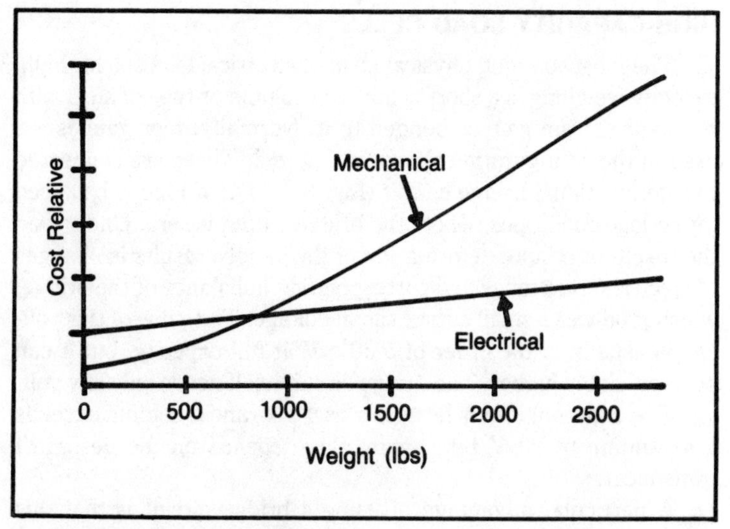

Fig. 8-1. Electronic weighing is more cost effective than mechanical weighing when the mass is very large.

The main advantages of electronic load cell weighing are

- □ The output is in the form of an electrical signal; this offers far greater flexibility than mechanical signals or pointer movements.
- □ Electrical load cells have no moving parts and can be hermetically sealed. Such cells are compact, robust, and normally need little or no maintenance.
- □ It is adaptable to situations or environments where mechanical weighing devices are impractical or impossible to use.
- □ The size of the load cell is largely independent of its load capacity. Larger and "stronger" elements are not needed to accommodate very large weights as in mechanical systems.
- □ Response is much faster than with mechanical systems; this enables it to be used under dynamic conditions, such as fluctuating loads, if necessary.
- □ Accuracy is at least comparable with mechanical weighing and may be better in particular circumstances. It does not follow that electronic weighing automatically gives greater accuracy, however. Under particular circumstances requiring high accuracy of measurement, mechanical weighers may achieve the accuracy required at lower cost. However, on a cost basis, electronic weighing is generally more favorable as the weighing capacity required increases.

HIGH-CAPACITY LOAD CELLS

The most common physical form of electrical load cell for high-capacity weighing is a short cylindrical column or tube of steel with resistance strain gauges bonded to it. Normally, four gauges are used in the configuration shown in Fig. 8-2. These are connected to a conventional bridge circuit (Fig. 8-3). The bridge is balanced for no-load conditions, where the bridge output is zero. Under load the resultant minute deformation of the gauges results in changes of electrical resistance and corresponding unbalance of the bridge, which produces a small output signal voltage. The value of this voltage is usually of the order of 5-30 mV at full capacity, but it can be appreciably higher. The bridge circuit itself needs only low voltage dc excitation; it may be as low as 4-5 V, and it seldom exceeds a maximum of 20 V but, again, this depends on the design of transducer.

A particular advantage of using a bridge circuit is that this renders the system inherently insensitive to temperature. Any

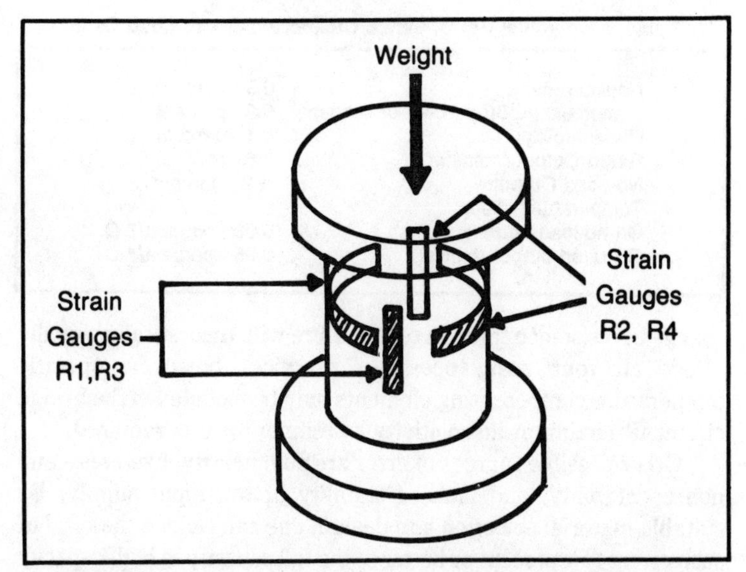

Fig. 8-2. Arrangement of strain gauges in an electrical load cell. The horizontally oriented gauges are designated R1 and R2; the vertical gauges, R3 and R4.

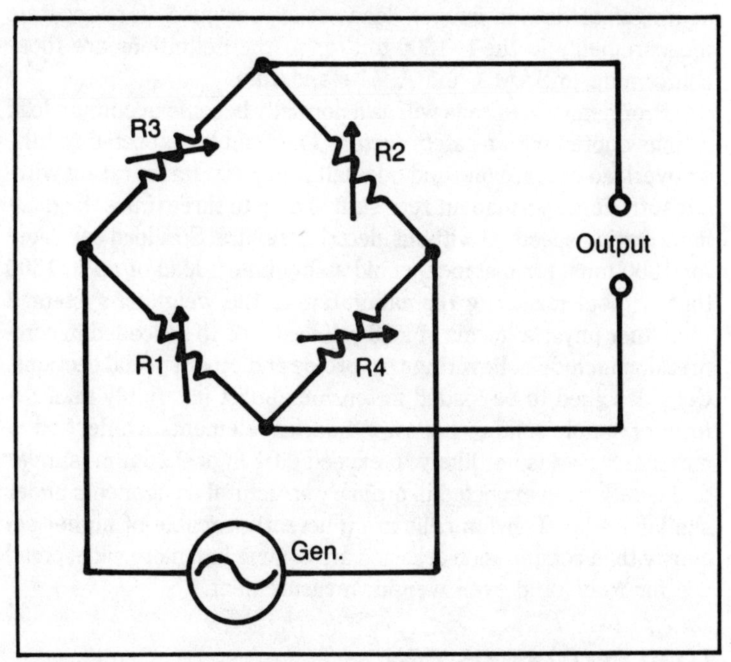

Fig. 8-3. Bridge circuit for interconnection of strain gauges shown in Fig. 8-2.

Table 8-1. Typical Performance Characteristics of Load Cells.

Nonlinearity:	0.50 percent
Hysteresis (at 50 percent of range):	0.2 percent
Repeatability:	0.1 percent
Rated Output (nominal):	1.6 mV/V
No-Load Output:	±2 percent
Temperature effect:	
On no-load output:	0.05 percent/° C.
On rated output (typical):	0.05 percent/° C.

change in resistance through temperature will, theoretically, equally affect all four resistances. In practice, however, specific temperature-compensating elements may be included in the bridge circuit if maximum insensitivity to temperature is required.

Other possible sources of error are nonlinearity, hysteresis and nonrepeatability, and lack of stability of the input supply. By suitable material selection and design, one can reduce the total of such errors to 1 percent or better of the full capacity in high-capacity cells; for lower-capacity cells it can be reduced even more.

Table 8-1 shows typical performance characteristics quoted for a range of (strain-gauge) load cells designed for accurate measurements in the 1–1000-ton range (the definitions are those conforming to SAMA and ASM standards).

Proprietary load cells will also normally become maximum load ratings quoted with a safety factor. They can be expected to take an overload of up to one and one-half times maximum rating without suffering a permanent zero shift and up to three times the maximum rating specified without electrical failure. So a load cell rated for 1000 tons, for instance, could safely take a load of up to 1500 tons without upsetting the calibration of the weighing system.

Other physical forms of load cell designed to be loaded in compression include hollow rings or toroids and other special sections. Cells designed to be loaded in tension almost invariably take the form of simple solid or hollow cylindrical elements. Deflection at maximum load is not likely to exceed 0.01 in or 0.25 mm (similar to the deflection expected in ordinary structural components under similar loads). Tension cells are inherently capable of higher accuracy than compression cells and are coming into more widespread use for force, and even weight, measurement.

LOAD CELL DISPOSITION

The simplest form of readout is to take the bridge circuit dc output (usually on the order of millivolts), amplify this, and then feed it directly to a moving-coil voltmeter. The voltmeter scale is then calibrated to indicate weight directly.

This is accurate enough for most general-purpose applications, but generally far less accurate than the performance of the load cell itself. To provide better accuracy, scientists developed a servo-potentiometric measurement system, using a self-balancing potentiometric indicator or recorder. In turn, this has now been replaced by its solid-state equivalent.

A typical schematic circuit is shown in Fig. 8-4. The principle of a servo-potentiometric indicator is that a voltage from a potentiometer is continuously compared with the input (that is, the voltage from the load cells). The potentiometer slider is driven by a small servomotor and fed via an amplifier from the "difference voltage" until the potentiometer voltage equals the input. At this point, called the *balance*, the position of the slider is an accurate measure of the unknown input voltage and is indicated by the pointer, which is coupled to it.

Voltage variations in mains are rendered harmless by the simple expedient of supplying the potentiometer from the same source as the load cells, so that the effects of fluctuations on the two cancel out.

The potentiometric indicator can be given dials of almost any size and can be made accurate enough to satisfy most requirements. When it is used with load cells, the power supply for these, the sum-

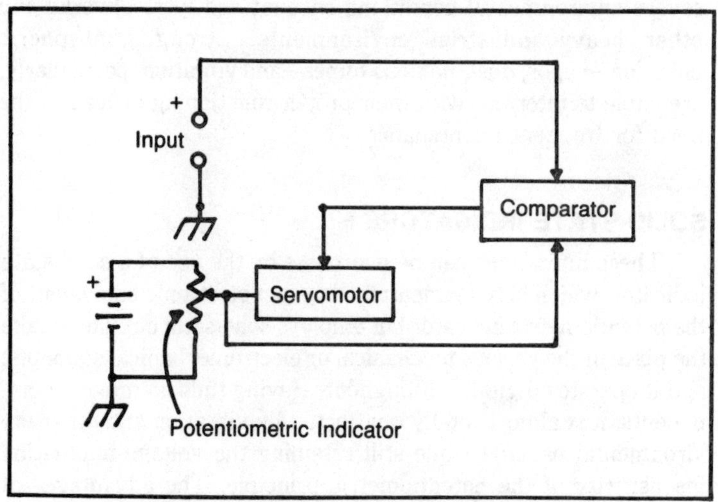

Fig. 8-4. A servo-potentiometric circuit for use with a load cell.

mating networks where several cells are used and the necessary calibrating circuits are always built into it.

The potentiometric indicator can be given multiple scales and measuring ranges, and microswitches, actuated by cams attached to the pointer spindle to make or break at any point of the scale, can be fitted into it, for initiation of a variety of alarm control or supervisory functions.

As an alternative to the circular-dial potentiometric indicator, a pen recorder, which marks a chart that may be either circular or in the form of a strip, may be used.

Where an indicator is used, often because it is usually more readable and accurate, a record may be obtained by another means. An encoding disc is coupled to the pointer and translates its angular position into a corresponding digital signal. This is then further processed in special electronic decoding circuits until it is in a form where it can actuate electrical printing (or adding) machines and also numerical readouts for digital conversion. The numerical readout has advantages over an analog pointer, because it is not subject to an operator's reading or interpretation error, and it can also be read at a greater distance.

Although potentiometric indicators and recorders are very satisfactory in many applications and have been developed to a state of high reliability and variety, they have one weakness. They are electromechanical devices and rely on moving components, such as potentiometers, drives, or motors, which may be affected by severe environmental conditions encountered in steelworks and other heavy industrial environments. Strong atmospheric pollution—vapor, dust, noxious fumes—and vibration, particularly, are liable to interfere with their proper functioning or lead to the need for frequent maintenance.

SOLID-STATE INDICATORS

These limitations can be overcome by the use of a solid-state indicator, which is fundamentally an exact electronic equivalent of the potentiometric indicator but employs solid-state circuits to take the place of the various mechanical or electromechanical elements, and it operates digitally throughout. Having thus no moving components, it is almost totally unaffected by vibration and other environmental hazards while still retaining the voltage-fluctuation insensitivity of the potentiometric principle. The advantages of digital output (which include its own built-in numerical weight display) are available without the need for costly conversion equipment.

HYDRAULIC LOAD CELLS

Although they are only of industrial interest a brief description will also be given of a modern electrohydraulic load-measuring system.

The load cell in this case is in the form of a shallow cylinder having its piston or platen bonded to the cylinder wall by a flexible rubber joint. Under the platen the space, which is circular in a compression capsule and annular in a tension capsule, is filled with a fluid such as water, which is incompressible for practical purposes if all the air is extracted.

The load cell is sensitive down to zero loadings. Under load the rubber joint is stressed in shear; it not only acts as a frictionless seal but also helps to constrain the platen to axial motion.

Also because of its construction, the perfect electrohydraulic load cell would be completely free from creep, because no metal parts are stressed to produce the signal. In practice, some small effect, typically 0.005 percent/5 mins., is measurable.

NONAXIAL LOADS

The electrohydraulic load cell is essentially free from errors introduced by nonaxial loading. However, the main advantage (compared to strain-gauge load cells) occurs when these nonaxial loads are transmitted. The electrohydraulic capsule, however, does not generate any signal but that due to true axial force.

A pressure sensor is sealed onto the body of the load cell. The sensor converts fluid pressure into an electrical output.

Various options on the performance of the sensor are available, together with a comprehensive range of recording methods, from simple and precise analog indication in pounds or kilograms to systems employing four-channel switching, summation, high/low alarm trips, peak hold, 0-10 mA, and BCD output.

Splash-proof enclosures housing the transmitter are available with the above facilities. Other systems, such as 16-channel signal conditioning units with auto- or manual scan employing printed output, are offered.

Differential and portable battery digital indicators are also available for use with modern hydraulic load cells.

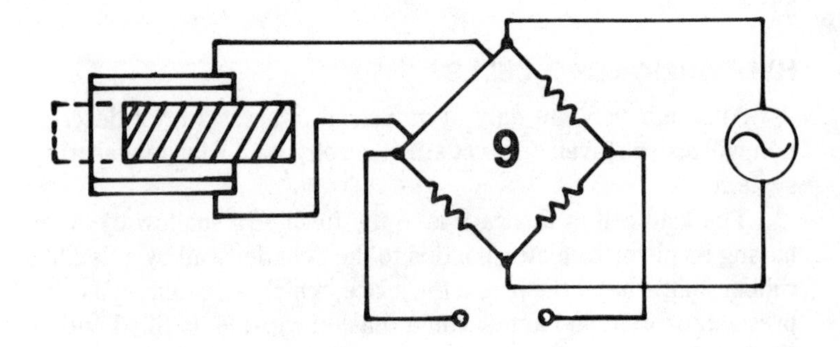

Magnetic and Inductive Transducers

The most elementary example of a magnetic transducer is the *magnetic compass*, which utilizes the force present in the earth's magnetic field to align a suspended or pivoted magnet (pointer) in a North-South direction. In other words, it uses magnetic force to produce mechanical movement in any misaligned position of the pointer.

Arguably, though, this is not true transducer action but a separate phenomenon. Nevertheless, magnetic devices working on magnetic field principles are widely used as true transducers; they are also used in other applications to produce controlled *movement* (whether or not you regard this as transducer action).

A simple example of the latter is the *magnetic follower*. If two bar magnets are mounted on separate spindles and placed close together, they will align themselves with opposite poles facing each other (Fig. 9-1). If one magnet is now rotated, the other will follow it. The two rotating systems are magnetically coupled: one being the driving system; the other, the driven system. If the driven system is braked by a load (such as appreciable friction), the driven magnet will lag behind the driving magnet by an amount proportional to the braking load. If this load is too high, the coupling effect will be overcome. The driven system will stall. Reducing the load will make it pick up again.

Magnetic drives of this type have numerous applications and can even work through a separating partition of nonmetallic mate-

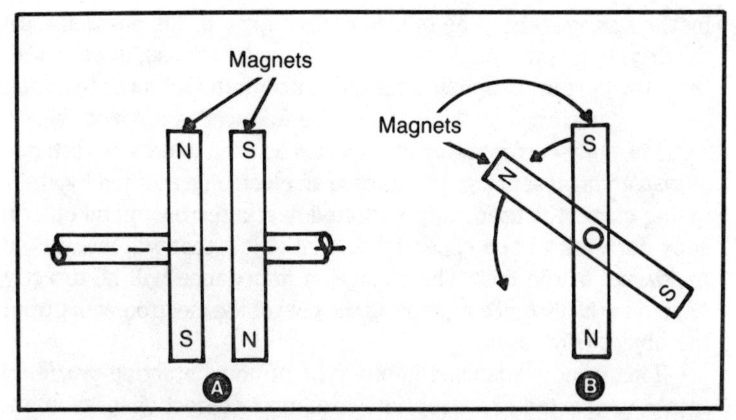

Fig. 9-1. Two magnets, placed adjacent to each other, will align with the opposite poles side by side (A). If one magnet is then rotated, the other will follow along (B).

rial. If the driving magnet is faced with a metallic material, however, its rotation will induce eddy currents in the metal object. If this object is in the form of a disk mounted on a spindle, these eddy currents will effectively couple the two systems, causing the second to rotate with the first (Fig. 9-2). This is *eddy-current coupling.*

This principle has been widely used in mechanical speedometer drives where the magnet is driven at "wheel" speed, the follower being an aluminum disc (or, more usually, cup stated) mounted on a spindle and lightly spring loaded to resist rotation.

The coupling thus pulls the follower (disc or cup) into sympathetic rotation until the eddy-current coupling force is overcome

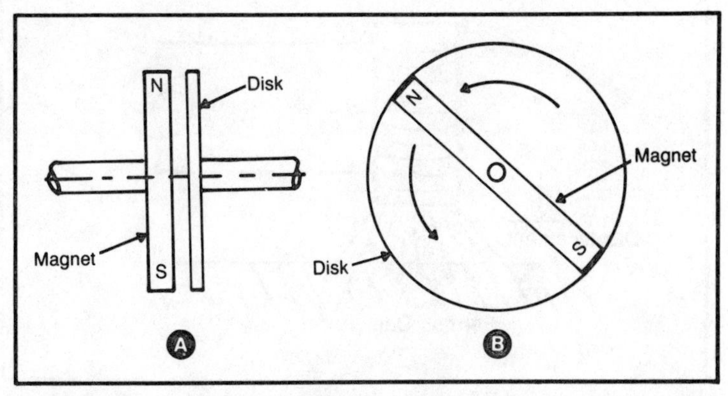

Fig. 9-2. Side view (A) and broadside view (B) of an eddy-current-coupled magnetic system.

by the spring force. The follower then stays in this position until the driving magnet speed changes. This will increase, or decrease, the eddy-current coupling strength, causing the follower to adjust to a new position. In this respect the follower is a speed *sensor.*

The eddy-current principle can also be used to provide distance or *displacement* sensing. In this case an electromagnet fed by alternating current is used. If it is located near a ferrous metal object, eddy currents will be generated in that object that will modify the *impedance* of the coil. The change in impedance will be directly related to the distance separating the pole of the electromagnet from the object (Fig. 9-3).

This principle is used in one type of noncontacting *proximity sensor*, connected to appropriate circuitry to detect changes in impedance in the coil and to display this in terms of displacement (distance) from the object. Other types of noncontacting proximity sensors work on capacitance effect or interruption of fluid (usually air) flow.

The use of magnets (especially permanent magnets) as sensors/transducers is attractive because they are simple devices. An example of a magnetic flowmeter is given in Fig. 9-4. Here two permanent magnets are mounted in the center of a pipe carrying fluid flow (A). Between them is a float containing a further magnet. With the magnet polarities shown in the diagram and no flow, the intermediate magnet will be suspended in the *null* field (midway between the two fixed magnets if they are of equal strength).

When there is fluid flowing through the pipe, the pressure on

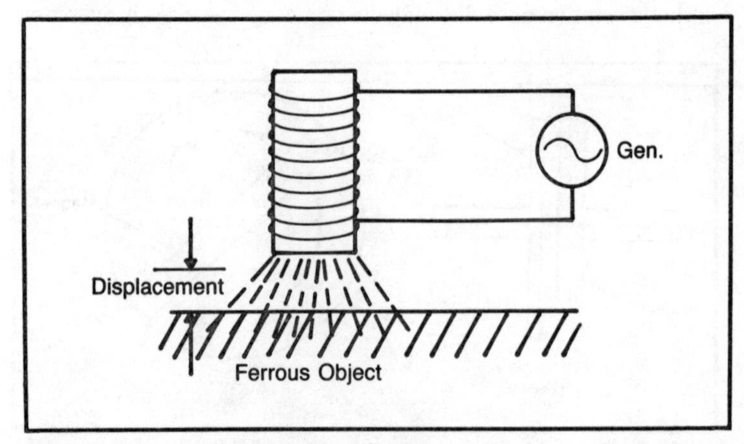

Fig. 9-3. When a coil supplied with alternating current is brought near a ferrous object, the coil impedance is altered.

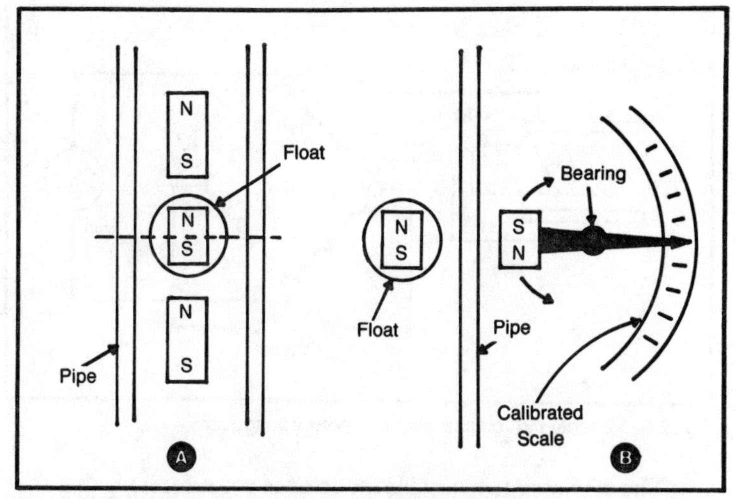

Fig. 9-4. A magnetic flowmeter. At A, condition of no fluid flow; at B, method of measuring fluid flow by means of magnetic coupling.

the float will lift it upwards away from the null field or static position by an amount proportional to the velocity of the flow. The actual displacement of the float will thus be a measure of the flow velocity.

To measure this movement, you can use a follower magnet mounted outside the pipe (which must be nonmetallic in this case) to move a pointer over a scale, as shown in Fig. 9-4B. This principle is used in certain proprietary flowmeters.

The majority of true transducers working in electromagnetic principles employ variable *induction* to generate the response required: for example, to render physical displacement in terms of a corresponding electrical signal. Main types are the *variable-inductance transducers*, the *differential transformer* (described separately in Chapter 10), and certain digital transducers (see Chapter 11).

The variable-inductance transducer is based on an ac coil with a movable iron core, armature, or diaphragm (Fig. 9-5). Movement of this core (armature or diaphragm) varies the inductance of the coil, converting motion into an electrical signal. If the transducer is used as one element of a bridge, and if the bridge is excited by a fixed-frequency, constant-amplitude signal, any change in inductance in the transducer will upset the balance of the bridge, providing an ac signal output proportional to the amount of displacement of the core.

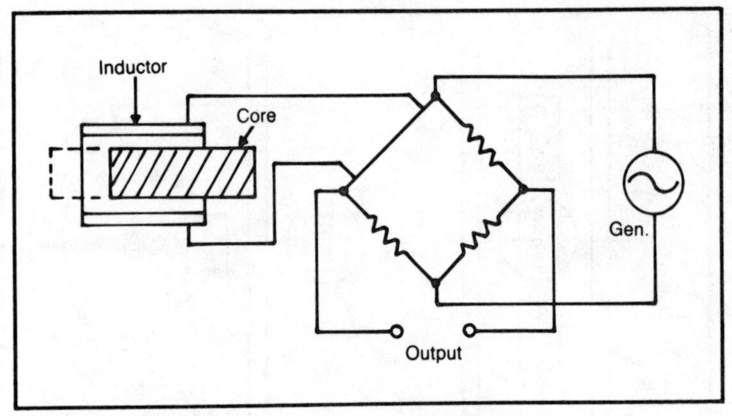

Fig. 9-5. Variable-inductance displacement transducer.

The advantage of variable inductance transducers is that they usually exhibit a high signal-to-noise ratio and provide continuous resolution. They do, however, require ac excitation, and they can provide spurious readings if mounted close to magnetic objects.

HALL-EFFECT SENSORS

When a conductor carrying a current is located in a magnetic field, a difference in potential is generated between the opposed edges of the conductor in a direction perpendicular to both the current and the magnetic field. This is known as *Hall effect,* and the potential difference generated is known as the *Hall voltage*. This principle is used in magnetically operated *position* sensors and *current* sensors. Hall-effect sensors, operated by a magnet in a plunger, are also used in mechanically operated solid-state switches.

Hall-effect sensors are used in a variety of instruments and equipment, including musical instruments, computer peripherals, home appliances, medical instruments, telephone equipment, office machines, farm machinery, copy machines, laboratory equipment, vending machines, and electronic games.

Other applications of Hall effect include cam, lever, or shaft positioning; cylinder positioning; length measurement; linear or rotary motion detection; sorting; current sensing; ignition timing; limit sensing; potentiometer and tachometer sensing.

A typical Hall-effect sensor looks like an electrical component with three or four ends. It may be unipolar or bipolar, the latter having a plus (South pole) maximum operating point and a minus (North pole) minimum release point. It may also incorporate a flux concentrator, concentrating flux in the sensory area.

Three-lead devices provide one output, and four-lead devices two outputs: that is, one output increasing with an increase in gauss, and one output decreasing with an increase in gauss (differential outputs). Positive gauss separates the South pole of the magnet facing the sensory area. Negative gauss represents the North pole of the magnet facing the sensory area. Leads may be identified by numbers; for example,

1 = − supply
2 = output
3 = + supply

or, for differential outputs,

1 = − supply
2 = 02
3 = 01
4 = + supply

or they may be color coded, such as

red = + supply
black = − supply (ground)
grey = output

A typical circuit in which a Hall-effect sensor is used is shown in Fig. 9-6, where the supply voltage is about 4-18 V, depending on the construction of the individual device.

To use Hall-effect sensors, you must know their characteristics in order to know their *operating* point and *release* point over their intended working range (also the supply voltage required). Thus, for a Honeywell 513SS16, for example, the maximum operating point is specified as 330 gauss and the release point as 85-305 gauss over a working temperature range of −40 to 100° C.

In this example, to ensure reliable operation, at least 330 gauss must be presented to the sensor. The gauss level must then be reduced to less than 85 gauss to guarantee that the sensor will release.

Therefore, when selecting a Hall-effect sensor, you must know the flux density (in gauss) measured at the chip to be able to select a device with the best characteristics for a particular application

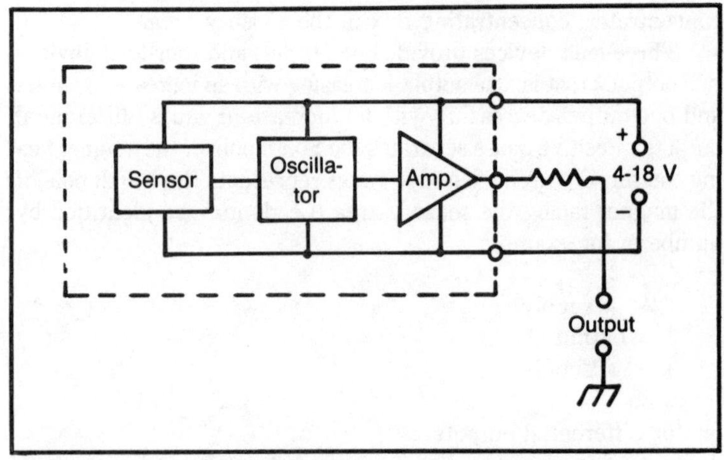

Fig. 9-6. A circuit for detecting position, using a Hall-effect sensor.

and to select the best magnet to use with it.

☐ *Head-on*—The target is centered over the point of maximum sensitivity and is moved "head-on" to the sensor, then backed off.

☐ *Slide-by*—The target is moved across the face of the sensor at a specified distance.

☐ *Rotary*—A rotating target, such as a ring magnet, provides an alternating pattern of on-off actuation.

☐ *Vane*—The target, a ferrous metal vane, slides through (or rotates through) the gap between magnet and sensor.

OUTPUT CIRCUITS

Two types of output circuits are common to solid-state circuitry: current sinking and current sourcing. The names are derived from the location of the load in the output circuit.

☐ *Current sourcing* (open emitter): The load is connected between the output and ground. The load is isolated from the supply voltage when the sensor is off. When the sensor turns on, current flows from the power supply, through the sensor, and into the load. The sensor supplies a "source" of power to the load. A current-sourcing output is normally low, but it goes high when the sensor is on.

☐ *Current sinking* (open collector): The load is connected between the power supply and the sensor. In this circuit the load is isolated from ground when the sensor is off. When the sensor turns on, the circuit is complete, and current flows from the supply, through the load, through the sensor, and then to ground. The sensor switches or "sinks" the output current to ground. A current-sinking output is normally high, but it goes low when the sensor is on.

MAGNETIC VELOCITY TRANSDUCERS

Magnetic/inductive-type transducers can be used to measure *linear velocity*. The usual form of such a transducer is two series-connected coils wound over a long, thin tube of nonmagnetic material. A free-floating permanent magnet lightly fitting the core of the tube is used as an armature. Motion of the armature in either direction produces a signal directly proportional to velocity as the magnet traverses the length of the coils. This signal is self-generated (that is, no external electrical supply is required) and can be read off by a standard high-impedance voltmeter having sufficiently high speed of response.

To design a simple project find a suitable cylindrical magnet of small diameter and reasonably long length. Look for an Alnico 5 magnet of suitable proportions. Then find a rigid plastic tube that is just large enough to let the magnet slide easily through it. Alternatively, wind a tube from gumstrip or similar material around a suitable size of mandrel (e.g. doweling). Make the tube length about 1.5 times the magnet length. Wind the two coils about each side of the midpoint of the tube, using an equal number of turns for each coil. Connect the two coils in series (end of coil 1 to beginning of coil 2). Connect the beginning of coil 1 and the end of coil 2 to a high-impedance voltmeter.

The question of providing a suitable "drive" for the armature depends on the application; that is, the armature must be linked to the motion involved in such a manner that it gives the armature a linear motion in or through the coil tube. Also, you will need to first measure a known linear velocity in order to calibrate the voltmeter readings in terms of velocity. This relationship will be linear over only a proportion of the length of the tube. (See Chapter 10 for more information on this particular subject.)

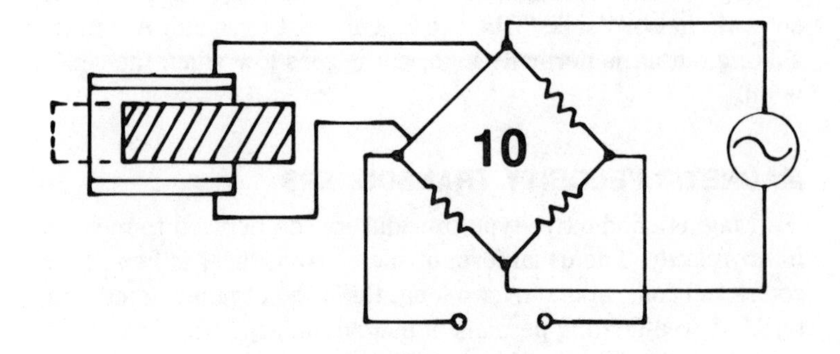

The Differential Transformer

The basic form of a *differential transformer* is three coils wound on a common bobbin, where the center coil is the primary, and the two end coils are the secondaries. Aligned in the center of the bobbin is an iron core that can be displaced linearily (Fig. 10-1). The position of this iron core relative to the primary and secondary coils controls the flux linkage into the secondaries and, hence, the voltages induced in them.

Connecting the secondary coils differentially, that is, start-to start, as in Fig. 10-2, the induced voltages in secondaries S1 and S2 with the core in the central position will be equal in plane and magnitude and cancel each other out. There will thus be no voltage appearing across the output terminals.

If the core is now displaced, the voltage induced is one secondary will increase, and that induced in the other secondary will decrease. The result will be a difference signal, the output of which is in linear proportion to the core movement.

In fact, it will be linear only over a proportion of the core movement. Output voltage will fall away as the core approaches the end(s) of the coil system due to reducing magnetic field at these points (Fig. 10-3). When used as a practical measuring device to give accurate linear response, a long coil assembly is needed to provide a relatively short linear range. This may be quite acceptable in many applications, and it represents the basic design of a linear variable displacement transformer (LVDT) for general use, measur-

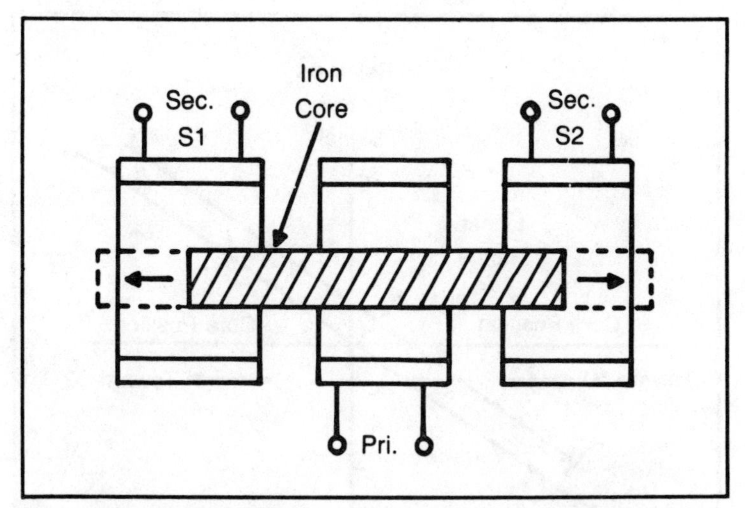

Fig. 10-1. A differential transformer. The core is movable. This is a linear device (LVDT).

ing smaller displacements (short strokes). To achieve linear output with longer strokes, you can use various modified forms of coil windings, although they are inevitably more expensive to produce. These are:

Balanced Linear-Tapered Secondaries (Fig. 10-4). In practice, a complicated form of winding where the secondary coils are wound over the primary and tapered in section from center to end. Provided the two secondary coils are perfectly symmetrical,

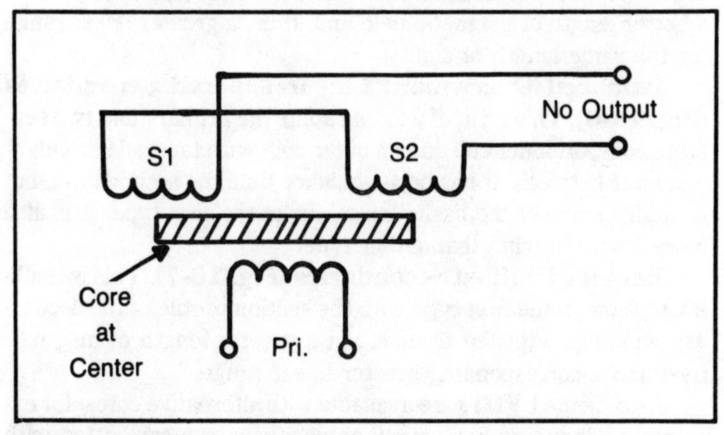

Fig. 10-2. When the core is centered and the secondaries connected as shown here, the output is zero.

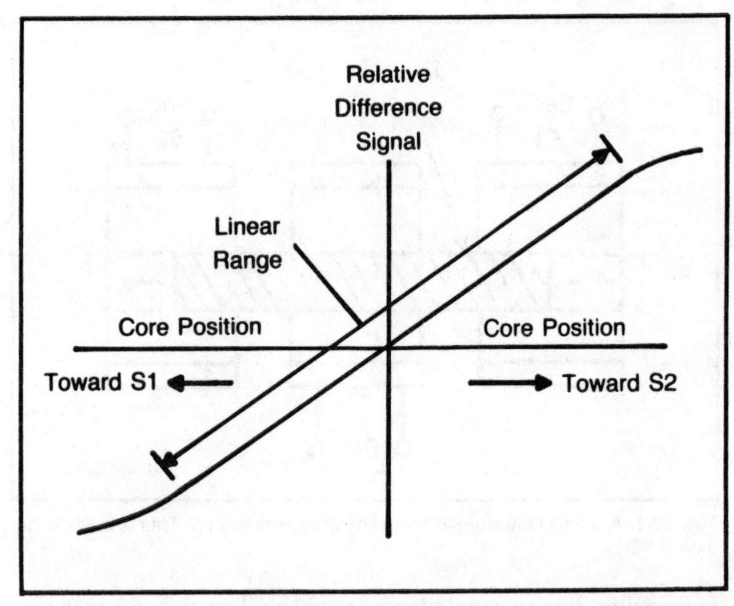

Fig. 10-3. The variation in output from a differential transformer is linear only within a certain core-displacement range.

good balance is achieved, but the magnetic field is still not maintained to the ends of the coils. Thus, linearity of output is not maintained over the whole of the possible stroke.

Overwound Linear-Tapered Secondaries (Fig. 10-5). This is an even more complicated form of winding to reproduce in practice, and it can exhibit unbalance. However, it does provide a better length of magnetic field and, thus, a greater linear range for the same length of coil.

Balanced Overwound Linear-Tapered Secondaries (Fig. 10-6). This form of winding splits the primary into two separate coils, an inner coil and an outer coil, with tapered secondary windings between. It has better balance than the second type, but it retains some of the basic limitations of the first type. It is also more costly to wind than either type.

Balance Profiled Secondaries (Fig. 10-7). This is really a variation on the first type with the section profile of the secondary windings adjusted to maintain a greater length of magnetic field and a corresponding greater linear range.

Note. Some LVDTs are available with alternative cores: for example, a plain core for general applications or a core fitted with low-friction sleeves (usually PTFE rings). The latter eliminate core skewing and would be the preferred choice where alignment is difficult or vibration is present.

INSTRUMENTATION

The linear variable-displacement transformer itself is only a signal generator requiring interfacing to an electrical circuit to provide indication or readout in analog or digital form. Such associated instrumentation can be relatively expensive. Because it is a transformer, ac excitation is essential. If a dc signal is required, ac output must be converted to dc when demodulation of the transformer output is required.

Two basic methods of achieving this are:

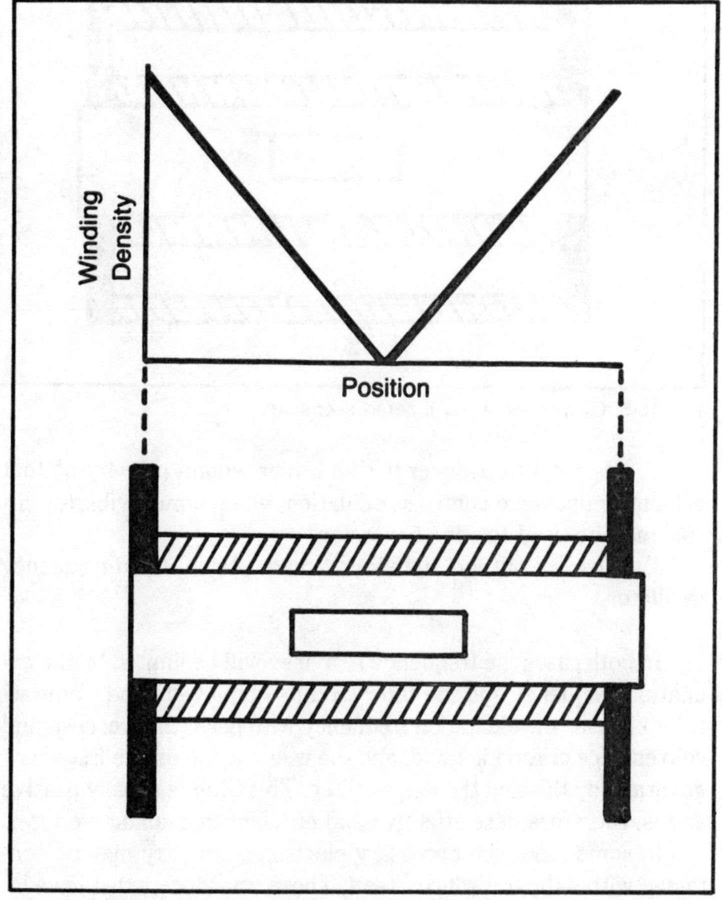

Fig. 10-4. Balanced linear-tapered secondary.

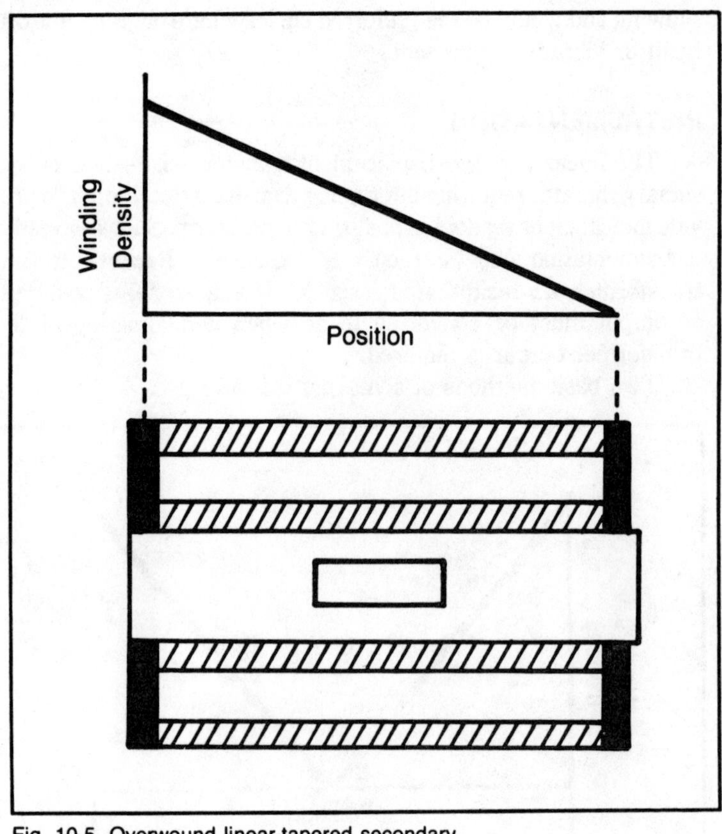

Fig. 10-5. Overwound linear-tapered secondary.

☐ Use of a transducer with a bifilar wound primary so that primary inductance controls oscillation, when a multivibrator circuit may be used for dc conversion.

☐ The ac output signal is "read" by a fixed-frequency oscillator.

In both cases the frequency response will be limited by the excitation frequency and the filter circuits employed. The manufacturer chooses the excitation frequency with performance, cost, and convenience criteria in mind, and the user has the usable frequency governed by this and the output filter. This filter is usually passive and is, therefore, less effective and efficient than an active filter.

In some cases the necessary electronic circuitry may be contained within the transducer itself. Those transducers that have integral dc input-dc output electronics are, in the main, available for excitation from 10-, 12-, and 24-V stabilized supplies. The use of electronics limits the temperature range; −20° C. to +100° C. is typical, with thermal coefficients of 0.04 percent to 0.01 percent of full range per degree Celsius for both zero and span.

The dc transducer with integral electronics contains a considerably larger number of components than its ac counterpart. Cost is also relatively high, but it is usually less than a full ac system.

In the case of ac transducers, excitation can be at any voltage level up to 26 V (rms), depending only upon the impedances of the windings, at frequencies from 50 Hz to 20 kHz. The operable temperature range is much wider than the dc type, being typically −55° C. to +150° C., but the temperature effects are difficult to define. For any particular transducer the thermal coefficients tend

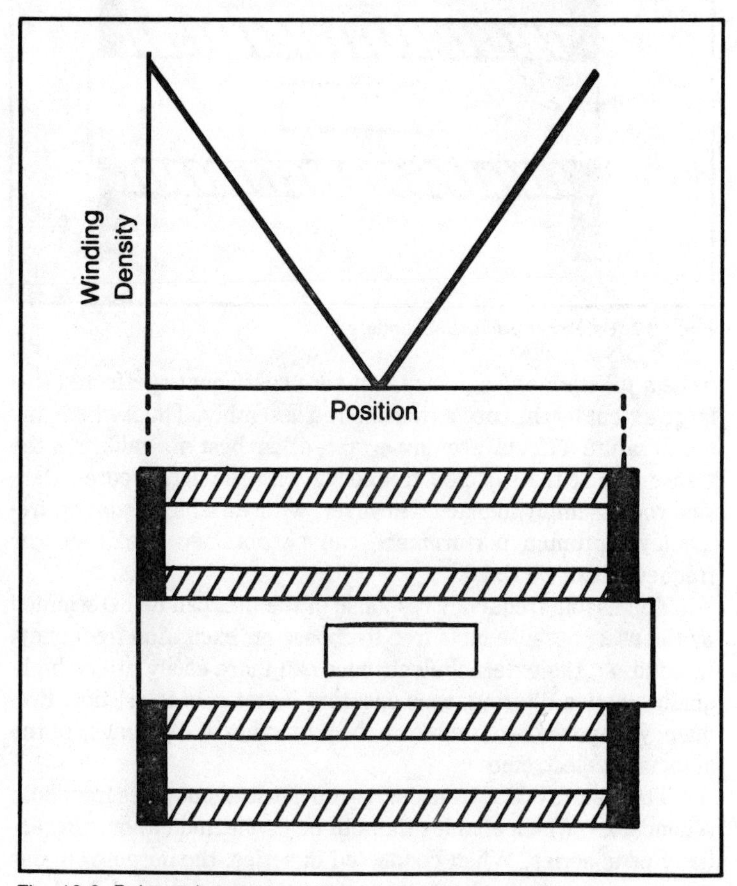

Fig. 10-6. Balanced overwound linear-tapered secondary.

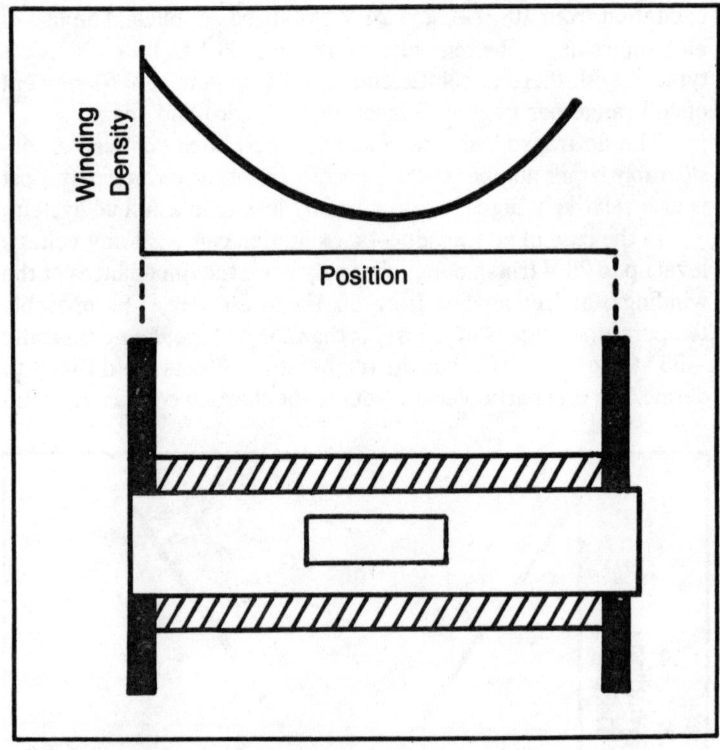

Fig. 10-7. Balanced profiled secondary.

to be a function of frequency, but zero coefficient is effected to a large extent by the core/extension rod assembly. Thus, where any temperature effects are important, it is best to calibrate the transducer at the selected frequency with the actual core/extension rod assembly in place. However, with variable excitation frequency, optimum performance can be obtained, and then the frequency can be fixed.

The usable frequency response of the unit can be determined by the user, because he is free to choose his excitation frequency. In addition, the external electronics can more easily utilize high-quality, active filtering, thus ensuring better characteristics. Frequency response can, therefore, be a function of the quality of the associated electronics.

The majority of ac transducers provide lead-out wires from both secondaries, which enables them to be connected either differentially or in series. When connected in series, the output remains constant, regardless of core position within the designed displacement range. This can be used as a self-test facility. Many users are unaware of this interesting facility, and it is most useful in applications where the integrity of the transducer is essential, and in some systems the connections are automatically switched to check the prime signal source. If, for example, the insulation were to break down on any turn in either of the secondaries or the primary, the output would change the value.

LVDT GAUGE HEADS

A typical LVDT gauge head consists of an LVDT whose core is connected to a spring-loaded probe shaft having a removable tip. The probe shaft is guided in a sleeve bearing that is retained in a case that also encloses the LVDT coil windings. The case is often threaded externally to simplify mounting.

There are several possible ways to classify LVDT gauge heads, but the most meaningful classification is according to performance capabilities. This leads to three general categories of LVDT gauge heads: economy, precision, and ultraprecision. Considerations of bearing precision and linearity are the principal variables from one class to another.

ECONOMY GAUGE HEADS

The *economy* classification is applied to gauge heads that are designed to give a reasonably good level of performance at moderate cost. Economy gauge heads incorporated a lower-cost LVDT having a linearity of better than 0.5 percent of full range. The shaft is loosely fitted in a sleeve bearing, typically of nylon, and is free to rotate. Shaft loading is accomplished by an external helical compression spring. The cases of economy gauge heads are usually not threaded, so they require an external mounting clamp.

The loose bearing fit and rotatable shaft may cause core skewing and core rotation. These contribute to a somewhat larger repeatability error, typically 0.00254 mm (0.0001 in.), than more precise types of gauge heads. The typical life of economy gauge heads exceeds 5 million cycles with no significant degradation in performance. Economy gauge heads are not recommended for operation in severe environments because special sealing and protection techniques are required.

PRECISION GAUGE HEADS

The majority of LVDT gauge heads fall into the *precision* category. Precision gauge heads are characterized by good linearity, typically 0.25 percent of full range, and excellent repeatability, typically 0.00127 mm (0.00005 in.). They incorporate more precise bearings that have a closer fit to a nonrotating shaft. In spring-loaded units the spring is located internally. Most precision gauge heads have an externally threaded case. Precision gauge heads incorporate either ac or dc LVDTs.

The life of a precision gauge head is comparable to that of an economy type but depends on operating environment to a greater extent.

ULTRAPRECISION GAUGE HEADS

An *ultraprecision* LVDT gauge head combines a special ac LVDT with a honed-and-lapped, selectively fit sleeve bearing or ball bushing to produce a linearity of up to 0.05 percent of full range and a repeatability of 0.0001016 mm (0.00004 in.). Such gauge heads have the lowest error from shaft play and core skewing. The nonrotating shaft can be spring loaded to produce tip loads from 10 grams to several pounds. The close fit of the probe shaft to the shaft bearing may cause bearing friction that can shorten the useful life of an ultraprecision gauge head. For this reason the operating environment must be carefully controlled to prevent dust, dirt, moisture, oil, and similar contaminants from affecting the bearing of an ultraprecision gauge head.

PNEUMATICALLY ACTUATED GAUGE HEADS

In a pneumatically activated gauge head the probe spring is deleted and low-pressure air is introduced to the end of the core opposite the probe shaft. The probe-shaft load force is proportional to applied air pressure. By changing the applied pressure, the probe contact force can be varied as needed to suit a variety of gauging requirements. Once the actuating pressure is set, usually by means of an external pressure regulator, the probe contact force remains constant; it does not vary with shaft position. This gives the effect of a zero-rate compression spring.

The probe shaft may be retracted between gauging cycles by applying a low vacuum—127 to 254 mm (5 to 10 in.) of mercury—instead of pressure. Alternatively, the probe shaft may be spring biased in the retracted position, and the air pressure can be released to retract the probe. However, this method loses the constant force effect of a zero-rate spring.

Pneumatic actuation is normally used with certain types of ultraprecision gauge heads. For this reason the applied air must be entirely dry and free of oil. It should also be filtered with a 10-μm (nominal) filter to prevent the introduction of contaminants into the bearing. Some LVDT gauge heads have a bleed orifice on the connector to facilitate low-pressure regulation. Typically, the probe-shaft force is 0.00175 gram per kg/cm^2 (0.4 oz per lb/in.^{2}) of applied air pressure.

LEVER- OR FINGER-PROBE GAUGE HEADS

In some gauging applications the dimension being checked is not directly accessible to the probe of an ordinary LVDT gauge head. For such measurements LVDT gauge heads incorporating lever- or finger-style probes are useful.

There are two types of finger-style gauge heads. One has a probe with a flexural pivot. A rotational (cosine) error is inherent in gauge heads using a flexural-pivot probe. In addition, the core of the LVDT also cocks, introducing other errors. For these reasons the parallel-flexure probe is preferred for most precise measurement applications.

ENVIRONMENT EFFECTS

The environmental considerations, with some restrictions, for an LVDT also apply to the LVDT gauge head. Economy and precision gauge heads can be operated over temperature ranges from −7° C. to 93° C. (−20° F. to 200° F.). The operating temperature of ultraprecision gauge heads should be restricted to the range from 4° C. to 60° C. (40° F. to 140° F.). The local environment should be clean and free from fluids or particles that could contaminate or otherwise affect the gauge-head bearing. Low relative humidity is desirable. In applications where cleanliness is difficult to maintain, a suitable environmentally protected gauge head should be used. No attempt should be made to lubricate the bearing of a gauge head.

Mechanical vibrations do not generally pose a problem under ordinary conditions and may even provide useful dither that reduces the effects of static friction. As with any standard LVDT, gauge heads are magnetically shielded and are not particularly affected by proximate magnetic fields or materials.

ADVANTAGES OF LVDTS

The LVDT has characteristics that make it a very useful (and, often, first choice) transducer for a wide variety of applications. Some of its features are unique to the LVDT. The following notes are quoted from Schaevitz.

Frictionless Measurement. Ordinarily, there is no physical contact between the movable core and coil structure, which means that the LVDT is a frictionless device. This permits its use in critical measurements that can tolerate the addition of the low-mass core but cannot tolerate friction loading. Two examples of such applications are dynamic deflection or vibration tests of delicate materials and tensile or creep tests on fibers or other highly elastic materials.

Infinite Mechanical Life. The absence of friction and contact between the coil and core of an LVDT means that there is nothing to wear out. This gives an LVDT essentially infinite mechanical life. This is a paramount requirement in applications such as the fatigue-life testing of materials and structures. The infinite mechanical life is also important in high-reliability mechanisms and systems found in aircraft, missiles, space vehicles,and critical industrial equipment.

Infinite Resolution. The frictionless operation of the LVDT combined with the induction principle by which the LVDT functions gives the LVDT two outstanding characteristics. The first is truly infinite resolution. This means that the LVDT can respond to even the most minute motion of the core and produce an output. The readability of the external electronics represents the only limitation on resolution.

Null Repeatability. The inherent symmetry of the LVDT construction produces the other feature, null repeatability. The null position of an LVDT is extremely stable and repeatable. Thus, the LVDT can be used as an excellent null-position indicator in high-gain, closed-loop control systems. It is also used in ratio systems where the resultant output is proportional to two independent variables at null.

Cross-Axis Rejection. An LVDT is predominantly sensitive to the effects of axial core motion and relatively insensitive to radial core motion. This means the LVDT can be used in applications where the core does not move in an exactly straight line (for example, when an LVDT is coupled to the end of a Bourdon tube to measure pressure).

Extreme Ruggedness. The combination of the materials used in an LVDT and the techniques used for assembling them result in an extremely rugged and durable transducer. This rugged construction permits an LVDT to continue to function even after exposure to substantial shock loads and the high vibration levels often encountered in industrial environments.

Core and Coil Separation. The separation between LVDT core and LVDT coil permits the isolation of media such as pressurized, corrosive, or caustic fluids from the coil assembly by a nonmagnetic barrier interposed between the core and the inside of the coil. It also makes the hermetic sealing of the coil assembly possible and eliminates the need for a dynamic seal on the moving member. Only a static seal is necessary to seal the coil assembly within the pressurized system.

Environmental Compatibility. An LVDT is one of the few transducers that can operate in a variety of hostile environments. For example, a hermetically sealed LVDT is constructed of materials such as stainless steel that can be exposed to corrosive liquids or vapors. Because it is hermetically sealed, the same LVDT can also be used in a hazardous location containing flammable vapors or particles if the external connections to the LVDT are made in an approved manner. LVDTs constructed with the proper materials and techniques can operate at cryogenic temperatures immersed in media such as liquid nitrogen or liquid oxygen. Other appropriately constructed LVDTs are available for operation at elevated temperatures (1100° F. or 600° C.) and in nuclear reactors with high radiation levels (10^{20} NVT total integrated flux). LVDTs are available to operate continuously in fluids pressurized to 3000 psi (210 bars). Suitably designed LVDTs can be used in various combinations of these hostile environments.

Input/Output Isolation. The fact that the LVDT is a transformer means that there is complete isolation between excitation input (primary) and output (secondaries). This makes an LVDT an effective analog computing element without the need for buffer amplifiers. It also facilitates the isolation of the signal ground from excitation ground in high-performance measurement and control loops.

ROTARY VARIABLE-DIFFERENTIAL TRANSFORMERS

A modified form of differential transformer can be used for measuring angular rather than linear displacement. This type is known as a rotary variable-differential transformer or RVDT.

A typical RVDT is illustrated in Fig. 10-8. The primary and secondary windings are wound symmetrically on a coil form (stator).

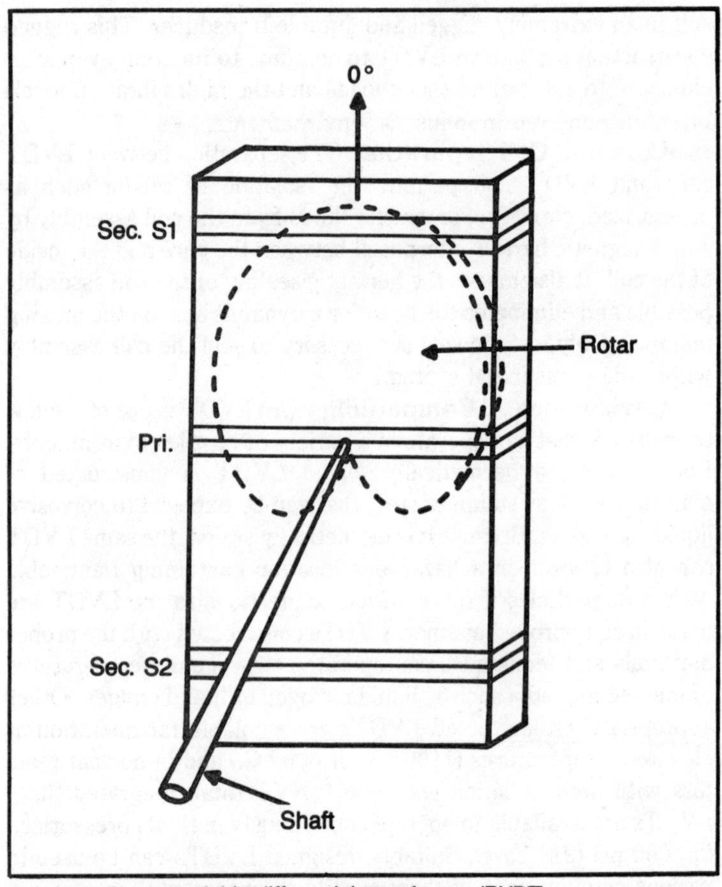

Fig. 10-8. Rotary variable differential transformer (RVDT).

A cardioid-shaped cam of magnetic material (rotor) is used as a core. The input shaft goes across the middle of the coil form at the plane of winding symmetry and is fastened to the cardioidal core. The shape of the rotor is carefully chosen to produce a highly linear output over a specified range of rotation.

The output curve of a typical RVDT is shown in Fig. 10-9. There are two linear operating ranges 180° apart for any RVDT, but only one is calibrated by the manufacturer. The factory-calibrated linear region is identified by markings on both the shaft and the RVDT case for the nominal zero-degree shaft position (null angular position) of the shaft.

The input shaft is usually supported by precision ball bearings to minimize friction and mechanical hysteresis. Because the bearing loads are ordinarily so low, the life of an RVDT transducer is comparable to the life of an LVDT. An important feature of the RVDT is the absence of any contact brushes in its construction. Also, like the LVDT, and RVDT is not affected by mechanical travel.

CHARACTERISTICS OF THE RVDT

Although the RVDT is a continuous (360 °) rotation device, the range of most linear operations for a typical RVDT is only about ± 40 °; operating range is better than 0.5 percent of full range. However, the linearity over smaller angular displacements is correspondingly improved. Thus, if an RVDT is used to measure a small angle of displacement, say ± 5 °, the linearity is found to be better than 0.1 percent of full range. The practical upper limit of angular measurement with an RVDT is about ± 60 °.

The resolution of the RVDT is theoretically infinite. Resolution to very small fractions of a degree is commonly achieved in practice.

APPLICATIONS

RVDTs are available in both ac and dc types; dc units may con-

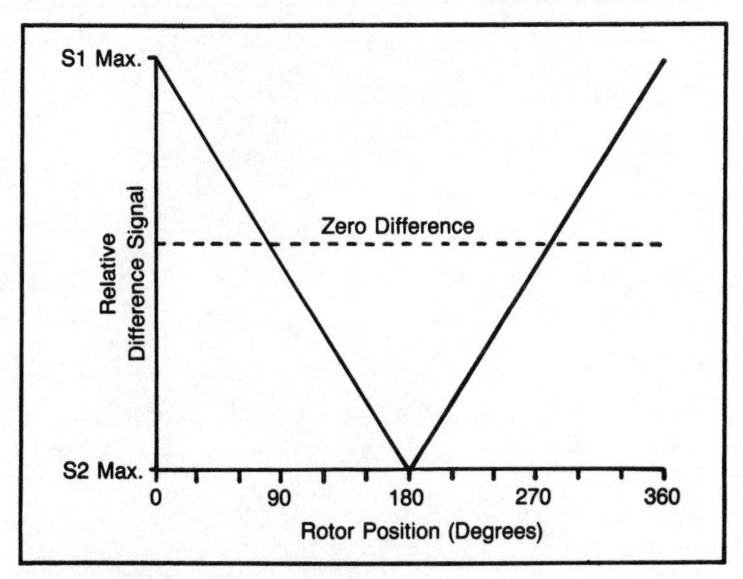

Fig. 10-9. Output curve of a typical RVDT. Only one-half (180 °) of the rotation range is needed.

tain integral thick-film signal conditioning. Both types are used extensively to measure angular position and offer significant advantages over other types of angular position transducers. The foremost advantage is the lack of physical contact between rotor and stator. Another major advantage is the absence of brushes and slip rings. The only limitation on mechanical life is the shaft bearing, but the bearing loads under normal operating conditions are so insignificant that the operating life of an RVDT is comparable to that of an LVDT.

Another advantage of the RVDT is its truly infinite electrical resolution. The bearing play is usually so negligible that an RVDT exhibits practically no mechanical hysteresis.

RVDTs having a separable rotor and stator can be used in applications where the rotor shaft must be located in a bearing assembly separate from the stator housing. The input shaft of the RVDT can be sealed against pressure by a simple O-ring, if required, but sealing the shaft can add significant friction to the transducer. Complete hermetic sealing of the coil assembly is also posssible.

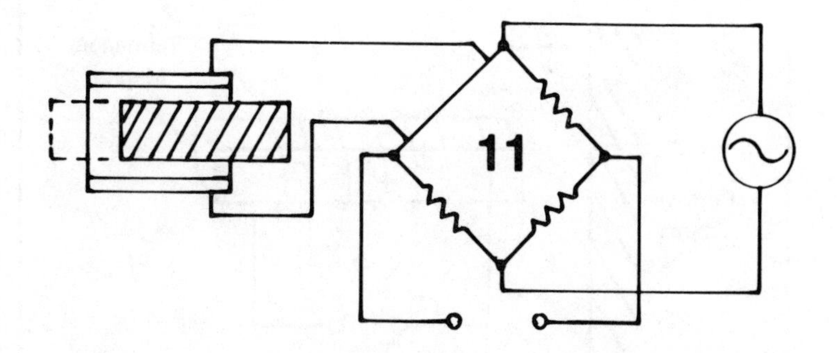

Digital Transducers

Digital transducers are devices that produce a pulse-form electrical signal (digital signal) corresponding to a train of mechanical events. The simplest type is the *magnetic pickup*, consisting of a coil wound round a permanent magnet, one pole of the magnet being located close to a ferrous object (Fig. 11-1). Movement of the ferrous object will induce a current in the coil. If a number of ferrous objects move past the coil, then a corresponding train of current pulses will be generated in the coil. Thus, if the ferrous object is a toothed wheel mounted on a rotating shaft, each revolution of the shaft will generate N pulses of electrical signal, where N is the number of teeth in the wheel (Fig. 11-2). If these pulses are counted and divided by N, this give an exact figure for the rate of revolutions of the shaft (the magnetic pickup in this case being used as a revolution counter). If the wheel has 60 teeth, the pulse frequency generated in the pickup will be the same as the rotational speed of the shaft in rpm. But, of course, the wheel does not need to have exactly 60 teeth. In practice, larger wheels with a smaller number of teeth (the preferred choice would be 30 teeth or 15 teeth) generally give better performance.

Good results can also be obtained by using pickoffs with shafts having a single indentation or projection to provide a single output pulse per revolution. Shaped cams may also be detected. Where the member to be monitored is nonferrous, the introduction of mild steel studs to provide the appropriate configuration is usually

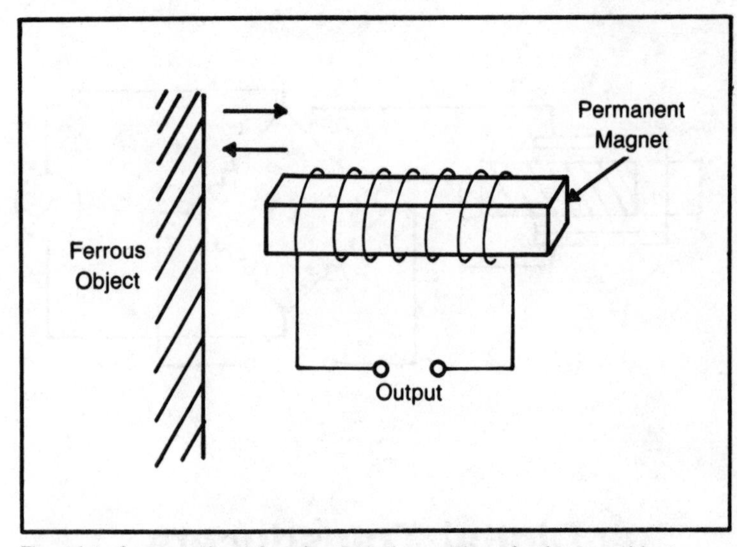

Fig. 11-1. A magnetic pickup for detecting motion of a ferrous object.

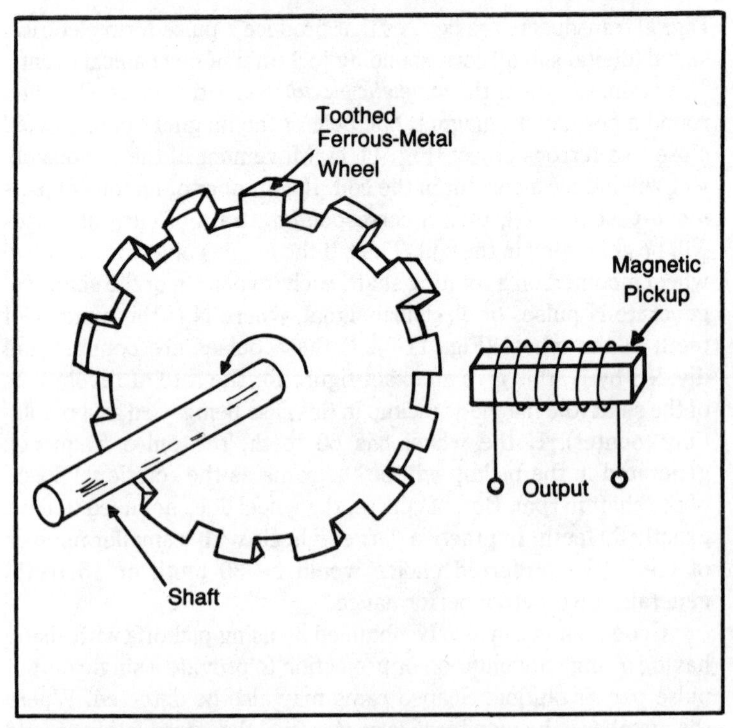

Fig. 11-2. A magnetic pickup can be used in conjunction with a toothed wheel for measuring the angular speed of a rotating shaft.

satisfactory. Moving permanent magnets in place of ferrous masses will provide a substantially higher output amplitude from the pickoff, and this is often of great value at low speeds, or where large air gaps are essential.

Apart from having no physical contact with the component being monitored (in this case the rotating shaft), the electrical information derived has the advantage of being in digital form suitable for working a digital display or readout with no need for analog-to-digital conversion.

SIGNAL OUTPUT

The strength of the signal generated, or *pulse amplitude,* depends on a number of factors, some dependent on the geometry of the installation and others on the design of the pickup. The most significant factors are:

☐ The *speed* at which the magnetic flux is interrupted or modulated. To a first-order approximation, the output amplitude is directly proportional to the speed or passage of ferrous objects past the pickoff pole piece but is attenuated by coil and iron losses that predominate at high frequencies. The output of magnetic pickoffs continues to rise with speed to frequencies in excess of 10 kHz; in some cases, to beyond 15 kHz.

☐ The *distance* separating the transducer polepiece and the moving ferrous object. To a first-order approximation, the output amplitude is inversely proportional to this distance. For very low-speed applications a separation of only 0.127 mm (0.005 in.) may be necessary, whereas for high-speed applications a separation of up to 2.540 mm (0.100 in.) or more may be satisfactory.

☐ The *mass* of metal modulating the flux. The relationship between the mass of iron passing the transducer pole piece and the output amplitude is complex and is, to some extent, related to the fourth factor. In general terms, and within the constraints of configuration, the output is greater for a greater mass of moving iron.

☐ The *configuration* of the iron circuit. The design of modern magnetic pickoffs is a compromise to achieve a high output voltage over a wide speed range in a small physical size. It is necessary that the active pole of the pickoff be small so that the sensing area is reduced to a level where the transducer may be used effectively with small gear wheels.

If the pole pieces are of circular section, then to generate maximum output it is desirable that one moving object should clear the active area before the next object enters it. When used with gear wheels, the best effects will be obtained if the tooth form is rectangular, the tooth width being equal to, or greater than, the pole-piece diameter, the gear width being at least twice the pole-piece diameter. The gap between teeth should also exceed the pole-piece diameter. However, the provision of these optimum conditions is seldom necessary in practice, and the use of standard dp or similar gear wheels usually gives entirely acceptable results. It should be borne in mind, however, that in installations where the gap between teeth is substantially less than the pole-piece diameter, the output of pickoffs will be well below optimum.

Diametrically split gear wheels should be avoided because they will almost certainly produce magnetic discontinuties that will affect performance.

LIMITATIONS OF MAGNETIC PICKUPS

The main limitation of a magnetic pickup is that there is a lower limit of speed monitoring where the output falls to a level where the signal-to-noise ratio is too poor to decode accurately. To a large extent this is also dependent on the sensitivity of the input circuits of the associated electronics. For low-speed measurement, therefore, preamplification and/or circuits with a sensitivty of not less than 1 V are usually required.

Another significant limitation is that it is difficult to provide a very high number of pulses per revolution without the use of large-diameter wheels having a large number of teeth, or stepup gearing to drive the wheel being monitored from the machine under investigation.

ROTARY DIGITIZERS

Rotary digitizers are another form of transducer generating electrical signal pulses when monitoring shaft rotation. They are much more sensitive than magnetic pickups and, in addition to measuring shaft rotational speed (working as pulse tachometers), can also accurately signal incremental shaft movements. In the latter application they are normally known as incremental shaft encoders, shaft angle encoders, or incremental pulse transducers.

In fact, there are two distinct forms of shaft encoders: the *incremental* encoder, and the *absolute* encoder. The incremental encoder or *rotary pulse generator* produces a train of electrical pulses resulting from angular rotation of the input shaft and interfaces with a digital counter to provide information giving the magnitude of angular rotation. The absolute encoder produces a number of electrical logic level outputs in parallel in binary form. It is a considerably more complex and expensive device.

An incremental or rotary pulse generator that produces a *single* train of pulses can be used for both shaft position and angular velocity measurements. It does not distinguish between different directions of angular movement. For that reason it is used only as a revolution counter. However, the addition of a second pulse train that changes its phase relation to the first pulse chain when the direction of rotation is reversed enables the transducer to detect *direction* of rotation via a suitable logic circuit. Such circuits can present this information in various ways, depending on actual requirements, and rotary digitizers of this type are known as dual-pulse rotary pulse generators (RPGs).

The following is a description of how they work, based on the industrial hardware developed and manufactured by Trump-Ross (a division of Datametrics Corporation).

Early forms of the RPG included electrical commutator type devices that caused interruptions of electrical power. Magnetic principles were also utilized in a number of ways, some of which included variable inductance techniques and reed switches operated by permanent magnets.

When light-sensitive devices, such as photoelectric cells, became available, it became a simple matter to drill a number of holes in a circle and use these as shutters to interrupt the electric power. In recent years light-sensitive devices have become extremely efficient and compact, resulting in the miniaturization of the once ponderous RPG. Advances in photographic reproduction techniques have enabled the disks that are attached to the RPG shaft to carry extremely fine-line patterns.

Typical low-cost RPGs have line densities up to 2540 on a 2-in. diameter circle. More sophisticated and expensive encoders sometimes pack in as many as 30,000 lines on a 2- or 3-in. circle.

The obvious benefit of using the optoelectronic approach is the elimination of any form of electrical contact, thus enabling the RPG to provide high reliability and long product life. In addition, the lack of contact eliminates any possibility of radio-frequency noise generation and contact surface wear. The use of very high-frequency response circuitry and the obviously very high speed of light itself enable the RPG to produce electrical pulses at frequencies approaching 1 million pulses per second.

THE DUAL-PULSE RPG

The more standard dual-channel RPG, which is designed to produce bidirectional-type pulse trains, must switch the incoming dc power in a form that is most acceptable to the logic circuit receiving the electrical signals.

A common electrical configuration is that which uses +5 Vdc as the power supply (this voltage level is normal to most logic systems), and the RPG switches the voltage level alternately from +5 V to a level approaching 0 V. The duration of the switched-on power is normally one-half of an angular pulse increment. If we call the +5 V output level *logic* 1 and the voltage level approaching zero as *logic* 0, we can associate the output signal more readily in terms understandable to the logic designer.

The electrical pulse train has a conventional configuration where the logic 1 level has the same angular duration as the logic 0. The 1:1 ratio of these states is known as the *symmetry* of the signal and is a closely controlled parameter of an RPG. The number of switched cycles per revolution of the input shaft of the RPG is known as the *line count* and is interchangeable with the term *resolution*. The degree to which the occurrence of a discrete pulse coincides with the incremental angular displacement of the shaft is known as *pulse accuracy*. The term *stability* is the extent of repeatability, or consistency, of the signal configuration under actual operating conditions. For a dual-channel RPG the relation of the second pulse train to the first is of considerable importance. In order to provide the maximum discrimination by the logic system, which is detecting the direction of rotation, it is necessary to place the switching point (the leading or trailing edge of the square wave) at a position that equally divides the logic 1 or logic 0 states of the other channel, or one-quarter of a full square-wave cycle. This is known as 90 electrical degrees of phase shift, or *Quadrature*.

THE DISK AND THE SIGNAL

The effective use of the rotary pulse generator does not depend on an intimate knowledge of the function of the device. For all practical purposes the system designer is only concerned with his two means of communication with the unit. The first, and quite obvious, requirement is to provide some form of mechanical connection to the input shaft in order to cause the shaft to rotate in proportion to the mechanical motion of the machine. The second, and equally obvious requirement, is to connect the electrical output of the unit to the control black box.

Experience, however, indicates that a more precise knowledge of the internal functioning of the RPG will assist the systems designer, and the ultimate user, in a more satisfactory selection and a greater understanding of the application and handling of the device.

All of the Trump-Ross rotary pulse generators operate with the use of some form of light source that shines through a transparent disc. A series of radial lines is printed on one side of the transparent disc, producing a complete circle of alternately transparent and opaque sectors (Fig. 11-3).

The light shining through the disk is optically corrected to produce a parallel beam. The technical term for light correction is *collimation*.

Because most RPGs have a relatively high number of lines on the disc, it is generally quite difficult to shine the light through one slit only and get a satisfactory electrical signal from the light-sensitive device receiving the light. It is normal, therefore, to utilize some form of grating that, when placed close to the encoder disc, will cause the light to be shuttered through many lines. This larger amount of light can then be utilized to produce a substantial and reliable signal (Fig. 11-4).

The distance between the grating and the disc should be as small as is physically possible in order to reduce the effects of poor collimation and stray light.

Conventional rotary pulse generators normally use tungsten-

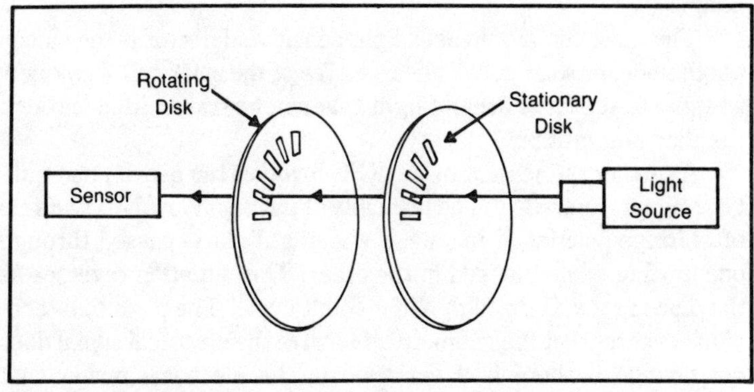

Fig. 11-3. A pair of rotating disks can be used to measure angular speed.

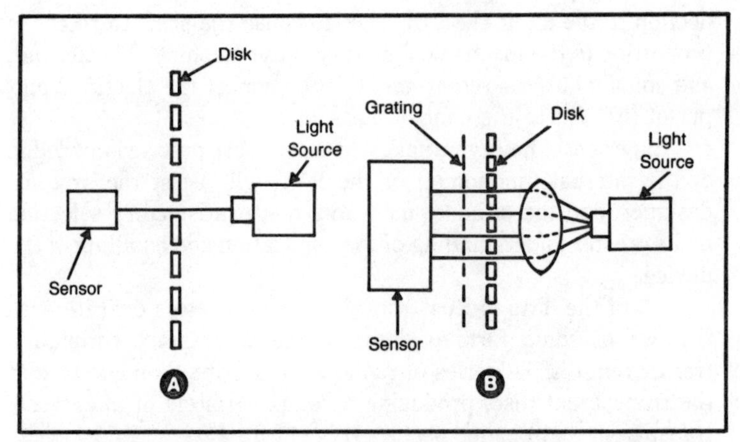

Fig. 11-4. At A, a narrow-beam light source must be used with a single rotating disk. At B, the addition of a grating allows a wider beam to be used.

filament lamps as the light source. The filament is generally very small, essentially appearing as a point source of light. The tungsten lamp, when operated at a voltage level that is significantly less than the manufacturer's rated voltage, has a useful life far in excess of the rated life of most RPGs. All Trump-Ross products have a built-in voltage reduction that assures a normal lamp life of 20 years or more.

For excessive shock and vibration, solid-state light-emitting diodes (LEDs) are used. The LED appears to have an almost infinite life when properly used. However, the lower energy output, coupled with sensitivity due to temperature variations and difficulties in collimating the light output, tends to present problems in matching the electrical performance of the tungsten-filament lamp.

The most commonly used light-sensitive detector is the silicon photodiode, or solar cell. The large area of the solar cell is uniquely suitable to the large area of light that can be transmitted through the disc and grating.

A further refinement of the RPG involves the use of push-pull-type signal generation. The principle of push-pull involves two sets of gratings positioned in such a way that light is passed through one grating while blocked in the other. This situation reverses as the disc is rotated through half a count cycle. The push-pull technique ensures that the geometrical form of the electrical signal does not change if there is a variation in the electrical power that energizes the signal source.

ELECTRONIC INTERFACE

High integrity is required from the electronic interface, the performance of which must be independent of variations of power-supply voltage, line resistance and capacitance, temperature changes, and other external factors that may tend to affect the electric signal generated by the RPG. Typically this is met by state-of-the-art solid-state circuitry.

SELECTING THE RIGHT RPG

The rotary pulse generator is an integral part of a widely varied range of digital control systems. Each application has its own peculiar set of conditions, and it is wise to match the physical and electrical configuration of the RPG to meet its particular environment. The instrument-type application requires an instrument-like RPG. The unit on a rolling mill, for example, must be heavy in construction in order to conform to the general atmosphere of the application.

The physical interface of the RPG to the machine or the element on the machine to which the shaft is connected is extremely important in order to protect the device.

Proper alignment of the fixing holes and shafts and the use of very flexible, but angularly stiff, couplings is mandatory. Electrical connections should be made, by the proper use of rugged disconnects and armored conduit, insensitive to the roughest treatment.

The RPG user must be prepared to receive equipment into his plant that may be unfamiliar to receiving personnel. Many machine-tool and heavy-equipment manufacturers experience "failures" of RPGs that may be traced to gross mishandling by personnel who are more familiar with large machine components. The incoming inspection department of these companies can systematically ruin RPGs by improper testing procedures.

Rotary pulse generators, like any other device, have a limited life. The life expectancy may be determined by the life of the power source, such as the lamp filament; or, on the other hand, the mechanical life may be the limiting factor, particularly if the shaft of the RPG is too heavily loaded.

The user must satisfy himself that his is buying a product that has a normal life expectancy in excess of five times the required life.

There is no substitute for experience. Life expectancy can best be judged by the history and experience that the RPG supplier can provide, plus the experience of the other users of the product.

Rotary pulse generators are frequently quite sophisticated and represents a complex electromechanical package. With the advent of solid-state devices, the resulting miniaturization allows the manufacturer to package much of the amplification and switching circuitry within the unit. The net result is a highly desirable package, which, unfortunately, has a higher potential for failure.

The occasional failure of an RPG necessitates, in most cases, some repair or replacement of the unit. Repair work is rarely possible in the field, and the unit must then be returned to the manufacturer. The speed with which the manufacturer can turn the unit around will materially affect the machine's down time.

Most manufacturers will warrant their product for one year and will be able to present plenty of evidence to show that their product should have a long life expectancy. The user would be foolish, however, to rely blindly on the reliability of the unit, particularly if the device is used in a critical application.

Installation of RPGs without a reasonable backup quantity of spare units is very shortsighted. The economics of such a policy can look very bleak when a highly complex, large machine is down for the want of a transistor.

(We gratefully acknowledge Trump-Ross for extensive quotations from their literature on the subject of RPGs.)

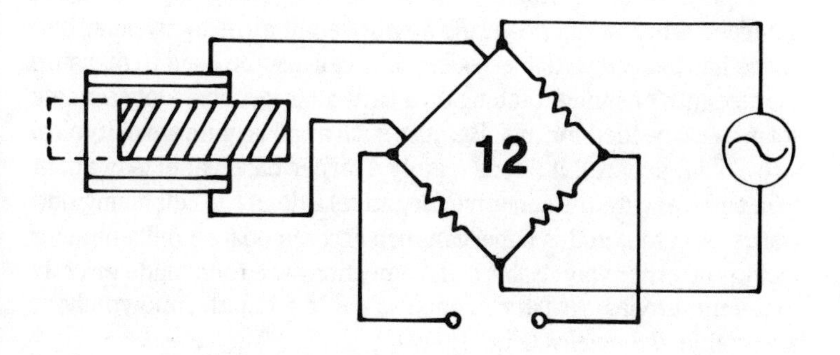

Photoelectric Devices

Photoelectric transducers are devices capable of converting light energy into electrical energy. In this respect only a limited number of photoelectric components comply with such a definition—that is, *photovoltaic* devices, which specifically *generate* electricity when light falls on them. Other photoelectric devices are *photoconductive* only, meaning that a charge of illumination or light falling on them produces a change in their conductive properties in an electric circuit. Thus they work as transducers only when part of an electrical circuit.

One thing all photoconductive devices have in common is that they are basically diodes, even the phototransistor that combines the photoconductivity properties of photodiode with the amplification properties of a transistor (except that the "diode" part is a photoelectric device).

A photovoltaic device, on the other hand, works like a battery and is generally called a cell or *photocell.* This is the type used in "self-powered" light meters. Like batteries they can be connected in series to generate a higher voltage (but still normally in the low-voltage range) or in parallel to generate higher current (in the low-current range). Top performer in this range is the solar cell, which in average sunlight may be capable of generating a photovoltaic potential of the order of 510 mV or more per cell with an output current of 3 mA into a 100-Ω load. It is also capable of being connected in series and/or parallel; a multiple cell arrangement is called a *solar battery*.

Apart from obvious applications for powering electrical devices—where even powering a lightweight airplane by solar batteries has proved possible—solar cells can also be used to measure the strength of sunlight falling on a particular location. Only a single solar cell is needed for this. Because such a cell is quite small (typical size 1/2 in. square), it needs mainly a larger base panel (say about 5 in. square) with the sensitive (negative) side of the cell facing outwards. A cutout in this panel can then accommodate a milliammeter facing the other way. Solar cell connections are then made directly to the meter via a resistor in one lead. Add a handle, and you have a portable *radiometer* (Fig. 12-1).

You can work out the resistor value and the meter range required in this way. In strong, average sunlight you can expect the cell to generate about 500 mV. Suppose you are going to use a 0-10-mA meter. A 500-mV signal thus has to develop a current of 10 mA through the meter for full-scale deflection. From Ohm's law the corresponding circuit resistance required is

$$R = \frac{\text{volts}}{\text{amps}}$$

$$= \frac{500 \times 10^{-3}}{10 \times 10^{-3}}$$

$$= 50\ \Omega$$

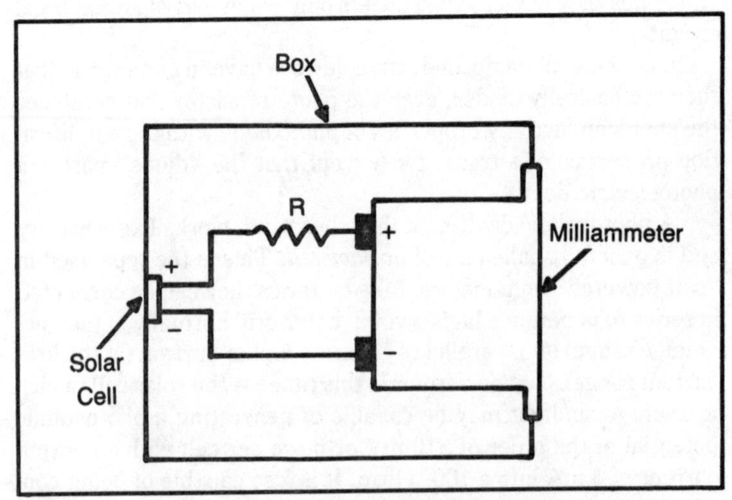

Fig. 12-1. A simple radiometer for measuring light intensity.

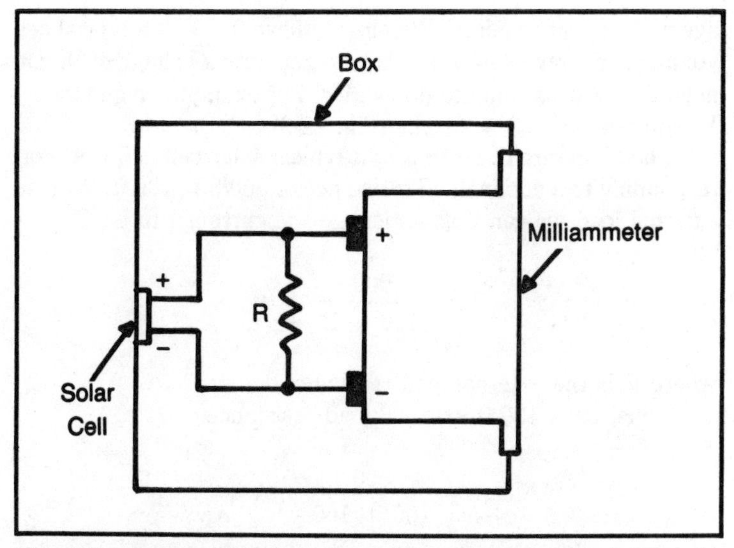

Fig. 12-2. An alternative wiring scheme for a radiometer using a meter with high internal resistance.

The resistance of the meter itself will be less than this, probably only a few ohms. So try a 39 Ω or 47 Ω resistor connected in series, and see if this gives near full-scale deflection. If not, use a smaller value for the resistor. Alternatively, if you can only get low readings on the meter in strong sunlight, this is because the effective meter resistance is higher than expected. In that case try connecting a low resistance (say 10 Ω or less) directly across the meter terminals instead of in series with it (Fig. 12-2).

Incidentally, a typical short-circuit current for a solar cell is 0-5 mA. Thus, if you used a 0-5 mA meter that had negligible resistance, you would not need any resistors in the circuit; the solar cell could be connected directly to the meter. However, the meter would still have some resistance, so to achieve full-scale deflection you would almost certainly need to short the meter with a small valve resistor connected across it.

MAKING A SOLAR BATTERY

A solar battery is a true transducer, which converts solar energy directly into electrical energy, just like a battery. Its effectiveness as a battery depends on the strength of the sunlight present and the number and arrangement of the cells used.

The starting point in the design of a solar battery is the voltage and current required. We already have 0.5 V as a typical cell voltage in average sunlight. Thus, to generate a voltage of V, you need 2 × V cells connected in series. For example, to generate 3 V, you need 2 × 3 = 6 cells (Fig. 12-3).

The short-circuit current of a typical solar cell is 5 mA, corresponding to a nominal *cell* resistance of 500/5 = 100 Ω. With an external load we can thus anticipate the current I to be

$$I = \frac{500}{100 + R} \text{ mA}$$

where R is the external load resistance.

Thus, for a 100 Ω external load resistance,

$$I = \frac{500}{100 + 100} \text{ mA}$$

$$= 2.5 \text{ mA}$$

Suppose we want to design the solar battery to work a 3-V device that has a load resistance of 25 Ω and needs to have a 12-mA current flowing through it to work properly. To generate 3 V, we need six cells connected in series, but the *current* through an external load of R = 50 Ω will only be

$$I = \frac{500}{100 + 25} \text{ mA}$$

$$= 4 \text{ mA}$$

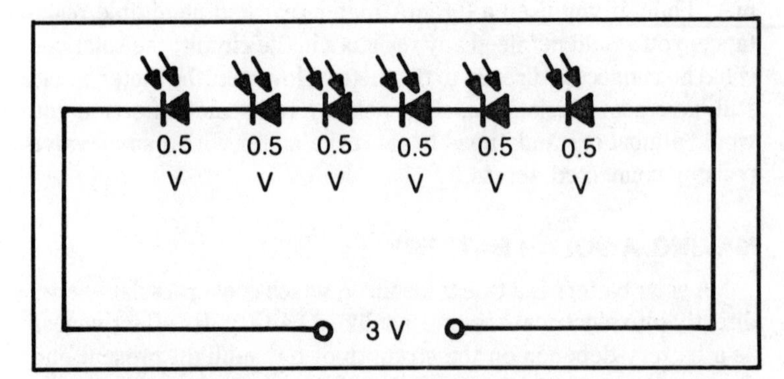

Fig. 12-3. Series connection of photovoltaic cells results in higher voltages than can be obtained with a single cell.

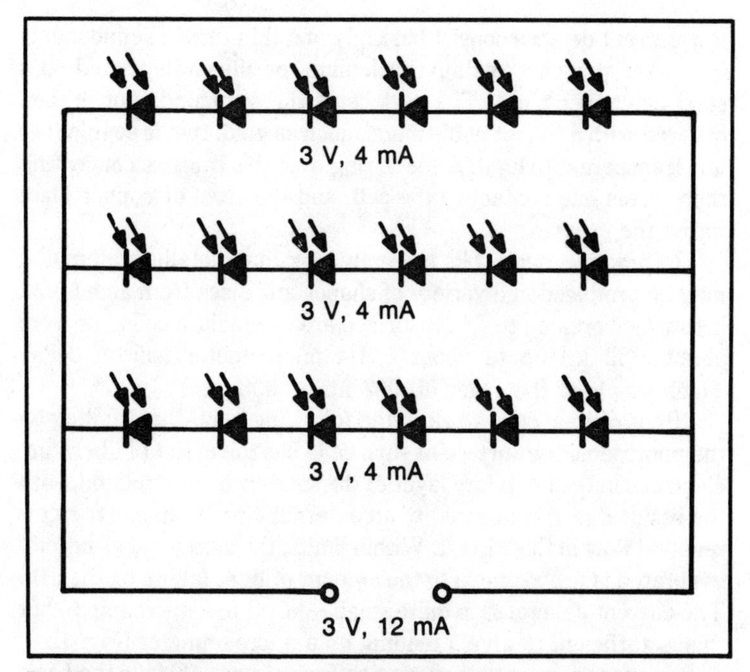

Fig. 12-4. A series-parallel array of photovoltaic cells for obtaining 3 V at 12 mA.

To boost this current to the required 12 mA, we need 12/4 = 3 rows of cells connected in parallel (Fig. 12-4). The total voltage generated is still the same (3 V), but each row contributes 4 mA of current, for a total of 3 × 4 = 12 mA. We now have the solar battery configuration required to power the device. Remember that when connecting them up and mounting them the sensitive side (negative side) of each cell must face directly towards the sun.

For maximum effect such a battery array should be mounted to point along the meridian (due south) and tilted away from the zenith at an angle of approximately the local latitude (Fig. 12-5). This is, of course, only possible with a fixed installation. In a solar battery that is an integral part of a moving device, the optimum arrangement is normally with cells facing virtually upwards and mounted on top of the device. Solar cells for powering an electric, motor-driven model airplane, for example, would need to be mounted on the top surface of the wing.

OTHER PHOTOVOLTAIC CELLS

Photovoltaic cells of the type used in light meters and other low-current devices consist basically of a thin film of semiconductor material such as gallium, selenium, or silicon deposited on a steel or copper plate. The surface of the semiconductor is then covered with a film of noble metal, such as gold, that is so thin that it is transparent to light. A metal ring over this transparent coating then forms one contact to the cell, and the steel or copper plate forms the other.

In practice, such cells normally look like metallic wafers and may be produced in a variety of shapes and sizes from about 3/32 in^2 surface area up to 12 in^2 surface area (selenium cells); or from about 3/32 in^2 up to about 1 1/4 in. diameter (silicon cells). Thickness is of the order of .002 in. to .006 in.

Photovoltaic cells work in the following way. Light falling on the photosensitive surface of such cells has the effect of liberating electrons in the boundary layer of the sandwich construction, with the result that if connected to an external circuit, an electric current will flow in that circuit. Within limits, the amount of electricity generated is proportional to the amount of light falling on the cell. The current generated is quite small—only a few microamps—but this is sufficient to give a reading on a microammeter (Fig. 12-6) or to operate a sensitive moving-coil movement. This is the basis of the light meter. All you need is a photovoltaic cell and a microammeter connected together, and you have a light meter capable of

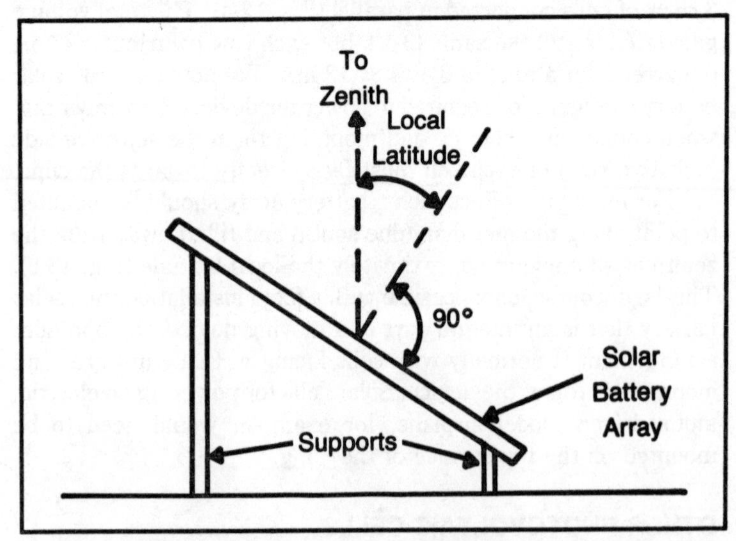

Fig. 12-5. Optimum positioning of a solar battery array. The array should face due south and be slanted from the zenith by approximately the local latitude.

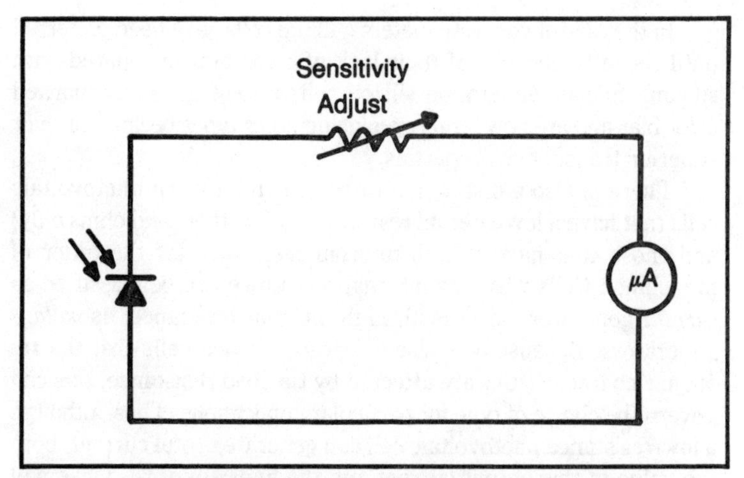

Fig. 12-6. A light meter for measuring low levels of illumination.

measuring the quantity of illumination falling on the face of the cell. It can only be calibrated, however, with reference to the specific performance curve for that cell, or, more simply, by comparing readings against those obtained from a standard light meter, such as a camera exposure meter.

Strictly speaking, provided the cell is not overilluminated so that the cell is saturated, the microammeter reading will be almost directly proportional to the level of illumination. Most light meters, however, have a scale where the divisions get closer together at the higher levels of illumination. This is either a feature of the design of the poles of the permanent magnet used with the moving coil movement in the meter, or it is a direct result of saturation with increasing illumination. Often, both factors are responsible.

The polarity of the current developed will always be the same for a given device, but the actual polarities of selenium and silicon photocells are different. Thus, with a selenium cell the base or back of the cells is positive. With a silicon cell the sensitive face is the positive electrode.

Many photovoltaic cells show different responses for the same level of illumination produced by different light sources. Thus, although a selenium-steel cell will give similar readings for daylight and tungsten-bulb light at the same levels of illumination it will underindicate the light from discharge lamps. A copper/oxide-copper cell, on the other hand, will overindicate daylight with respect to tungsten light and underindicate with discharge lamps (but not fluorescent lamps).

In the case of converse meters gallium cells have been preferred until recently because of their lack of color bias (compared with silicon). Second-generation silicon cells do not have any marked color bias and are now largely replacing other types because of their superior transducer properties.

There is also a distinction to be drawn between photovoltaic cells that have a low internal resistance (a few thousand ohms only) and those that have a high internal resistance (of the order of megohms). Cells with low internal resistance can be classified as *current* generators; cells with high internal resistance, as *voltage* generators. Because it is also a feature of such cells that the response characteristics are affected by the load resistance, this can govern the choice of type for particular applications. Thus, although a low-resistance photovoltaic cell can generate useful current, both the value of this output current and the linearity of response will fall off with increasing external resistance. Thus, any meter movement employed in working the cell in a practical circuit, for example, must have a very low resistance. A high-resistance cell, on the other hand, even though generating a much smaller current, is less affected by load resistance and can develop a useful voltage across a high load resistance, which can then be amplified as necessary.

PHOTOCONDUCTIVE CELLS

Photoconductive cells are quite different in their working properties, for they need to be supplied with an external source of electricity. The effort of varying the level of illumination falling on the cell is to give it a variable electrical resistance, the effective resistance changing in proportion to the light level.

Earlier photocells were of two main types: cadmium sulphide cells responsive to visible light, and cesium sulphide cells highly sensitive to light in the infrared spectrum. For general use they have largely been replaced by *photodiodes* and *phototransistors*. With both of these (photodiodes in particular), current charges realized are small compared with photoconductive cells, so in practical circuits they normally have to be used with one or more stages of amplification.

PHOTODIODES

If any semiconductor diode is reverse biased and the junction illuminated, the reverse current flow will vary in proportion to the amount of light. This effect is utilized in the photodiode, which has a clear window through which light can fall on one side of the crystal and across the junction of the p and n zones.

In effect, such a diode will work in a circuit as a *variable resistance*, the amount of resistance offered by the diode being dependent on the amount of light falling on the diode. In the dark the photodiode will have normal reverse working characteristics; that is, it will provide almost infinitely high resistance with no current flow. At increasing levels of illumination, resistance will become proportionately reduced, thus allowing increasing current to flow through the diode. The actual amount of current is proportionate to the illumination only, provided there is sufficient reverse voltage. In other words, once past the "knee" of the curve (Fig. 12-7), the diode current at any level of illumination will not increase substantially with increasing reverse voltage.

Photodiodes are very useful for working as light-operated switches (and can be made sensitive to infrared as well as visible light). They also have quite a high switching speed, so they can be used for counter circuits counting the interruption to a light beam.

A basic photodiode light switch circuit is shown in Fig. 12-8, where the diode, together with the resistor, controls the current

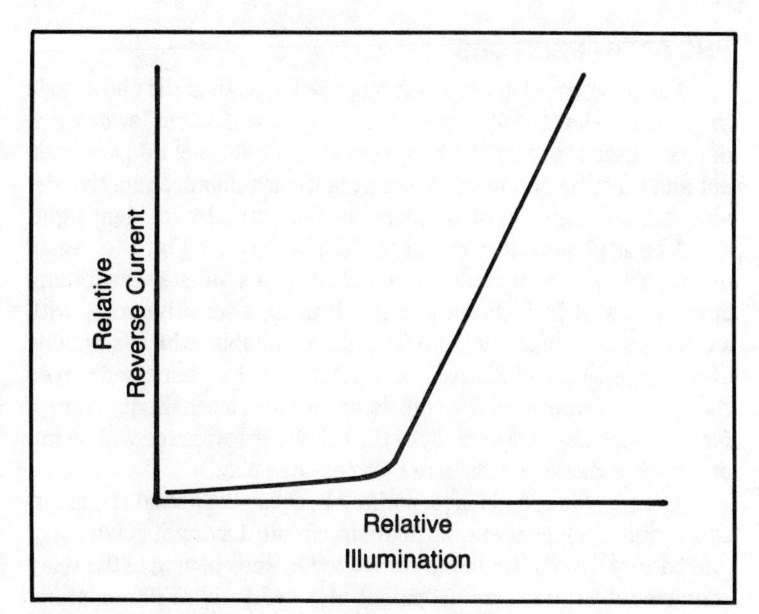

Fig. 12-7. The reverse current through a photodiode is a function of the relative illumination falling on it.

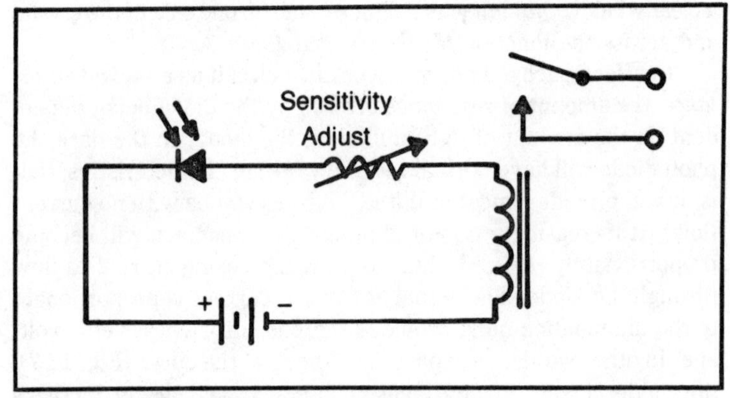

Fig. 12-8. A simple photodiode-actuated relay circuit.

flowing through a relay coil. The relay pulls in when the diode is illuminated and drops out when no light falls on the diode. The sensitivity of the circuit can be adjusted if a variable resistor is used in the primary circuit. No amplification is needed in this circuit because any battery voltage less than the breakdown voltage of the diode can be used to power the relay directly.

PHOTOTRANSISTORS

The phototransistor is much more sensitive than the photodiode to changes in level of illumination, thus making a better "switching" device where fairly small changes of level of illumination are present and must be detected. It works both as a photoconductive device and an *amplifier* of the current generated by incident light.

A basic phototransistor light switch is shown in Fig. 12-9, again using a relay (which could be replaced by a solid-state switching circuit). If a OCP71 phototransistor is used, a sensitive relay with a coil resistance of about 2000 Ω would be suitable, which is capable of pulling in at about 2 mA. The variable resistor connected across the base and emitter of the transistor provides a sensitivity control for adjusting the pull-in point of the relay. Performance will be improved if a diode is connected across the relay.

No reset function is incorporated because the actual alarm circuit is quite independent of the main circuit, having its own separate battery, the circuit being completed by the opening of the relay contacts when the relay drops out at a low level of illumination. This alarm circuit will be switched off again as soon as illumination is restored to the phototransistor, and thus the bell will ring only during the period the light beam is interrupted—just momentarily if anything passes through the light beam.

The sensitivity of this circuit can be greatly improved by adding a second conventional transistor to provide one stage of amplification following the phototransistor (Fig. 12-10). A second potentiometer (R2) is included to adjust the bias applied to the base of the second transistor, which in turn affects the relay current and, thus, the pull-in of the relay. In practice, R2 is adjusted with the phototransistor shielded or covered up so that the relay does not pull in with circuit switched on. The phototransistor is then uncovered, and R1 is adjusted to set the level of illumination at which the relay pulls in.

ANNUNCIATOR PHOTORELAY

The circuit shown in Fig. 12-11 employs a conventional photoconductive cell with straightforward amplification, giving a simple design, again with a minimum of components. Virtually any type of relay can be used with a coil resistance between 1 Ω and 10 Ω and that can pull in at about 3 mA or less. A potentiometer can be used to adjust the relay current to the required operating level. Alternatively, with a relay coil resistance of 5000 Ω, this poten-

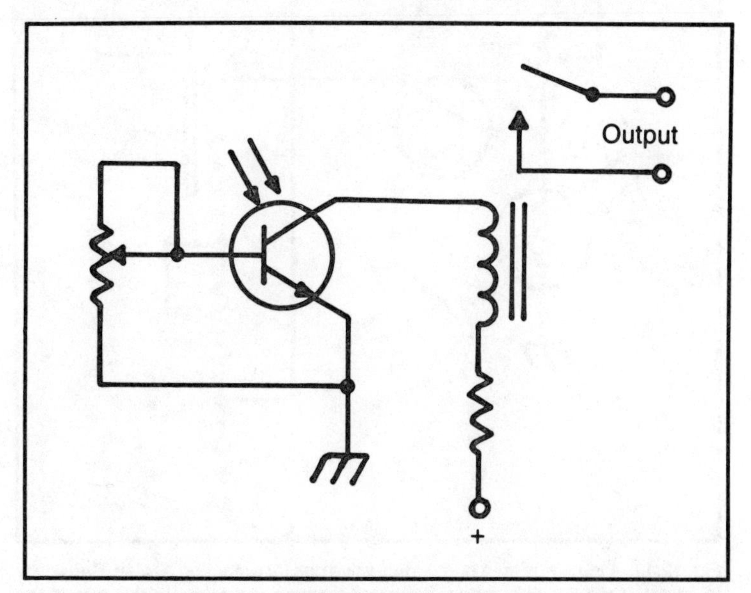

Fig. 12-9. A phototransistor-actuated relay circuit. The relay pulls in when the illumination increases to a level determined by the setting of the potentiometer.

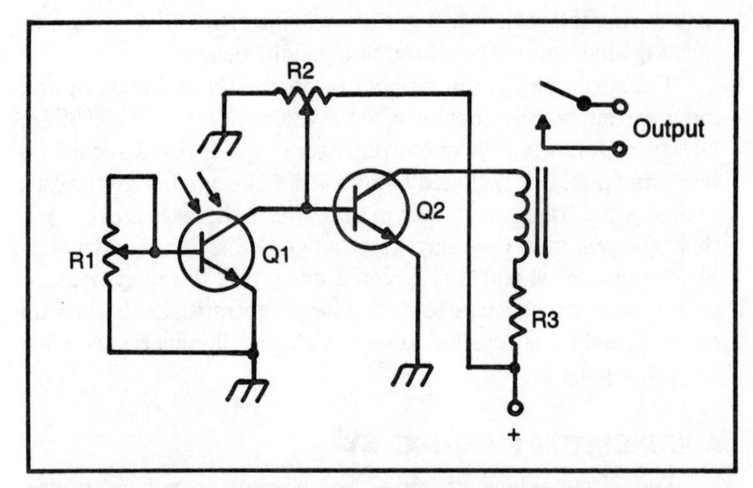

Fig. 12-10. A more sensitive phototransistor-actuated relay circuit. Transistor Q2 amplifies the output from phototransistor Q1. Sensitivity is adjusted via R1 and R2. Resistor R3 limits the current through the relay.

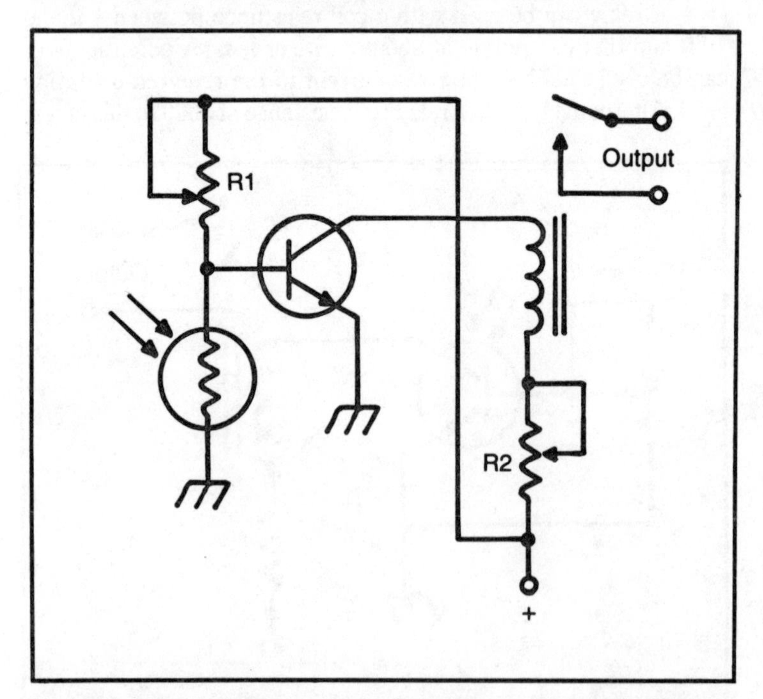

Fig. 12-11. A circuit using a photoconductive cell for actuating a relay. Sensitivity is adjusted via R1; R2 limits the current through the relay. In this circuit the relay is open with illumination and closes when the light intensity falls below a certain level.

tiometer can be dispensed with. A sensitivity control for the whole circuit is provided by R1, which can be adjusted to establish the pull-in of the delay at the desired level of illumination. Ideally, this should be with as much of R1 left "in" as possible (corresponding, that is, to a fairly high level of illumination) so that the current drawn when the light is on is very low. The circuit can thus be left set with very little drain on the battery. When the light is interrupted or removed from the photocell, the current will rise to between 3 to 5 mA, causing the relay to pull in and closing the contacts to complete the alarm circuit.

ANNUNCIATOR RELAY WITH "HOLD"

The circuit just described can, with slight modifications, be made to work with a "hold" function. That is, when the relay drops out in response to an interruption of the light, it continues to hold out until the circuit is reset.

This is accomplished by completing the circuit through the relay contacts, as shown in Fig. 12-12. In this case only a single battery is required, and the circuit is reset by a pushbutton switch. It is similar to the burglar alarm circuit previously described.

The circuit works as follows. With the photocell illuminated, R1 is adjusted so that the relay is not pulled in. When the light level falls and the photocell resistance rises, the balance of the potential-divider circuit shifts sufficiently for the current, amplified by the transistor, to operate the relay. The relay contacts now change over, removing the photocell from the circuit and providing a large positive bias on the base of the transistor. This maintains the current through the relay to keep it held in. At the same time the changeover of the relay contacts completes the alarm circuit. Pressing the reset button de-energizes the relay and restores the photocell to the circuit once more.

EXTRA-SENSITIVE ANNUNCIATOR RELAY

The addition of a second stage of transistor amplification to the type of circuit already described produces a circuit with extreme sensitivity to light changes. Such a circuit is shown in Fig. 12-13, where the potentiometer R4 controls the overall sensitivity by tapping off a voltage drop applied to the base of the second transistor as bias.

This again is a current-rise circuit: an almost negligible current is drawn from the battery when the photocell is strongly il-

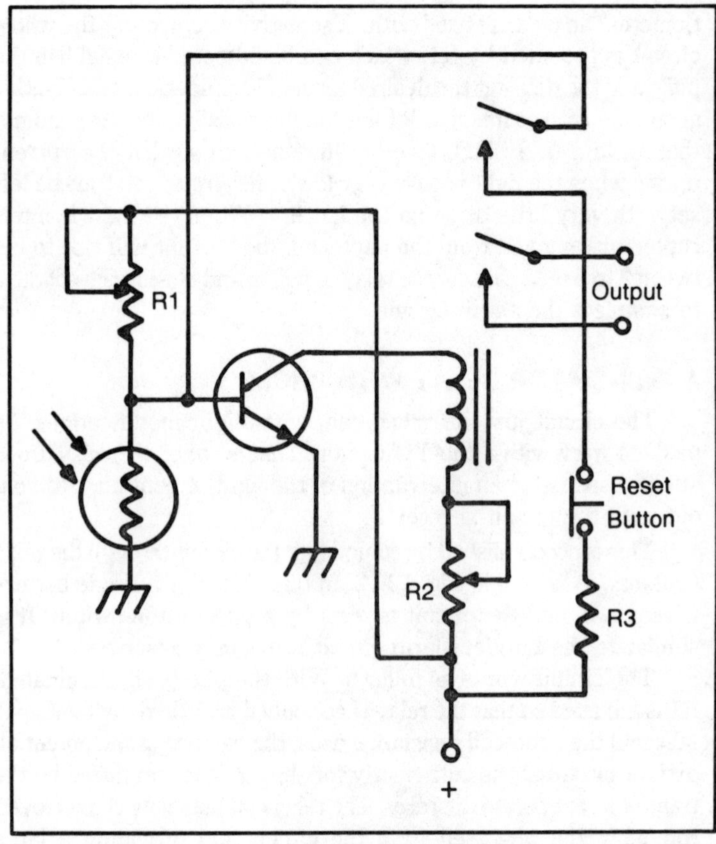

Fig. 12-12. A variation of the circuit of Fig. 12-11, incorporating a "hold" function.

luminated, and the relay is not operated. A fall in illumination produces a change in circuit conditions, which triggers the amplifier circuits so that sufficient current is then passed to pull the relay in. The current drain on the battery then rises from a few microamps to the order of 5 mA.

THE OPTOISOLATOR

A phototransistor and a light-emitting diode (LED) may be combined in a single envelope, such a device being known as an *optoisolator*. In this case the LED provides the source of illumination to which the phototransistor reacts. It can be used in two working modes: either as a *photodiode* with the emitter of the transistor part left disconnected; or as a *phototransistor*. In both cases the operation is governed by the *current* flowing through the LED section.

Further variations on this device are the *opto-Darlington-Isolator* and the *opto-triac-isolator*. The former comprises, typically, an optically coupled, gallium-arsenide, infrared-emitting LED and an npn silicon photo-Darlington transistor in a six-pin IC package, capable of collector-current switching from 100 mA (off-state) to 100 mA (on-state). The opto-triac-isolation is essentially similar, with a triac replacing the transistor pair, and capable of being worked directly from household ac voltage.

PHOTOELECTRIC CONTROLS

Photoelectric controls are used in numerous industrial and commercial applications for sensing, detecting, counting, and similar functions. They are commonly interfaced with logic capabilities to cope specifically with individual applications. Most photoelectric controls consist of a light source/photoreceiver combination providing a signal to a control *base*, which then amplifies this signal and applies the signal logic to transform it into a usable electrical component.

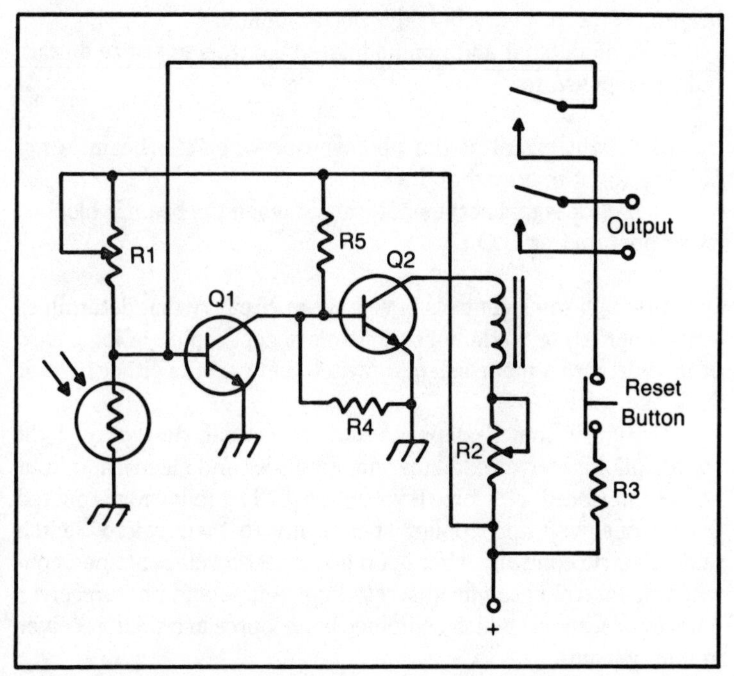

Fig. 12-13. A more sensitive version of the circuit in Fig. 12-12.

There are two main types of controls. The *self-contained* control includes the light source, photoreceiver, and the control base function, and the *modular control* uses a light source/photoreceiver combination or reflective scanner separate from the control base. Self-contained retroreflective controls require less wiring and are less susceptible to alignment problems, whereas modular controls are more flexible in allowing remote positioning of the control base from the input components and are more easily customized.

Modulated LED controls respond only to a narrow frequency band in the infrared. Consequently, they do not recognize bright, visible ambient light.

Nonmodulated controls respond to the intensity of visible light. Therefore, to maintain control reliability, they should not be used where the photosensor is subject to bright ambient light, such as sunlight.

Controls typically respond to a change in light intensity above or below a certain value or threshold response. However, certain plug-in amplifier/logic circuits cause controls to respond to the rate of light change (transition response), rather than to the intensity. Thus, the control responds only if the change in intensity or brightness occurs very quickly, not gradually.

Both modulated and nonmodulated controls energize an output in response to

☐ a light signal at the photosensor when the beam is not blocked (light operated or L.O.)

☐ a dark signal at the photosensor when the beam is blocked (dark operated or D.O.)

Although some controls have built-in circuitry that determines a fixed operating mode, most controls accept a plug-in logic card or module with a mode-selector switch that permits either light or dark operation.

Therefore, much depends on the actual design of light source/photoreceiver combination, amplifier and electrical output device interfaced with logic level circuitry. The following is quoted from Honeywell and applies specifically to their Micro Switch photoelectric controls, where you can select a self-contained control or a control base plus discrete light source and photoreceiver units (or a scanner, which combines light source and photoreceiver in one housing).

SCANNING TECHNIQUES

There are several scanning techniques (ways to set up the light source and photoreceiver to detect objects). The best technique to use is the one that yields the highest signal ratio for the particular object to be detected, subject to scanning distance and mounting restrictions.

Characteristics of the objects to be detected that have a bearing on scan technique include whether the objects are (1) opaque or translucent; (2) highly, or only slightly, reflective; (3) in the same position or randomly positioned as they pass the sensor. Detecting change in color is a special consideration.

Scanning techniques fall into two broad categories: *through scan* and *reflective scan.*

THROUGH SCAN (Fig. 12-14)

In through (direct) scan the light source and photoreceiver are positioned opposite each other, so light from the source shines directly at the sensor. The object to be detected passes between the two. If the object is opaque, direct scan will usually yield the highest signal ratio and, therefore, should be your first choice.

As long as an object blocks enough light as it interrupts the light beam, it may be skewed or tipped in any manner. As a rule of thumb, object size should be at least 50 percent of the diameter of the photoreceiver lens. To block enough light when detecting small objects, special converging lenses for the light source and photoreceiver can be used to focus the light in a small bright spot (where the object should be made to pass), thereby eliminating the

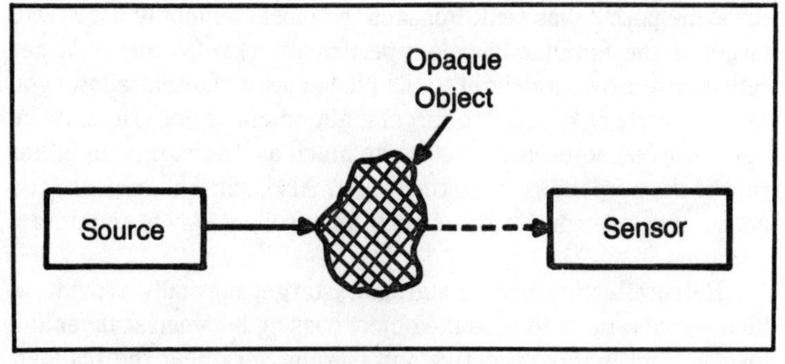

Fig. 12-14. Through (direct) scanning. The source is located some distance from the sensor.

need for the object to be half the diameter of the lens. An alternative is to place an aperture over the photoreceiver lens in order to reduce its diameter. Detecting small objects typically requires direct scan.

Because direct scan does not rely on the reflectiveness of the object to be detected (or a permanent reflector) for light to reach the photosensor, no light is lost at a reflecting surface. Therefore, the direct-scan technique lets you scan farther than reflective scanning.

Direct scan, however, is not without limitations. Alignment is critical and difficult to maintain where vibration is a factor. Also, with separate light source and photoreceiver, there is additional wiring, which may be inconvenient if the application is difficult to reach.

REFLECTIVE SCAN (Fig. 12-15)

In reflective scan the light source and photoreceiver are placed on the same side of the object to be detected. Limited space or mounting restrictions may prevent aiming the light source directly at the photoreceiver, so the light beam is reflected either from a permanent reflective target or surface, or from the object to be detected, back to the photoreceiver. There are three types of reflective scan: *retroflective scan, specular scan,* and *diffuse scan.*

RETROREFLECTIVE SCAN (Fig. 12-16)

With retroreflective scan, light source and photosensor occupy a common housing. The light beam is directed at a retroreflective target (acrylic disc, tape, or chalk) which returns the light along the same path it was sent. Perhaps the most commonly used retro target is the familiar bicycle-type (tricorner) reflector. A larger reflector returns more light to the photosensor and thus allows you to scan farther. With retro targets, alignment is not critical. The light source/photosensor can be as much as 15 degrees to either side of the perpendicular to the target. Also, since alignment need not be exact, retroreflective scan is an excellent way to counteract vibration.

Retroreflection from a stationary target normally provides a high signal ratio as long as the object passing between scanner and target is not highly reflective and passing very near the scanner. Retroreflective scan is a preferred technique to detect translucent objects and assures a higher signal ratio than is obtainable with

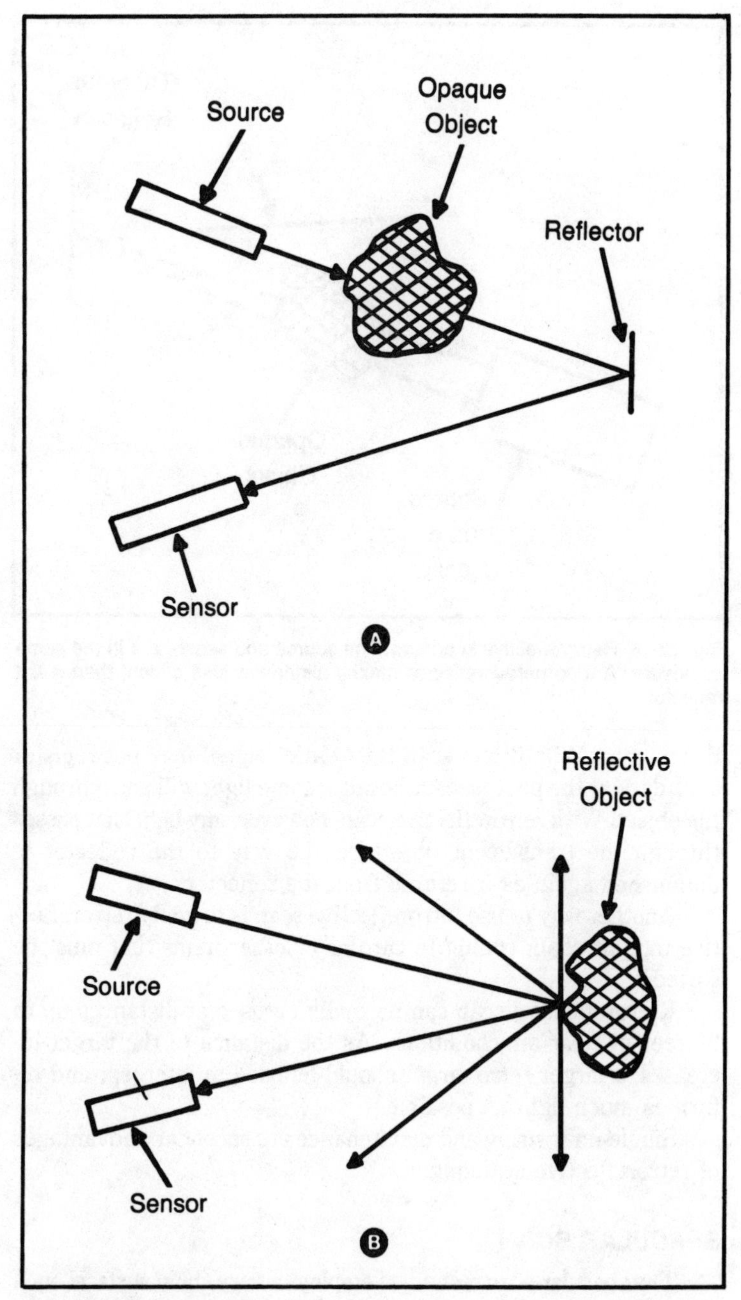

Fig. 12-15. Reflective scanning. At A, a permanent, fixed reflector is used. At B, the object itself reflects light from the source.

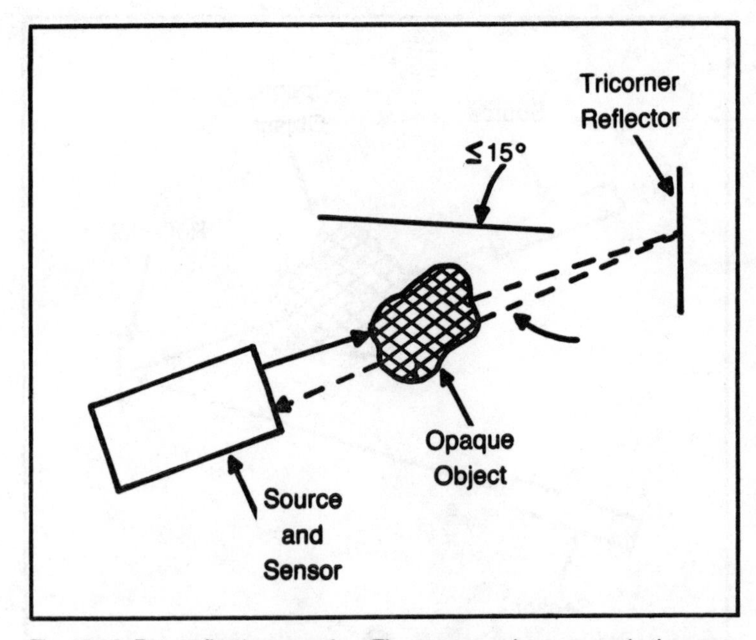

Fig. 12-16. Retroreflective scanning. The source and sensor are in the same enclosure. A tricornered reflector makes alignment less critical than a flat reflector.

direct scan. With direct scan the "dark" signal may not register very dark at the photosensor because some light will pass through the object. With retroreflective scan, however, any light that passes through the translucent object on the way to the reflector is diminished again as it returns from the reflector.

Another way to use retroreflective scan is to apply retroreflective tape or chalk coding to cartons or other items that must be sorted.

Retroreflective scan can normally be used at distances up to 30 feet in clear air conditions. As the distance to the target increases, a larger retro target should be used to intercept and return as much light as possible.

Single-unit wiring and maintenance are secondary advantages of retroreflective scanning.

SPECULAR SCAN

The specular-scan technique employs a very shiny surface, such as rolled or polished metal, shiny plastic, or a mirror, to reflect light to the photosensor.

With a shiny surface the angle at which light strikes the reflecting surface equals the angle at which it reflects from the surface. Positioning of the light source and photoreceiver must be precise (mounting brackets that fix the light source/photoreceiver relationship are available), and the distance of the reflecting surface from the light source and photoreceiver must be consistently controlled. The size of the angle between light source and photoreceiver determines the depth of scanning field. With a narrower angle there is more depth of field. With a wider angle, there is less depth of field. In a fill-level detection application, for example, this means that a wider angle between light source and photoreceiver enables you to detect fill level more precisely.

Specular scan can give a good signal ratio when required to distinguish between shiny and nonshiny (matte) surfaces, or when using depth of field to reflect selectively off shiny surfaces of a certain height. When monitoring a nonflat shiny surface with high and/or low points that fall outside the depth of field, these points will appear as dark signals to the photosensor.

DIFFUSE SCAN

Nonshiny surfaces such as kraft paper, rubber, and cork reflect only a small amount of light directly. Light is reflected or scattered nearly equally in all directions. In diffuse scan the light source is positioned perpendicularly to a dull surface. Emitted light is reflected back from the target to operate the photoreceiver. Because the light is scattered, only a small percentage returns. Therefore, scanning distance is limited (except with some high-intensity modulated LED controls), even with very bright light sources. It is often difficult to get a sufficient signal ratio with diffuse scan when the surface to be detected is almost the same distance from the sensor as another surface (for instance, a nearly flat or low-profile cork liner moving along a conveyor belt). When you are trying to distinguish between the reflection from two unlike surfaces, signal ratio can be improved considerably where the unlike surfaces have contrasting colors.

Diffuse scan is used in registration control and to detect material (corrugated metal, for example) with a slight vertical flutter, which might prevent a consistent signal with specular scan. Alignment is not critical in picking up diffuse reflection.

COLOR DIFFERENTIATION

In distinguishing color, as in registration mark detection, contrast is the key. High contrast (dark color on light, or vice versa) provides the best signal ratio and control reliability. Therefore, if possible, plan early to use bright, well-defined, contrasting colors in your operation.

Diffuse scan is normally used to detect color change. The chart gives some of the common color combinations that must be distinguished in registration control, plus the most suitable type of photosensor and scan technique.

When the background is clear (transparent), the best method is to detect any color mark with direct scan. When the background is a second color, contrasts such as black against white usually assure sufficient signal ratio (difference between dark and light signals) to be handled routinely with diffuse scan. Red, or a color that contains much red pigment (yellow, orange, brown) on a white or light background is a special case. You should use a photoreceiver with a cadmium sulphide (CdS) cell to detect red marks because it makes red appear dark on a light background.

A retroreflective scanner with a short-focal-length lens (but without a retro target) can be used to detect registration marks. It is placed near the mark and is actually used in the diffuse-scan technique. If you use a retro scanner to detect marks on a shiny surface, cock the scanner somewhat off the perpendicular to make certain you pick up only diffuse reflection. Otherwise the shiny surface of the mark could mirror-reflect so brightly it would overcome the dark signal a CdS cell normally gets from red. This would mean a light signal from both background and mark. In detecting colors a rule of thumb is to use diffuse (weakened) rather than specular (mirror) reflection.

SENSITIVITY ADJUSTMENT

Most photoelectric controls have a sensitivity adjustment to determine the light level at which the control will respond.

Conditions that could require the sensitivity to be adjusted to less than fully clockwise (maximum) include

- ☐ Detecting translucent objects
- ☐ High-speed response
- ☐ High cyclic rate
- ☐ Line voltage variation
- ☐ High electrical noise atmosphere

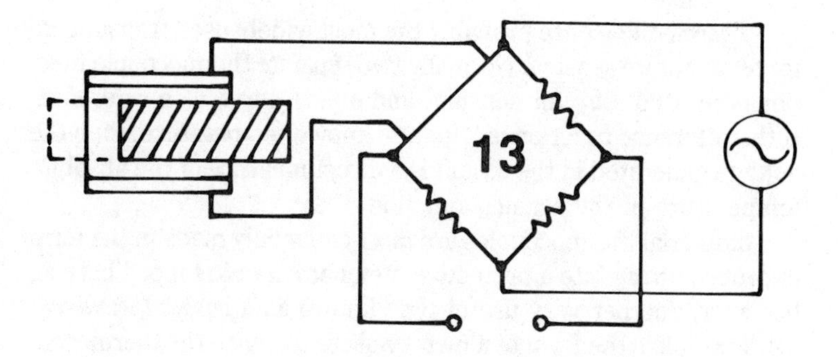

Sensors

Sensors are devices that change a parameter into a form that can be measured or recorded easily. In this sense they can be classified as *transducers,* although not all types of sensors convert one form of energy into another. Thus, there are (true) transducer-type sensors and nontransducer-type sensors. An example of the latter is a mechanical switch operated by a physical movement to "sense" the end of that movement and apply an electrical signal to indicate or control it. It could reverse the direction of current to a dc motor powering the movement, for instance, making the movement reverse when it reached a limiting position. This is consistent with the main use of sensors in providing an electrical signal output that is usually the simplest way of measuring or recording a parameter required.

TEMPERATURE SENSING

Temperature can be sensed and indicated, or recorded, by a thermometer. A simple thermometer, however, can only provide a *visual* indication of temperature. To convert heat energy (temperature) into a corresponding electrical signal that can be indicated, or recorded, on a meter, you need a time transducer. The most usual type in this case is the *thermocouple* (see Chapter 14). Alternatively, heat energy (temperature) can be converted into mechanical energy to operate a pointer movement directly, in which case another type of transducer—the bimetal strip (see Chapter 15)—is used.

Thermocouples are probably the most widely used transducers for temperature sensing. Normally two separate thermocouple functions are used: one for sensing, and one to serve as a reference. If the reference function is kept at a known temperature, then the voltage generated in the circuit is a direct measure of the absolute temperature of the sensing function.

Industrial thermocouples are most commonly made in the form of probes fitting into a protective metal or ceramic tube. There is, however, another very useful type known as a *gasket thermocouple.* This takes the form of a metal washer to which the thermocouple wires are bonded. This kind of thermocouple can be attached to existing bolts or studs like an ordinary washer, even under spark plugs for measuring cylinder-head temperatures.

THERMISTOR AND RESISTANCE PROBES

Another type of transducer used for temperature sensing is the *thermistor* or *resistance probe.* Actually there are two distinct types, but they work on the same principle. The material used for the probe undergoes a change in electrical resistance proportional to any change in temperature. Thus, measuring this change of resistance in an electric circuit (as a change in voltage or change in current) gives a direct measure of the change in temperature involved.

In the case of the thermistor and the resistance probe, the change in resistance is opposite in effect. With a thermistor the resistance decreases with increasing temperature, but with a resistance temperature screen the resistance increases with increasing temperature. The reason is that two quite different materials are involved. The thermistor uses a semiconductor material as the sensory element. The resistance temperature sensor uses a metal sensory element.

Thermistor probes are normally in the shape of heads, thin rods, disks, or washers. They are particularly sensitive to temperature changes and are thus capable of providing enough signal strength to operate a meter in a simple circuit without amplification. A typical thermistor probe may show a resistance change from 75,000 Ω at −40° C. to only 40 Ω at 150° C., or something like 400 Δ per degree ° C. temperature change (although not necessarily in strictly linear proportion).

On this basis we could anticipate a typical resistance at 0° C.

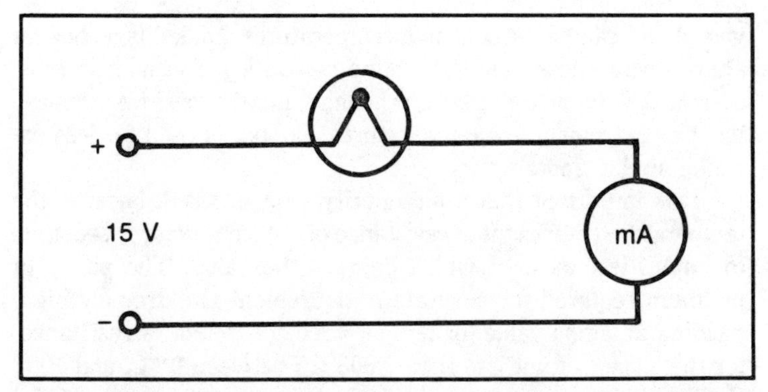

Fig. 13-1. A thermistor in conjunction with a power supply and a milliammeter can be used to measure temperature.

of say, 60,000 Ω, and a resistance of 100° C. of 16,000 Ω. In a simple meter circuit, as in Fig. 13-1, with a 15-V supply, this would give a circuit range from 15/60,000 = 0.25 mA to 15/60,000 = 0.93 mA. Thus, a 0-1-mA meter would cover temperature measurement from 0° to 100° C. comfortably with good scale spacing and no signal amplification necessary.

Resistance temperature probes are considerably less sensitive but, being of metal, can be used for much higher temperatures than thermistors. Their upper temperature limit is two or three times higher than the typical limit for thermistors. They are also more suitable for use as sensors of very low temperatures.

The resistance/temperature characteristics of metallic wires are expressed by their temperature coefficient of resistance, which is not a constant. It normally tends to increase with increasing temperature; the change (increase) in resistance becomes more marked. However, some metals shown an opposite effect. Alloys of nickel and manganese, for example, have equal and opposite nonlinearities in resistance-temperature characteristics. Thus, combining the two in a sensor can provide a substantially linear change of overall resistance with temperature.

Typical performance one might expect from a nickel-manganese sensor is a resistance of the order of 300 Ω at 80° F., falling to about 200 Ω at some very low temperature, say around -400° F. The actual resistance change is thus quite small—only about one-fifth of an ohm per degree Fahrenheit.

In practice, resistance temperature sensors are normally based on a single metal of high purity. Platinum is the usual choice because it does not oxidize or corrode (except in the presence of carbon gases) and can be used at high temperatures. Nickel is a cheaper alternative for less exacting commercial duties. Special materials, such as 0.5 atomic-percent iron-rhodium, may be used for measurement of extremely low temperatures. Copper is used for less extreme applications.

It is important that a high-purity wire be used, because the temperature coefficient of resistance of any conductor is sensitive to impurities, most of which depress the value. The purity of platinum required for accurate measurement is extremely high, yielding an alpha value (or temperature coefficient of resistance) for the sensor of not less than 0.003925 between 0° C. and 100° C. The purity of platinum wire is, in fact, normally expressed by its alpha value.

The relationship between variations in resistance of a wire with temperature and the reference scale (gas scale) can be expressed mathematically as

$$tg - tp = C\left(\frac{tg}{100}\right)^2 - \frac{tg}{100}$$

where tg = temperature on the gas scale
tp = temperature on the platinum-wire scale
C = a constant depending on the purity of the wire (for pure platinum the value is 1.5)

Differences between the two scales are simply corrected by calibration.

The typical resistance temperature probe uses platinum, nickel, or copper based in a metal or ceramic enclosure. Because it is a large surface area relative to its mass, it is mainly suitable for measuring temperatures of an area rather than a point.

Such elements are connected to a compensated bridge circuit for signal readout. Figure 13-2 shows a wheatstone bridge circuit (A) and a ratio-arm bridge circuit (B).

CERAMIC TEMPERATURE SENSORS

Certain ceramic materials have an electrical resistance that is largely unaffected by temperature until it reaches a particular value where the material undergoes a crystalline change and the resistance increases abruptly. The temperature at which this occurs is the Curie point, where the resistance rises within a span of a few degrees.

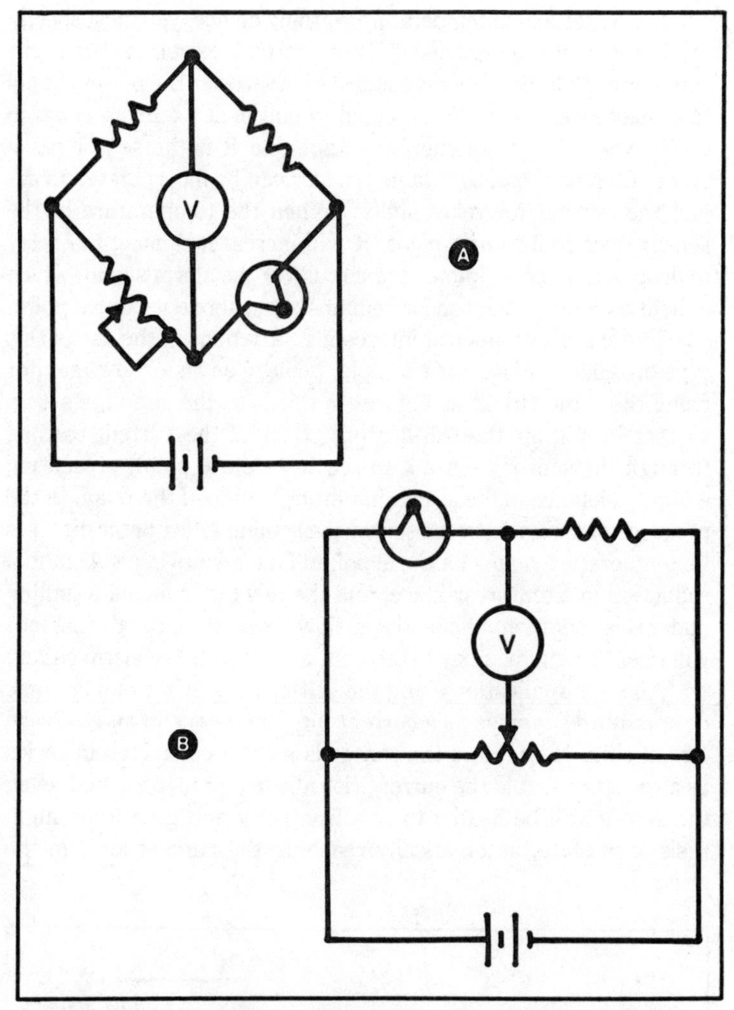

Fig. 13-2. Two types of bridge circuits that can be used with thermistors. At A, a Wheatstone bridge; at B, a ratio-arm bridge.

Such ceramic devices thus sense when a particular temperature (its Curie point) is reached, generating a substantial rise in resistance in an electrical circuit. By selecting the composition of the material and its treatment during processing, one can set the Curie point at any level between about 60° C. and 180° C. Also, the resulting resistance change can be controlled, that is, both the actual change in resistance and the small range of temperature (normally not more than 5° C.) over which this change occurs.

For use as a high-temperature warning device, you must specify the Curie point required and know its normal resistance. In the circuit in Fig. 13-3 the sensor is connected in series in the power supply to a relay selected and/or adjusted to pull in at a current equal to V_S/R, where V_S is the supply voltage and R is the sensor resistance. Alternatively, a variable resistor can be incorporated to adjust the current for relay pull-in. When the temperature of the sensor rises to its Curie point, R will increase, causing the relay to drop out. This completes the circuit for the alarm signal, which is held as long as the sensor remains at or above its Curie point.

There are a number of interesting variations in the use of this type of sensor. It can, for example, be used as an airflow monitor using the same circuit as before. In this case the circuit is set up so that in still air the self-heating effect of the current passing through the sensor causes it to reach its Curie point, presenting a high resistance in the circuit inhibiting pull-in of the relay. In the presence of airflow past the sensor the cooling effect of the air loses its temperature below its Curie point. This results in a substantial reduction in circuit resistance, and the relay is adjusted to pull-in under this condition. Should the airflow cease, the circuit resistance will rise, causing the relay to drop out and activate the alarm circuit.

Another application using the self-heating effect of a ceramic temperature sensor is as a current limiting device in an electrical circuit (Fig. 13-4). Here the sensor is merely connected in series as a resistor. Should the current rise above a predetermined level, the sensor will be heated to its Curie point and go into its high-resistance state, automatically reaching the current level in the circuit.

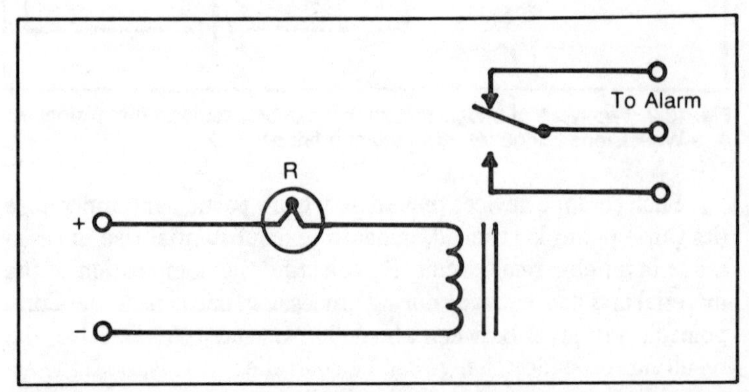

Fig. 13-3. A simple thermistor-operated alarm control for warning of overheating condition.

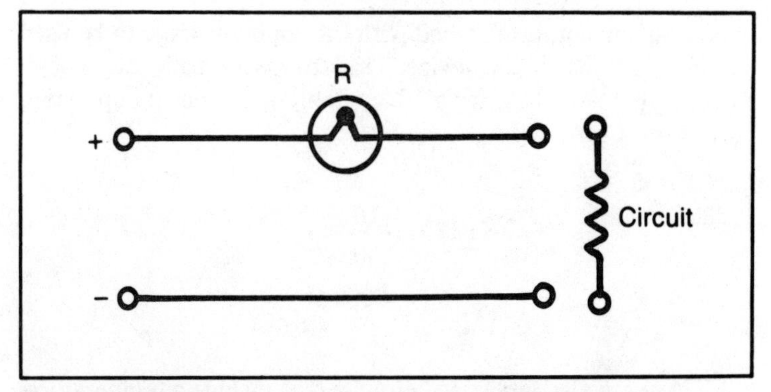

Fig. 13-4. A thermistor can be used for limiting the current in a circuit, preventing damage from overheating.

SOLID-STATE TEMPERATURE SENSORS

A simple solid-state temperature sensor can be based on a form of zener diode so constructed that its breakdown voltage is directly proportional to absolute temperature. A typical device of this type may have a working range of − 10 ° to + 100 ° C. with a linear output of the order of 10 mV per degree Celsius.

Characteristics of such a device will normally specify an output voltage at a specific temperature; a typical value would be of the order of 3 V at 25 ° C. This, over the temperature range of, say 0 ° to 100 ° C., the anticipated voltage range would be from 2.75 V (at 0° C. to 3.75 V (at 100° C.). Figure 13-5 illustrates a temperature-measurement circuit using such a device.

This device can be expected to have an operating current range from about 0.5 to 5 mA, or an optimum operating current of 1-2 mA. Here the value of the resistor R is selected to give a suitable

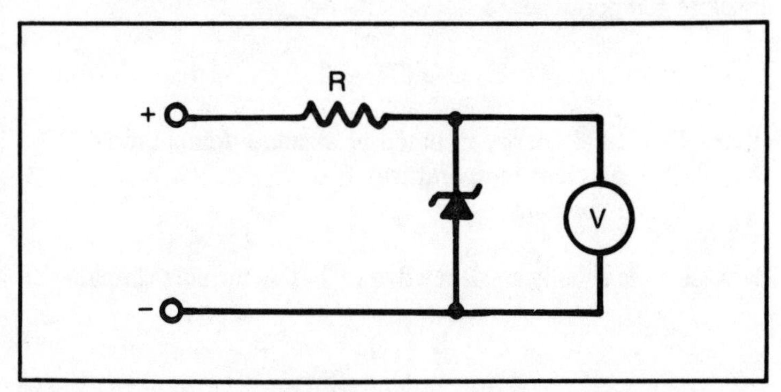

Fig. 13-5. A circuit using a solid-state sensor for measuring temperature.

operating current, established with the supply voltage to be used and the resistance of the device. The latter is normally low enough to be negligible. Thus, for a 10-V supply and a design operating current of 2 mA, say,

$$R = \frac{10}{.002}$$

$$= 5000\ \Omega$$

There are also integrated circuits that include a temperature-saving device of the type described together with a stable voltage reference and an op amp in the same IC chip. If the sensing device has the same characteristics (10 mV per degree Celsius), output voltage from the IC then goes negative at this rate with temperature increase.

The particular attraction of such an IC is that any temperature-scale factor can be realized with external resistors.

PYROMETERS

Pyrometers are another type of temperature sensor. Optical and radiation pyrometers depend on the intensity of the radiation, either monochromatic or total, emitted by a hot body. Optical pyrometers rely on visual observation of the indicator; radiation pyrometers use a receiver, such as a thermocouple, that reacts to the change in temperature produced by the radiation.

The principle of operation is based on the fact that the total energy by a blackbody is proportional to the fourth power of its absolute temperature, or

$$E = O\ (T^4 - T_0^4)$$

where E = total energy radiated at absolute temperature T
T_0 = ambient temperature
O is a constant.

Because T_0 is usually small relative to T, this formula simplifies to

$$E = OT^4$$

In the case of an optical pyrometer the intensity of the light emitted by the hot body is compared with that given by a standard hot body in the same wavelength, usually either by matching to a constant comparison lamp or by adjusting the brightness of a single lamp until the filament "disappears" (has the same color temperature as the emitted light). Optical pyrometers are usually fitted with a red monochromatic glass, which assists color matching and also protects the eye. They can be used to measure temperatures from about 100° C. and up but are normally only employed for measuring very high temperatures above the range covered by other instruments (about 2000° C. and above). They are generally less accurate at any temperature than the other types of thermometers described.

In addition to the well-established forms of optical and radiation pyrometers, other instruments include:

Infrared radiation pyrometers: generally more sensitive to lower temperature and can give accurate readings within 1 to 2 percent down to temperatures as low as 50° C.

Two-color pyrometers: operate on two wavelengths used on a ratio basis and eliminate the need for estimating the emissary effect of nonblackbodies (or coating the surface with black).

Color still anf cinematography: the optical density or color of the film provides a quantitive measurement of temperature.

Thermographic pyrometers: the ultraviolet content of the radiation is determined by its effect on a phosphor-coated screen, which can be analyzed quantitatively in terms of optical density. Largely, however, this method is mainly useful for displaying the temperature *pattern* over a surface.

HEAT-SENSITIVE MATERIALS

Finally, on the subject of temperature sensing, a number of materials can act directly as sensors by changing their color or appearance at a specific temperature or specific combination of temperature and exposure time. These are known as *thermographic materials.*

Simple examples are crayons or lacquer, which can be applied to a surface to produce a mark. Once a particular temperature is reached, the mark liquefies sharply, the change in appearance being easily noticeable. Similar materials are also available in pellet form, the "calibrated" temperature being indicated by the first signs of liquid appearing due to melting of the pellet.

Quite close "calibrated" temperatures can be provided; for example, steps of about 3 to 4 degrees Celsius over a range that may extend from about 45 ° C. to over 1000 ° C. are common. Accuracy of temperature rating can be within plus or minus 1 percent.

Temperature-indicating materials that liquefy at a specific temperature are usually thermoplastic or "reversible." That is, once having melted, they will solidify again to a "dry" mark when the temperature falls below the rated value. Other crayons or lacquers may undergo a permanent color change at their rated temperature and thus do not need monitoring or a continuous watch. They indicate that a specific temperature has been reached or exceeded. A series of marks with different temperature ratings could be used to indicate a peak temperature reached when the subject was not under observation.

Pyrometric cones and bars are another type of heat-sensitive material, mainly used for temperature-time or "heat-soak" indication in pottery kilns (although they can have other industrial applications). They are refractory materials cast in rigid shapes that are maintained until the cone or bar has absorbed a specific quantity of heat. At this point they will sag, and if heating continues, they will melt. A series of such cones, each with a different temperature rating, can thus be used to determine oven or bulk heating by direct observation or to record the amount of heating they have received during a firing cycle.

A variation on this technique is the use of special alloy "plugs" or "pellets," that undergo a permanent change in hardness, permeability, or some other physical property when exposed for a certain duration at a fixed temperature. After a given time they are removed, and the change in characteristics is measured, from which the degree of heat soaking they have received can be assessed.

LIGHT SENSORS

Light sensors or true optical transducers are based on both photovoltaic and photoconductive devices (see Chapter 12). Such systems can be extremely flexible, capable of detecting both visible and invisible light signals, and used to detect, count, stop, start, sort, measure, position, control, and perform any number of similar functions.

Light sensors are packaged as individual units or in various arrays for reading multiple-impulse patterns. Complete units with associated optics and electronics may be available as scanners, recorders, counters, or cartridges. In this case four basic elements are involved: a light source, an optional system, a photoelectric sensor, and necessary electrical processing equipment (although, as is common, the latter may be separated from the sensor package).

Light sources may consist of an incandescent lamp, a red lamp, or a LED (emitting visible light or infrared).

The *optical system* normally employs lenses, mirrors, and prisms, as appropriate, for straight-line optive paths. For more complex paths, or to simplify presentation of light at a point source, fiber-optic bundles can be used.

The *photoelectric sensor* used is normally one of the four following types, the choice largely depending on the application.

Photovoltaic cells: self-generating with a maximum output of the order of 0.5 V. Output is substantially linear with illumination and varies logarithmically with incident radiation. Types of cells include silicon, selenium, germanium, gallium arsenide, and indium antimonide, with silicon and selenium being the most active types. Silicon cells provide approximately 20 times more output than selenium cells and have a much broader range of frequency response. Selenium cells are more responsive in the visible-light range, although they are now outperformed by second-generation silicon cells. Selenium cells are also subject to fatigue with repeated exposure to high levels of illumination.

Photodiodes: operate in a switching mode.

Phototransistors: also operate in a switching mode but with amplification of signal or gain in proportion to the circuit of light falling on them.

Photo SCRs: capable of switching large amounts of power as the result of the absence or presence of light. This can result in considerably simplified electronic circuitry for on-off (digital) switching. Photo SCRs can switch as much as 500 mA at 200 V, with trigger threshold readily adjusted by bias resistance.

PROXIMITY SENSORS

Proximity sensors detect and/or measure the proximity of some object relative to a base position without actually touching the object. They are thus *noncontacting* sensors or probes in the mechanical sense. A variety of principles is used in the design of these sensors, including reflected light, radiated heat, reflected electromagnetic radiation (up to and including radar, which is a more extreme form of proximity sensor), pneumatic (airflow) and fluidic devices, and magnetic disturbance. Some are power sensors and some are true transducers.

Proximity sensors are described collectively in Chapter 25. Also see Chapters 5, 16, and 17. Photoelectric devices are widely used as proximity sensors (Chapter 12).

FIBER-OPTIC TACHOMETER

The *fiber-optic tachometer* consists of a flexible light guide connected to an electro-optic unit. The guide consists of mixed light-transmitting and light-receiving fibers terminating in a focusing optional head. Light from the electro-optical source passes down one set of fibers when it is focused and directed onto a target on the straight edge or rotor to be measured. The target consists of a reflecting strip, so that light is reflected back through the second set of fibers to the detection circuit in the electronic unit.

Each movement of the target past the end of the light probe (once per revolution) thus generates a corresponding return light pulse, which the electronic unit converts into an electronic pulse—usually a direct digital readout signal, the pulse frequency of which is directly related to shaft rotational speed.

This type of transistor has many advantages for revolution counting. Alignment problems are minimal because the transmitter and receiver are in the same head; also, the probe can operate over a wide range of distances from the target (typically from 1/4 in. to over 20 in. if necessary). The output pulse signal is noise free and of constant amplitude, making it easy to process. It can also work over a wide range of frequencies from 0 (100-percent work-space pulse on target) to about 6000 pulses per second (equivalent to a rotational speed of 360,000 rpm with a single target). The actual signal-pulse frequency generated can also be adjusted upwards for low speeds (low rpm) by using multiple targets on the shaft or rotor.

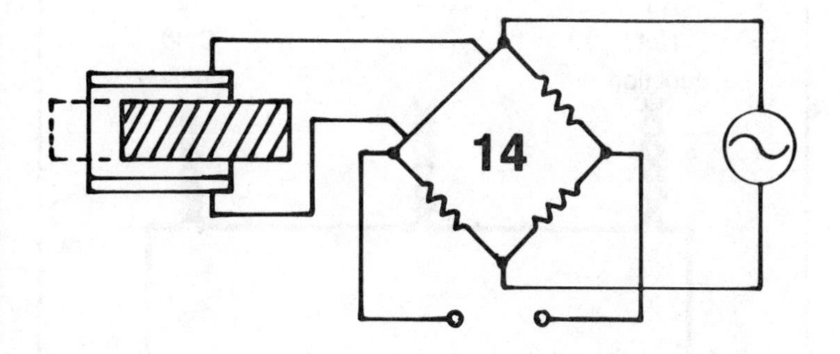

Thermocouples

A *thermocouple* is a transducer that transforms heat into electrical voltage. The construction involved is that of two pieces (usually wires) of dissimilar metals fused together at one end, with the other two ends separated. Heating the joined end will make the device work as a voltage generator. In this respect, and in geometry, it differs from the other transducer utilizing two dissimilar metals—the bimetallic strip—which transforms heat into mechanical movement (see Chapter 15).

PRINCIPLE OF THE THERMOCOUPLE

To work as a voltage generator, you must convert the thermocouple into a conductive loop (see Fig. 14-1). The joined end must then be at a higher temperature than the open ends, known as the *hot junction* and *cold junction*, respectively. The value of the voltage generated is then given by

$$\log V = A \log (T1\text{-}T2) + B[(T1)^2\text{-}(T2)^2]$$

where V is the voltage (or strictly speaking, the emf) generated in microvolts
T1 is the hot-junction temperature
T2 is the cold-junction temperature
A and B are constants depending on the metals forming the thermocouple

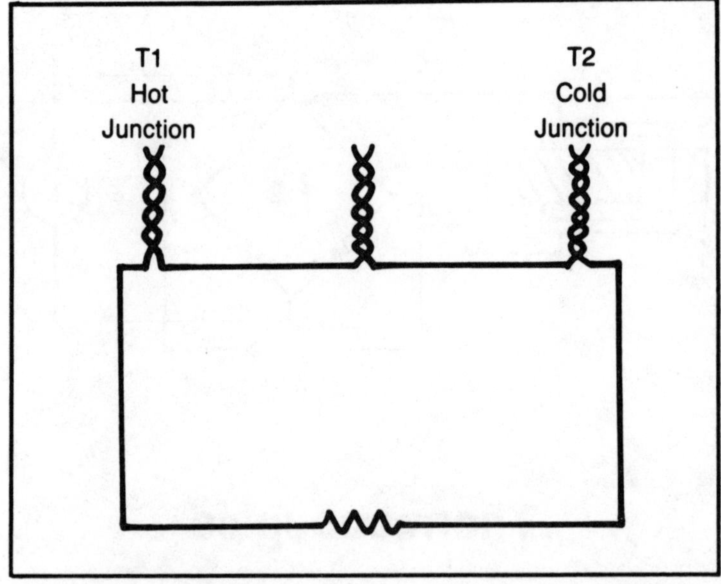

Fig. 14-1. A thermocouple is connected in a conductive loop, generating current through the load resistance.

A wide variety of different metal combinations will work as thermocouples, but some combinations are much better than others. These are the combinations used in making proprietary thermocouples. Typical examples are copper—constantan and platinum/nickel—platinum, representative of a base-metal combination and noble-metal combination, respectively. Here the values of the constants A and B are

copper-constantan -	A = 1.14
	B = 1.36
plantinum/nickel-platinum -	A = 1.22
	B = 0.36

CHARACTERISTICS OF THERMOCOUPLES

It is a general characteristic of thermocouple combinations that the values of the constant A are fairly similar, but the values of constant B are usually much larger for base-metal combinations than noble-metal combinations. As a result, base-metal thermocouples generate higher voltages for the same differences in temperature between hot and cold junctions. However, they are much more limited in *maximum* temperature they can withstand, so actual maximum voltages that can be generated are much lower.

For example, the maximum service temperature for a copper-constantan thermocouple is 400° C. (or 500° C. for intermittent use). At this temperature it will generate an output of about 20,000 mV. A platinum-platinum alloy thermocouple, on the other hand, may have a maximum service temperature of 1500°-1600° C., at which temperature it will generate an output of about 10,000 mV, still lower than the copper-constantan thermocouple. On the other hand, a palladium-gold/indium-platinum thermocouple has a maximum service temperature of 1200° C., at which temperature it will generate an output of 40,000-60,000 mV (depending on the alloy composition).

Thermocouples are seldom used as transducers as such but are used as *sensors* for temperature measurement. Examples of metal combinations used and their applications are summarized in Table 14-1.

CONSTRUCTION

Thermocouples are commonly constructed in the form of parallel lengths of wire joined at one end, usually by fusion welding, to form a hot junction. The wires are protected and insulated from each other by twin-bore refractory insulators, the whole thermocouple being supported and protected by a closed-end refractory outer covering or sheath. Instrumentation for measuring the emf output of the circuit is then connected to the two cold-end wires.

Proprietary types of thermocouples also include those with metal cladding for protection, insulated types, and others in flexible coil form and other shapes for specific applications.

It is equally possible to make thermocouples by joining two pieces of wire together by welding, choosing one of the metal combinations in the table. However, the availability of the right metal wires is strictly limited (most are produced only for thermocouple manufacture). Also, difficulty may be experienced in welding the ends together to produce a hot junction (no joint other than welding will be suitable). It makes sense, therefore, that if you want to use a thermocouple for a project, you should buy a ready-made proprietary article.

THERMOCOUPLE CIRCUITS

Only a simple loop circuit is needed to use a thermocouple as a "working" thermometer (Fig. 14-2). The main requirement is to

Table 14-1. Thermocouple Materials.

Combination	Output mV	Maxiumum service temperature	Remarks
Platinum: 10% rhodium-platinum Platinum: 13% rhodium-platinum	10,334 at 1064° C.	1500° C.	Used in a wide variety of industries and generally the most accurate type. Extremely stable (provided rhodium drift is avoided).
6% rhodium-platinum: 30% rhodium-platinum	5437 at 1064° C.	1600° C.	Developed for higher temperatures. 6% rhodium-platinum: 30% rhodium-platinum particular suitable for long life at elevated temperatures. Other rhodium-platinum proportions also used.
70% rhodium-platinum: 40% rhodium-platinum	1590 at 1064° C.	1700° C.	
Iron-gold: chromel		100° C.	Developed for measuring small temperature differences. Highly sensitive. Very suitable for cryogenic temperatures.
Iron: rhodium		—	Developed for cryogenic temperature measurement.
Indium: indium-rhodium-iridium	11,000 at 2000° C.	2100° C.	Developed for temperature measurement in range 1500-2000° C.
Copper: constantan	20,680 a 400° C.	400° C.	Low-cost, low-temperature thermocouples.
Iron: constantan	47,390 at 850° C.	850° C.	Low-cost, medium-temperature thermocouple.
Chromel: alumel	45,000 at 1100° C.	1100° C.	Low-cost, high-temperature thermocouple.
40% palladium-gold: 10% iridium-platinum	63,470 at 1064° C.	1200° C.	Developed for accurate measurement of temperature over the range 0-1200° C. Thermal emf comparable to constantan: iron.
46° palladium-gold: 12.5% iridium-platinum	43,440 at 1064° C.	1200° C.	Developed for accurate measurement of temperature over the range 0-1200° C. Noble metal alternative to chromel: alumel.

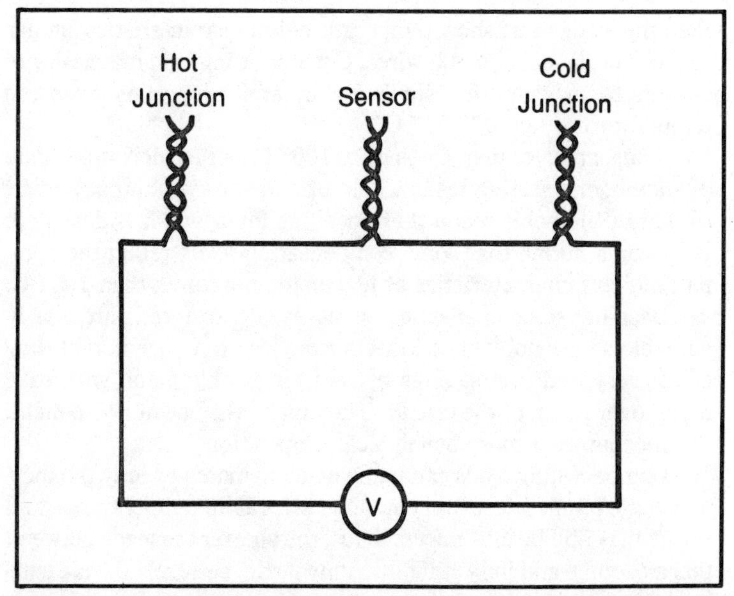

Fig. 14-2. A thermocouple temperature sensor. Either a voltmeter (shown here) or a milliammeter may be used.

match the voltmeter/millivoltmeter to the output range over the temperature at which the thermocouple is to be used. Employing a proprietary device for the thermocouple also means that you will be supplied with its output characteristics, from which you can calibrate the voltmeter/millivoltmeter scale in degrees of temperature (or simply refer to the characteristics to convert millivolt readings into temperature).

This simple type of circuit can, however, be highly inaccurate. The reason is that the leads connected to the thermocouple can themselves have a thermocouple effect. Effectively they extend the length of the thermocouple and thus affect the actual position of the cold junction. Ideally, the cold junction should be kept at a fixed temperature so that the voltage generated by the hot junction of the thermocouple at a given temperature is always the same.

To overcome this particular problem, you need *compensating leads* to transmit to the measuring instrument, with a minimum of error, a thermocouple signal from a cold junction that may be at either an unknown but fixed temperature or possibly at a temperature that is both variable and unknown.

When base-metal compensating leads replace what would otherwise be continuations of the noble-metal thermocouple limbs, then the leads must show temperature-emf characteristics similar to those of the nobel-metal wires. Compensating lead materials are chosen to provide this similarity in emf properties over the temperature range 0°-100° C.

Thus, at all temperatures up to 100° C., a thermocouple made of such compensating leads would provide an emf output similar to that of the noble-metal thermocouple for which it is designed. It is worth noting that some compensating leads retain the comparable emf characteristics at higher temperatures than 100° C.

Base-metal compensating leads are, therefore, particularly valuable where noble-metal thermocouples are in use in that they effect very real cost savings by replacing noble metal with base metal over much of the circuit. They make the use of noble-metal thermocouples a more economical proposition.

Compensating leads cannot be made to match exactly the thermocouple temperature-emf characteristics at all temperatures, and errors may still be introduced. Thus, the shorter the leads between thermocouple and indicating instrument, the better. If the measuring instrument is located some distance from the thermocouple, perhaps to remove it from a high-temperature area, then compensating leads are strictly necessary, with one exception. A few thermocouple combinations (6-percent rhodium-platinum/30-percent rhodium-platinum, for example) do not need compensating leads at all. Here ordinary copper leads can be used.

When compensating leads are used, two conditions must be met to prevent large errors in temperature measurement:

☐ The temperature of the noble-metal/base-metal junctions can vary but must not rise above the maximum operating temperature for the compensating lead (usually 100° C.).

☐ The noble-metal/base-metal junctions must each be at the same temperature. This should not be a problem because the two junctions are usually adjacent.

The first condition dictates the length of noble metal used for the thermocouple.

The noble-metal/base-metal junction is best made by small terminal connectors, but any of the standard methods of electrical connection are sufficient. Ideally, the junctions should have low thermal mass and be small, so that they can be kept close together without being in contact. This ensures that both joints are at the same temperature in service.

Table 14-2. Positive/Negative Characteristics of Thermocouple Materials.

Thermocouple	Compensating leads
10-13% rhodium-platinum/platinum	copper/0.6% nickel-copper
40% rhodium-platinum/platinum	copper/1.0% zinc-copper
10% iridium-platinum/40% palladium-gold	chromel/alumel
40% iridium-rhodium/iridium	15% nickel-copper/19% nickel-copper
copper/constantan	copper/constantan
iron/constantan	iron/constantan
chromel/alumel	chromel/alumel

Note also that different materials are needed for the positive leg and negative leg in compensating leads to match the positive-negative characteristics of the thermocouple materials. Examples are shown in Table 14-2, with positive first in each case.

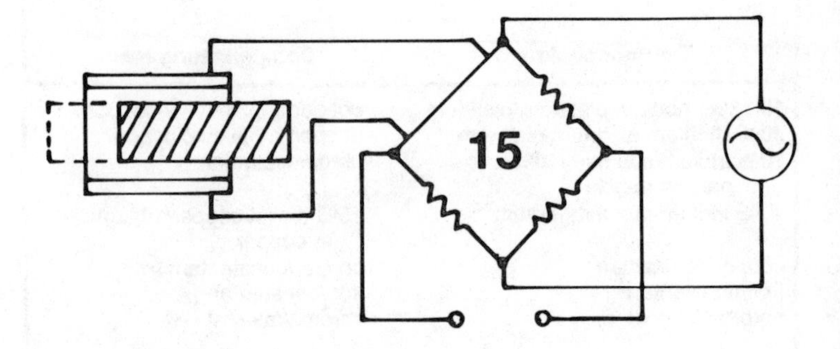

Bimetal Strips

Bimetal strips or *thermostat metals* comprise two (or more) layers of dissimilar metals bonded together.

When subjected to a change in temperature, the strip will bend by virtue of the difference in thermal coefficients of expansion of the metals. Basically, therefore, thermostat metals are simple transducers, converting thermal energy into mechanical energy.

Bimetal strips, also known as *thermometals,* may be used for the following functions:

Temperature indication: for example, in thermometers employing a bimetal strip sensor

Temperature control: for example, in thermostats

Control of functions with temperature change: for example, in automatic chokes on automobile engines

Compensation for length or force with changes in temperature: for example, in lever mechanisms

Control devices: for example, circuit breakers, which are heated by the load circuit and which trip if the current reaches overload values.

The first four functions are direct transducer reactions, where temperature change (heat) is converted into mechanical movement. The fifth function is an indirect application involving two separate stages of conversion: electric current into heat, and then heat into mechanical movement.

STRAIGHT-STRIP DEFLECTION

In the case of a narrow straight strip of bimetal (Fig. 15-1), the strip will bend to an arc of a circle, the radius of curvature of which is dependent on the difference in coefficients of expansion of the two metals, their thicknesses and elastic moduli, and the change in temperature. If you ignore the difference in elastic moduli (which is usually quite small with thermometals), you can use the working formula

$$R = \frac{(t_1 + t_2)^3}{6d\,T\,t_1 t_2}$$

where R is the radius of curvature
- t_2 and t_2 are the thicknesses of the two metals in millimeters
- d is the difference in coefficients of expansion of the two values
- T is the temperature change in degrees Celsius

The significance of this formula is that curvature is greatest

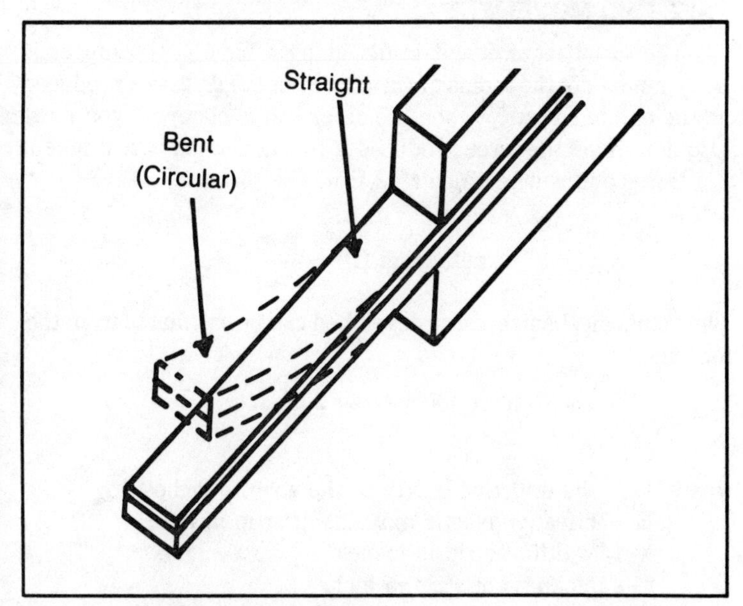

Fig. 15-1. A bimetal strip bends along a circular arc with a rise or fall in temperature from the straight condition.

when the two metal thicknesses are the same. Thus, for maximum mechanical movement, if you were making your own bimetal strip, you would bond two strips of the same thickness together from metals with the largest difference in coefficients of expansion.

In practice, ready-made thermometals (bimetal strips) would normally be used, where the curvature can be determined from the simple formula

$$R = \frac{t}{2KT}$$

where t is the total thickness of the thermometal, and K is a constant that is different for each thermometal.

The quantity 2K, known as the *flexivity*, is the change in curvature per unit temperature change per unit thickness. For proprietary thermometals the value of K is usually of the order of 7.3×10^{-5} but may be higher or lower with special types. If possible get the maker or supplier to specify the actual K value for the thermometal you are using. If not, use the value 3×10^{-6}.

FREE-END MOVEMENT

The usual requirement in designing a bimetal transducer is determining not the radius of curvature but the *deflection* produced at one end of the strip when the other end is clamped; you must also determine the force produced if this deflection is restrained.

Deflection can be calculated from the formula

$$\text{deflection (D)} = \frac{KTL^2}{t}$$

The actual mechanical force developed can be calculated from the formula

$$\text{force (ounces)} = \frac{4ETwt^2}{L}$$

where L = the effective length of the strip in inches
E = effective elastic modulus of strip in lb/in²
w = width of strip in inches
t = thickness of strip in inches

Typically the modulas of elasticity of thermometal will be of the order of 2.4×10^{-5} lb/in². Use this value of calculating force, unless you have a specific value for the thermometal chosen.

If we assume typical values for K and E, the force formula can be simplified to

$$F(\text{ounces}) = 550 \frac{Twt^2}{L}$$

where w, t, and L are in inches
T is temperature difference in degrees Fahrenheit

Note that in determining *deflection* or force developed by a bimetal strip, the *width* of the strip does not affect the result. In other words, the width used can be whatever is convenient. When it comes to determining the *force* developed, however, the width of the strip must also be taken into account.

Project 1. Design a high-temperature warning system or thermostat using a sample piece of thermometal available. The warning signal is to come on at a temperature of 90° F. Normal ambient temperature is 70° F.

The basic design is shown in Fig. 15-2, where deflection of the thermometal (bimetal strip) completes the lamp circuit when the temperature rises by 90° – 70° or 20° F. Thus T in the deflection formula is 20.

We assume that only a piece of thermometal is available; its specific properties being unknown. We therefore adopt a typical value for K of 7.3×10^{-5}. We can find the thickness of the strip (t) by actual measurement. Suppose it is 0.025 in.

This now leaves only L and D as unknown values. We can either allocate a value for length L and calculate the resulting deflec-

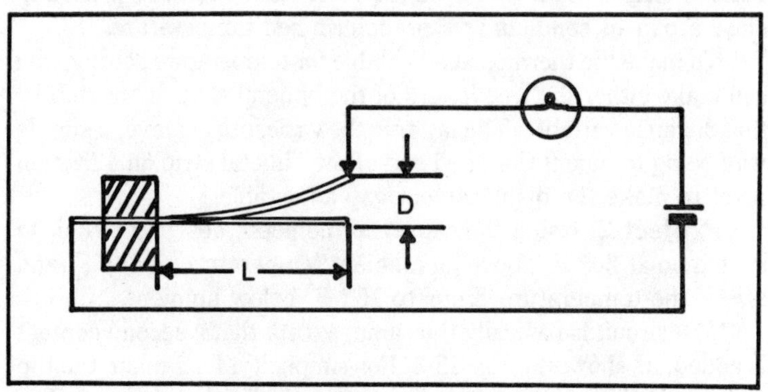

Fig. 15-2. A temperature-warning system using a bimetal strip.

tion, or we can calculate the value of L required to produce a given deflection. The latter is the best design practice, so let us set D = 0.1 in.

Rewriting the deflection formula as a solution for L, we get

$$L = \sqrt{\frac{Dt}{KT}}$$

Substituting values gives

$$L = \sqrt{\frac{0.1 \times 0.025}{7.3 \times 10^{-5} \times 20}}$$

$$= \sqrt{1.7}$$

$$= 1.3 \text{ in}$$

It now only remains to check that the strip deflects the right way with increasing temperature: that is, towards the contact, not away from it. The only way to do this is to try it and see. If it deflects the wrong way, simply turn the strip over.

Note that this circuit is a high-temperature warning, switching on at 90° F. It will remain switched off at any lower temperature. Check its actual switch-on temperature against a thermometer and a flow of hot air directed against it (from a hairdryer, for example). The actual contact position may need adjustment up or down, because the K value for the strip may not be the typical value used for the calculation.

Note: this example is also the basis for the design of a thermostat. In this case the mechanical movement (deflection) of the bimetal strip is separated from the electrical circuit and is used to close a pair of contacts at a predetermined temperature.

To make the thermostat adjustable for temperature setting, you can make either the free length of the bimetal strip or the deflection distance variable. The latter is the easiest to achieve, a simple way being to mount the fixed end of the bimetal strip on a friction pivot to make the displacement gap adjustable.

Project 2. Using the same thermometal, design a circuit to switch on at 20° F. above an ambient temperature of 70° F. and also if the temperature drops to 15° F. below ambient.

The circuit is basically the same, except that a second contact is added, as shown in Fig. 15-3. For simplicity, let's make the top contact the high-temperature contact with the same strip deflec-

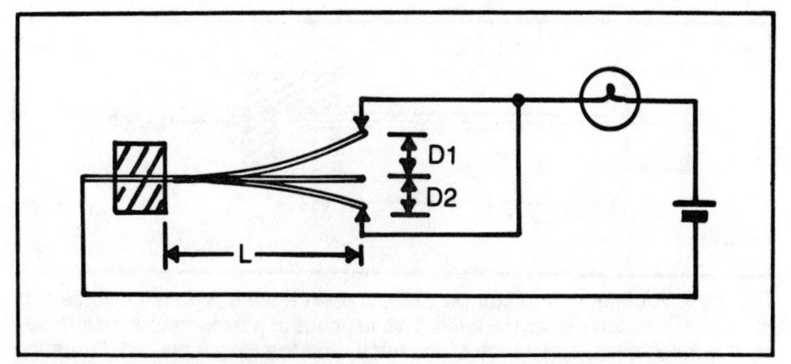

Fig. 15-3. Another temperature-warning system, giving an indication for temperatures outside a prescribed range.

tion and strip length as before. It now only remains to calculate the deflection of the strip downwards (D2) for a temperature drop of 15° F. below ambient. This will determine the position of the low-temperature contact.

Using the same formula, we get

$$\begin{aligned} D &= 7.3 \times 10^{-5} \times \frac{15 \times 1.3 \times 1.3}{0.025} \\ &= 7.3 \times 10^{-5} \times 1014 \\ &= 0.074 \text{ in} \end{aligned}$$

Project 3. Design a bimetal-strip system to operate a latch when the temperature rises by 10° F. The force required to operate the latch is 2 oz. Thermometal material available is 0.025 in thick. No other properties are known.

This is a purely mechanical system with the free end of the bimetal strip resting against the latch at normal temperature (Fig. 15-4).

The known rates are

$$\begin{aligned} T &= 10^\circ \text{ F.} \\ t &= 0.025 \text{ in} \end{aligned}$$

This leaves the length of strip (L) and the width of strip (w) to be used as unknowns.

The force formula is

$$F(\text{ounces}) = 550 \ \frac{Twt^2}{L}$$

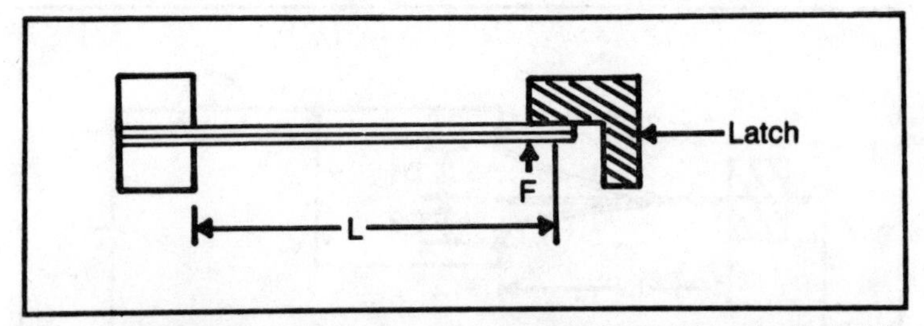

Fig. 15-4. A bimetal strip can be used to open a latch. When the force F is sufficient, the strip moves the latch. This happens at a temperature determined by the mechanical resistance of the latch, and the length and configuration of the bimetal strip.

Inserting the known values, we have

$$F = 2 = 550 \times \frac{10 \times w \times 0.025 \times 0.025}{L}$$

$$= 3.4375 \times \frac{w}{L}$$

$$\text{or } \frac{w}{L} = \frac{2}{3.4375}$$

$$= 0.58$$

In other words, we can use any combination of strip length and width where the length L is 1/0.58 = 1.724 × w. Let's choose 0.5 in as a convenient width to cut the strip. Then

$$L = 1.72 \times 0.5$$
$$= 0.86 \text{ in}$$

As a check, recalculate the force with these values:

$$F = 550 \times \frac{10 \times 0.5 \times 0.025 \times 0.025}{0.86}$$

$$= 1.999 \text{ oz.}$$

We should also check that the bimetal strip is not overstressed with this load applied. A formula for the safe maximum load (maximum force) is

$$\text{max. force (ounces)} = 40{,}000 \times \frac{wt^2}{L}$$

Check against the project calculation:

$$\text{max. force} = 40{,}000 \times \frac{0.5 \times 0.025 \times 0.025}{0.86}$$

$$= 14.5 \text{ oz.}$$

Our design load is well below this.

COILED BIMETAL STRIP

Bimetal-strip transducers can also be used in the form of coils, when the deflection will be in the form of angular rotation. Such coils will thus develop a torque with a change in temperature, the value of which can be calculated from the formula

$$\text{torque (oz in)} = 4700\ Twt^2$$

where T is the temperature difference in degrees Fahrenheit
w is the width of the strip in inches
t is the thickness of the strip in inches

This is an approximate formula, based on typical values for K and E for thermometals.

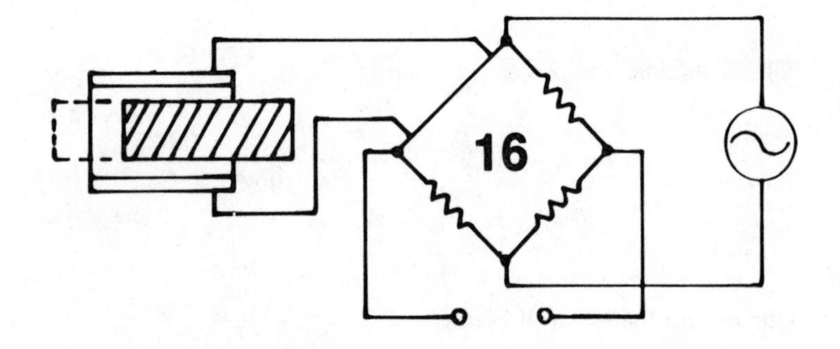

Infrared Techniques

Electromagnetic radiation covers a whole family of natural phenomena with the common characteristics of traveling with the speed of light. Radio waves, light, X-rays, and cosmic radiation are all part of this family, characterized by wavelengths with a different spectrum range. Light waves occupy only a tiny part of the overall range, bordered on either side by bands of invisible light: infrared with lower wavelengths than visible light, and ultraviolet with higher wavelengths than visible light. The infrared, with a spectrum range from about 1 μm to 1000 μm wavelength (one-millionth of a meter to one thousandth of a meter) is the type of radiation produced by *heat*.

Specifically, for measurement purposes, this spectrum is directed into *short-wavelength* infrared, with wavelengths from 1 μm to 5 μm, and *long-wavelength* infrared, with wavelengths from 5 μm to 15 μm.

The reason for choosing these two ranges is that with infrared wavelengths greater than about 15 μm the attentuation produced by passage through air becomes increasingly evident. The short-wave areas and the long-wave areas, on the other hand, provide minimal attenuation of infrared and are known as atmospheric windows. At even longer wavelengths—around 1000 μm or 1 mm which represents the extreme of the infrared spectrum (bordering on microwaves)—atmospheric attenuation again decreases.

In fact, infrared is not itself a transducer. It is electromagnetic radiation. Transducers associated with infrared techniques are *infrared detectors*, which provide thermal images for temperature measurement and heat detection within industrial and medical fields.

In its simplest form, such a detector, or true transducer, is a device, such as a photo sensitive diode, that generates an electric current when exposed to infrared radiation. This current is dependent on the intensity and wavelengths of the radiation and can be used in many different ways to produce an infrared picture or *thermal picture*.

The current produced by such a detector is very small and needs to be amplified and processed. Amplification increases the strength of the signal. Processing reduces the possibility of interference due to external sources and internal sources, such as electrical noise.

This particular requirement is complicated by the fact that semiconductor materials (say diodes) used as transducers always generate a certain amount of electrical noise, which is dependent on the temperature of the material. This is known as thermal agitation noise. Mounting the detector in a flask of liquid nitrogen at a temperature of $-196°$ C. will reduce this noise to a minimum. At this temperature, thermal agitation is very low. A further advantage of cooling is that it increases the cutoff range of the detector; that is, the wavelength range outside which the detector no longer responds to the radiation.

BLACKBODY CONCEPT

A perfect *blackbody* is an object that absorbs all incident radiation striking it and emits all of this energy in the form of transmitted radiation. No objects are perfect blackbodies, but the behavior of most objects can be related to perfect blackbody performance in terms of strength of radiation and wavelengths transmitted.

In the case of a body being treated, the radiation from a blackbody—or *spectral radiation emittance*, as it is called—takes the form of clearly defined curves, as in Fig. 16-1. These curves define the broad wavelength range over which radiation is emitted, which is also characterized by a distinct maximum peak. The actual shape of a curve and the peak value are dependent on temperature. Increasing temperature rapidly increases the peak radiation, and at the same time progressively shifts the peak wavelength.

Blackbodies radiating at a temperature of 800 to 900 degrees

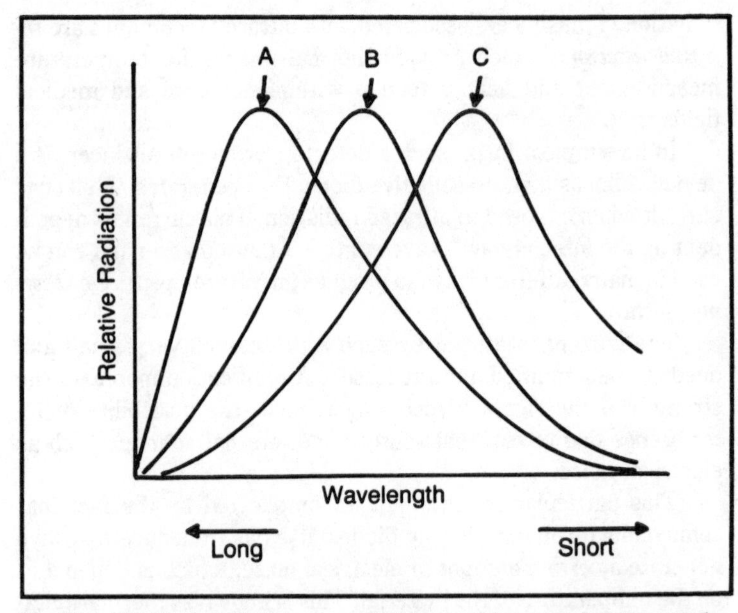

Fig. 16-1. Blackbody radiation curves. In this example, curve A represents the coolest temperature; curve B represents a warmer temperature; curve C, a still warmer temperature.

Kelvin (K) emit a very low level of radiation with wavelengths less than 1 μm and a peak wavelength of around 4 μm. The wavelength range for visible light is about 0.4-0.8 μm, so only minimum blackbody radiation appears at a wavelength of 0.8 μm. The major part is in the infrared spectrum. In a fully darkened room there would be just enough 0.8 μm radiation present to make the blackbody faintly visible as a dull red.

Note. Blackbody temperatures are always quoted in temperatures to the Kelvin scale and designated as K (not K°). 0 K corresponds to absolute zero temperature, which is −273° C. or −459° F. Thus, 0° C. or 32° F. correspond to 273 K; and 100° C. or 212° F. correspond to 373 K (from 0° C. upwards the K scale corresponds numerically to ° C. + 273).

The wavelength for maximum radiation from a blackbody is given by Planck's law:

$$\lambda_{max} = \frac{3000}{T}$$

where λ (wavelength) is in micrometers, and T is the temperature in degrees Kelvin.

Some interesting quick calculations can be made on this basis. The temperature of the sun, for example, is 6000 K. Considering the sun as a perfect blackbody gives

$$\max\lambda = \frac{9000}{6000} = 0.5\ \mu m$$

In other words, this lies towards the lower end of the *visible* spectrum range (0.4-0.8 μm upwards), so the sun appears yellowish in color. Emission in the infrared range (0.8 μm upwards) has a much lower intensity but is still readily detectable.

Stars cooler than the sun have peak radiation at a longer wavelength. We can, in fact, easily tell the temperature at which they would become largely invisible (that is, with a peak radiation wavelength of, say 1 μm). Using the same formula as above, we get

$$1 = \frac{3000}{T}$$

or

$$T = 3000\ K \text{ or } 2727°\ C.$$

Such a star would not be seen through an ordinary telescope but would readily be detectable by its infrared radiation.

EMISSIVITY

The total emitted radiation energy of a blackbody is given by

$$W = \sigma\ T^4 W/m^2$$

where σ is the Stefan-Boltzmann constant, 5.67×10^{-8}

No object is a perfect blackbody, but its relative efficiency in this respect can be related to blackbody radiation by its emission factor or emissivity. The law for nonblackbodies then becomes

$$W = 5.67 \times 10^{-8} \times E\ T^4$$

where E is the emissivity.

Emissivity can vary from 0 (a perfectly reflective surface) to 1 (for a perfect blackbody). It will also vary somewhat with temperature. Table 16-1 gives some typical emissivity values for familiar surfaces.

Digressing a little, it is interesting to note that the unclothed human body is a near-perfect blackbody as far as heat loss is concerned. At a temperature of 27° C. or 300 K, for example, heat loss is

$$\begin{aligned} W &= 5.67 \times 10^{-8} \times 0.98 \times 300^4 \\ &= 450 \text{ W/m}^2 \end{aligned}$$

Because the surface area of a man's body is about 2 m^2 this represents a radiated heat loss of nearly 1 kW.

GRAY BODIES AND COLORED BODIES

A body with a *constant* emission factor (emissivity) regardless of wavelength is called a *gray body*. A body with an emission factor that varies with the wavelength of the radiation is called a *colored body*. An example of a colored-body curve, compared with that of a blackbody, is shown in Fig. 16-2. Apart from having an emissivity of less than 1 at any wavelength, the principal difference between the two is that a colored body tends to behave like a blackbody at certain wavelengths, but a gray body does not.

THERMOGRAPHIC IMAGES

A thermographic image or *thermogram* is an image showing the heat radiation from a particular object at a particular temperature. In a complex object not all parts will be at the same

Table 16-1. Typical Emissivity Values for Surfaces.

Surface	Emissivity
highly polished metal surface (100° C.)	0.02-0.05
rough or oxidized metal surface (200° C.)	0.7-0.8
concrete (20° C.)	0.92
snow (−10° C.)	0.85
highly polished glass (20° C.)	0.94
ice (−10° C.)	0.96
human skin (30° C.)	0.98

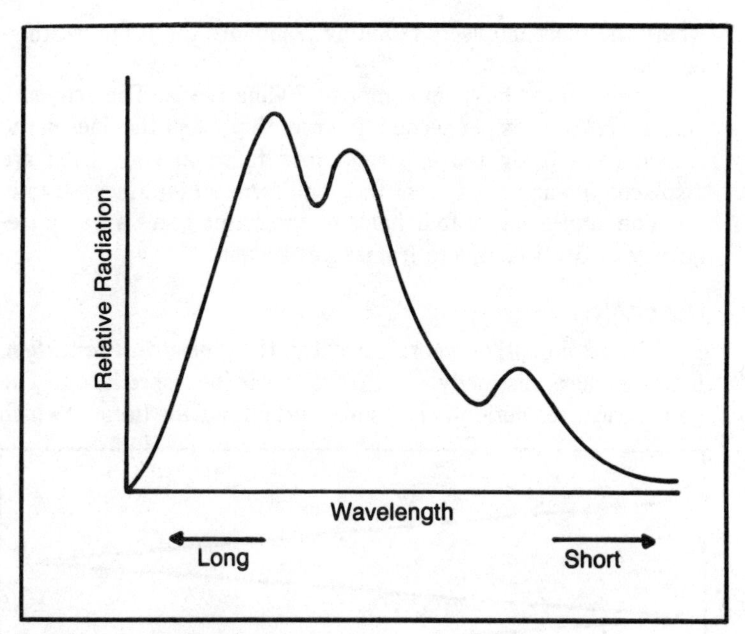

Fig. 16-2. A colored-body radiation curve.

temperature, so a thermogram also shows the heat distribution of the object. This is where a transducer/scanner unit is necessary to compile the thermogram.

The first requirement is an optical system to focus the radiation on the detector. The detector will only detect the strength of such focused radiation received. Thus, to produce a complete thermal picture the optical system must be flexible in the sense that it must allow the detector to sense the object point by point. In this way a complete thermal picture of the object can be presented, somewhat like a still TV picture.

There are several methods of making the detector scan the object. The most suitable is usually an optical system using both horizontal and vertical rotating prisms. One prism then deflects vertically and the other horizontally, scanning the object with a pattern like that shown in Fig. 16-3.

In the AGA scanner a horizontally deflecting prism rotates in the scanner at a speed of more than 18,000 rpm, a vertically deflecting prism at 180 rpm. Therefore, the detector constantly monitors different points, and the electrical current from the detector varies in strength with the radiation of the object at every point. This variable current is used to produce a picture in a gray scale where the most intensely radiating points are the brightest (hottest areas).

The pictures built up comprise 70-line fields. The scanning velocity is 25 fields per second. In order to prevent the lines in the screen from being too apparent, four fields in succession are displaced in such a way that they will form an interlaced frame.

The interlacing of four fields to one frame gives a frame frequency of 6 1/4 complete frames per second.

Resolution

The *resolution*, or, more correctly, the *geometrical resolution*, indicates how distinctly small details can be reproduced. The geometrical resolution is measured in milliradians (mrad), which

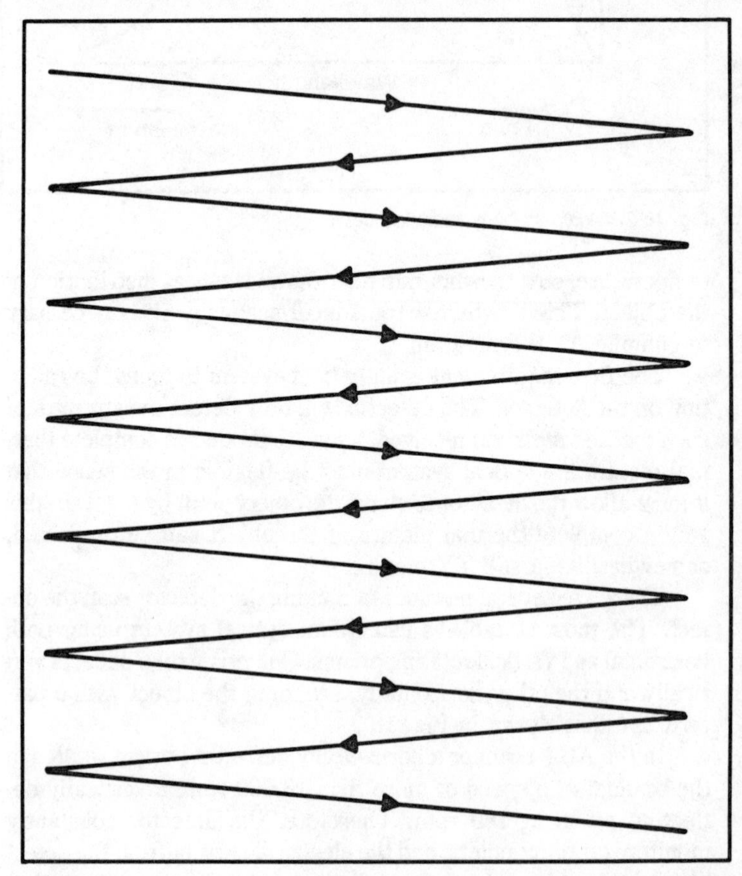

Fig. 16-3. Scanning pattern for infrared imaging. (The vertical scale is greatly exaggerated.)

is an angle measurement. The resolution is dependent on the size of the detector and the optical arrangement in the scanner. A lens with long focal length (telephoto lens) will give better resolution, for example, than a lens with a wide angle. This follows the principle that a smaller part of an object can be viewed using a telephoto lens than with a normal relatively wide-angle lens.

Aperture

To enable as much radiation as possible to be focused on to the detector, use a lens with a wide aperture. When viewing a very intensively radiating object (for example, a very hot oven), the radiation intensity will be too strong, and the system will be overloaded; in this case it treats all radiation as maximum, and the entire picture becomes white even if the picture controls are adjusted to their end positions. The system will not break down because of this, but it is overmodulated. This in itself can be overcome by using the aperture control. Then the radiation intensity can be reduced, and the detector will not be overloaded.

With the scanner set to the highest aperture number and without the use of external filters, objects of a temperature up to 800 ° C. can be viewed.

FILTERS

There are filters for a number of different purposes and applications, all aiming at increasing the scope of the thermal scanner.

For example, there is a high-temperature filter that both decreases the intensity of the radiation received from very hot objects and cuts off short wavelengths, in which the radiation from very hot objects is concentrated. This is used when registering the long-wave radiation from hot objects.

For special applications to measure the temperature on liquid (fluid) glass, having a temperature of 700 °-800 ° C., a problem arises. Glass is transparent to short infrared wavelengths, and therefore the temperature of the background tends to be measured instead of that of the glass. By using the filter it is possible to remove the very shortwave radiation. Therefore, the long-wavelength infrared radiation will be measured, and the glass will be completely opaque, so that we measure the temperature of the glass and not of the surroundings.

PRESENTING THE THERMAL PICTURE

The final thermal picture is presented on a picture tube similar to that used in television. The detector output signal is in analog form. After being amplified, it needs processing into a *video signal.*

All values of intensity between the signal's minimum and maximum values, respectively, form the dynamics of the tube. In a gray scale the dynamics range from dark black to brilliant white. The thermal resolution (the smallest variations in temperature that can be registered on AGA's equipment), is approximately 0.1° C. at room temperature.

It is of great importance that the thermal resolution of the system be dictated by the characteristics of the thermal scanner. What is missed by the thermal scanner cannot be replaced by refined processing of the signal.

The above procedure implies that objects of the same temperature, or at least what the scanner recognizes as the same temperature, should be reproduced in the same gray tone. In the processing of the video signal from the detector, it is possible, by marking the temperature chosen with saturated white pulses in the pictures, to add a function that will emphasize an area with the same temperature.

COLOR PRESENTATION

The human eye has a very high sensitivity when it comes to separating various color tones from each other. This is the explanation for using a color monitor instead of the usual black and white monitor when illustrating thermal pictures. The number of lines, and therefore the resolution, remains unchanged but the various levels of temperature are illustrated by different colors instead of gray shades. The colors are usually ranged such that blue illustrates the coldest parts, and yellow and white the warmest. This order can be changed if it for some reason should be better to increase the contrast between close parts in the object. Simultaneously, a range of colors is displayed on one side of the picture, indicating the various isotherms for reference purposes.

PICTURE ANALYSIS

AGA's thermal picture is built of a number of lines, 4 × 70, where 70 lines constitute a field and four fields fill a frame. The horizontal and vertical resolution is set by the size of the detector and is 100 points per line. However, lines and points are overlapping each other, so the true resolution results in 280 × 100 = 28,000 points.

When a *digital* rendition of the picture is made, the information is first quantized. The value of the signal is measured frequently. This is called *sampling*. A sample can be quantized to one out of, say, 24 values. The analog signal can take an unlimited amount of values; the quantizing causes the analog value of the sample to be replaced by the closest possible digital value (Fig. 16-4).

The values of all points, which together build up a picture, is stored in a memory, from which it can later on be fetched for further processing. This is normally done by a computer.

A line is divided into 100 points, where each point forms a picture element (pixel). Each point is quantized into one of 256 possible values. With 256 levels the quantizing differences are minimal, and the digital rendition is essentially equivalent to the analog rendition.

One of the advantages of digital processing is that, with the help of a digital computer, it is possible to "subtract" pictures from

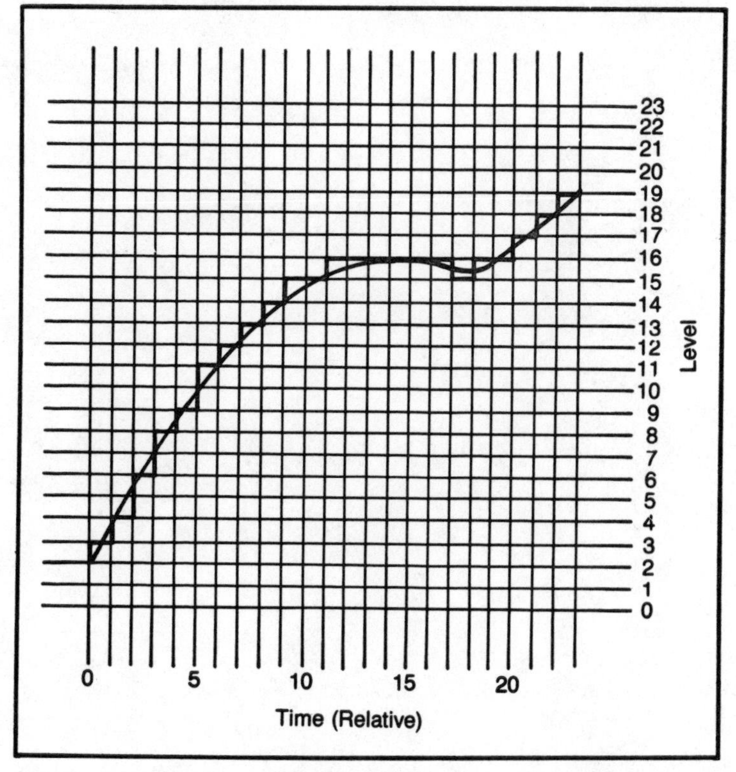

Fig. 16-4. An example of digitizing, or quantizing, an analog signal.

each other in order to procure variations that otherwise would be very hard to obtain. Furthermore, in this way, complex picture analyses can be carried out. For example, quick events can be analyzed, and the gradual change from picture to picture can be shown very clearly.

An example of very advanced picture processing is the evaluation of pictures taken from American space shuttles. The very violently heated front and wing leading edges on the space shuttle have been investigated when reentering the atmosphere. This was thermographed with the assistance of a modified AGA Thermovision® camera.

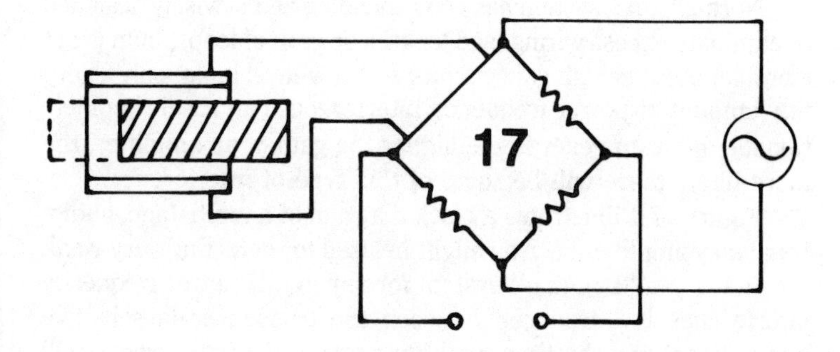

Interference

In any transducer system it is important that interference be minimized. There are numerous ways to do this; the most obvious and fundamental is good system design, employing optimum wiring techniques, shielding, and impedance matching. But there are certain cases in which interference may be a problem despite all of the commonly accepted layout and design practices. This chapter describes a few of the ways in which special interference problems can be dealt with.

WHAT IS INTERFERENCE?

We may say that, in general, interference is any energy that is present in a system and is not desired. Alternating-current "hum" is a good example of interference because it is almost never wanted. You have probably experienced this type of interference, for example, in a public-address system having inadequate shielding of the microphone cord. Two radio stations may interfere with each other; which station is "desired" and which is "interference" depends on your preference.

The method to be used for minimizing interference depends on the type of interference, the spectral characteristics, the location of the interfering source with respect to the system, and many other factors.

AC HUM

Normally, if a system has good shielding and is wisely designed to eliminate excess wiring lead lengths or ground loops, hum is not a problem. But in high-gain circuits with low-level input, only a very tiny amount of power-frequency interference can result in objectionable hum. In general, the higher the gain of an amplifier, the more likely there will be some of this type of interference.

Figure 17-1 illustrates a block diagram of a multistage, audio-frequency amplifier such as might be used for detecting very weak noises (as in a "bugging" system for spying). If power-frequency interference is introduced between the transducer (a sensitive microphone) and the first amplifying stage, the interference will be amplified by all of the succeeding stages. The result will likely be a severe hum in the output if the interference is input at point A or point B. An equal amount of interference introduced at point C or point D would cause less trouble. Clearly, wiring and shielding precautions are most important in low-level parts of a circuit.

Suppose we have a sensitive amplifier chain, such as that shown in Fig. 17-1, and have taken all normal precautions against hum, but nonetheless we have interference. There are various means we might use to alleviate the problem.

One method of reducing hum interference is to use a low-impedance input circuit. We might, for example, employ a bipolar transistor rather than a FET at the first stage; or we might connect an operational amplifier for a low-input-impedance condition. When the input impedance is low, electrostatic coupling to the outside environment is minimized.

Another method of reducing hum in an audio-frequency circuit is to use a band-rejection filter or high-pass filter. Filters may be installed at the input only, or between each stage, or between

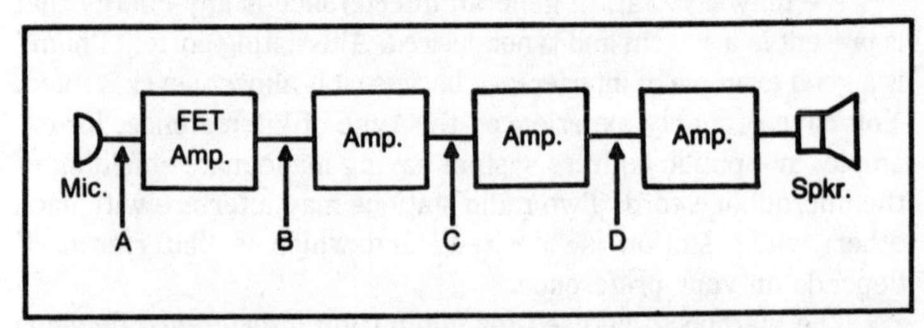

Fig. 17-1. Block diagram of an amplifier chain, showing various points at which interference might enter the system. The most severe problem would occur if the interference entered at point A; less and less severe trouble would occur at B, C, and D entry points.

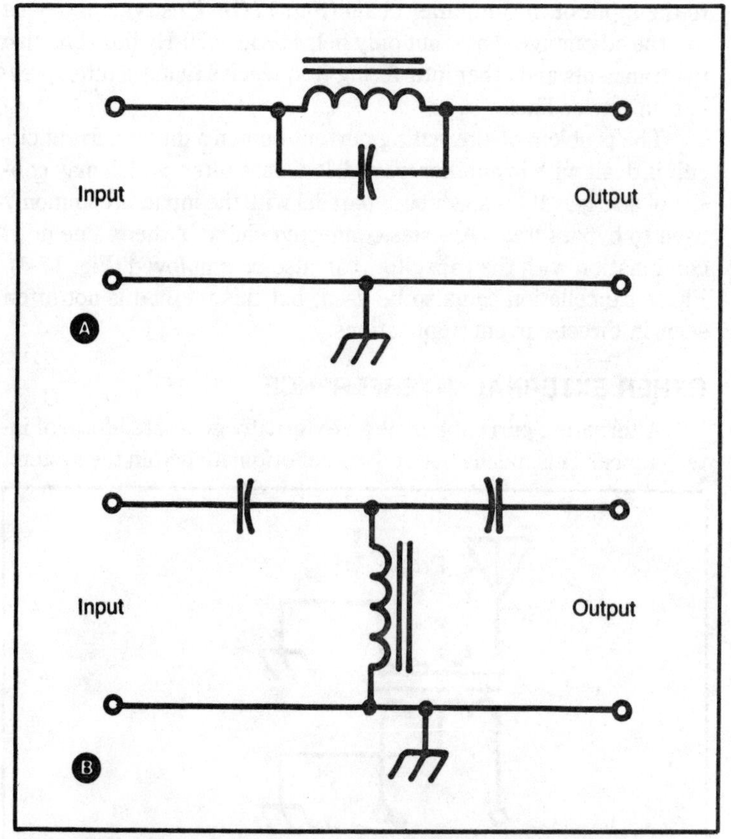

Fig. 17-2. Passive band rejection (A) and high-pass (B) hum filters.

selected stages. It is almost always advantageous to place a filter at the input, because amplified hum may cause reduced gain or nonlinearity of later stages in an amplifier chain. Figure 17-2 shows simple passive LC band-rejection (A) and high-pass (B) filters. The band-rejection filter is resonant at 60 Hz; the high-pass device is designed for a cutoff frequency somewhat above 60 Hz.

Still another method of reducing hum is to apply a small alternating-current signal, equal in magnitude but opposite in phase to the interference, somewhere along the amplifier chain. This can prove extremely effective in such devices as very-low-frequency (VLF) receivers and modulated-light devices. We simply construct a small probe that picks up some hum from the environment, connect it to a circuit that provides either zero phase or phase opposition (selectable) and variable gain, and apply the resulting output to the input of the amplifier chain (Fig. 17-3). This type of device has the advantage that it not only helps reduce 60-Hz hum but also the transients and other interfering frequencies that are often present on power lines.

The problem of alternating-current hum in a direct-current circuit is dealt with in another way. A low-pass filter, which may consist of a large-value capacitor in parallel with the input, is commonly used to bypass hum. A series-connected choke, either alone or in combination with the capacitor, can also be employed (Fig. 17-4). Phase cancellation can also be used, but this method is not often seen in direct-current applications.

OTHER EXTERNAL INTERFERENCE

Alternating-current hum is an externally generated form of interference. This means that it does not originate within the system,

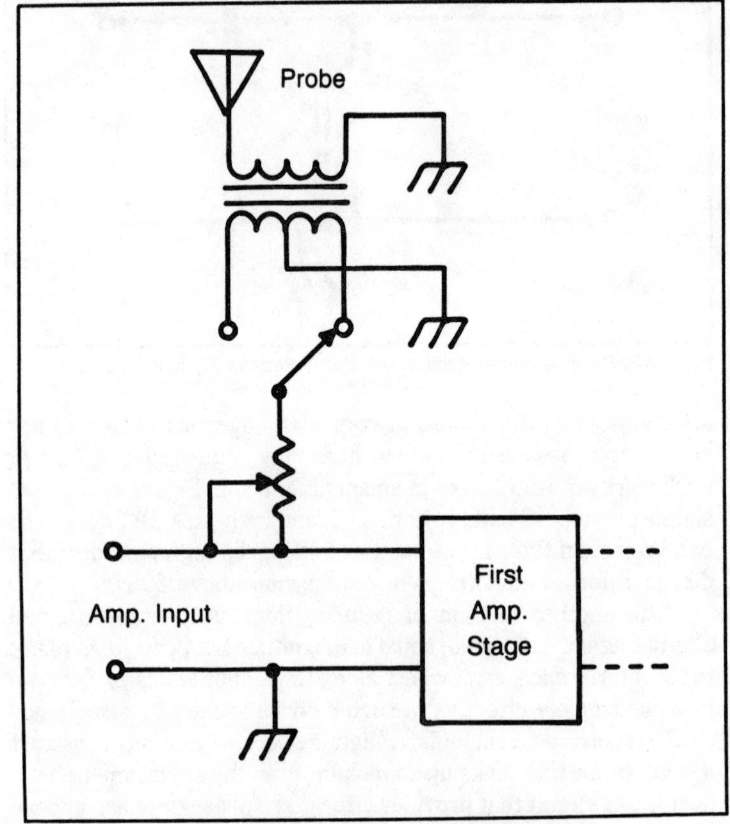

Fig. 17-3. Phase-cancellation circuit for reducing interference.

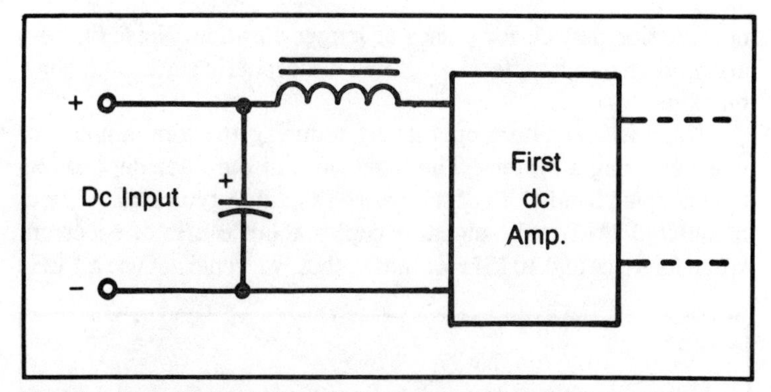

Fig. 17-4. Filter for eliminating alternating-current interference at the input of a direct-current circuit.

but it gets into the system from outside. (The ultimate source of hum is, of course, the power-plant generators.) There are other forms of externally generated interference that can affect the performance of a transducer system. Examples include sferics (atmospheric static), radio signals, and even cosmic noise. In some cases electric or magnetic fields, even the earth's magnetic field, can adversely affect a system. Let's look at some examples.

Sferics are present at all frequencies from subaudible to the visible-light and ultraviolet ranges. Most sferics are generated by lightning strokes, carrying momentary currents of hundreds of thousands of amperes. Sferics, unlike alternating-current hum, are a *random* form of electromagnetic energy. Sferics exhibit no evident pattern, such as a repeating wave cycle. The randomness of sferics makes this type of interference more difficult to deal with than hum because there is no specific kind of filter that is unfailingly effective against it.

Meticulous attention to shielding of amplifier stages and interstage wiring is essential if we are to minimize the interference caused by sferics. But sometimes, as in a radio receiver, the sferics enter the system along with the signal. Then we have no recourse except to use a "brute force" approach. Noise limiters, noise blankers, and narrowband filters are the most often-used methods of reducing interference caused by sferics.

The *noise limiter* circuit simply prevents interfering signals from exceeding the amplitude of the desired signal (Fig. 17-5). A *noise blanker* circuit shuts down the amplifier chain (at one stage) during a noise pulse (Fig. 17-6). The noise blanker can be extremely effective when interference pulses have short duration, but it does not function as well for pulses of longer duration. Noise limiters are more generally effective against atmospheric static than noise blankers.

Narrowband filters operate by reducing the total amount of noise entering a system. The bandwidth of the filter depends on the occupied bandwidth of the desired signal. A typical amplitude-modulated (AM) radio signal occupies about 6 kHz of spectrum space for voice and 10 kHz for music; thus, we could not use a 1-kHz

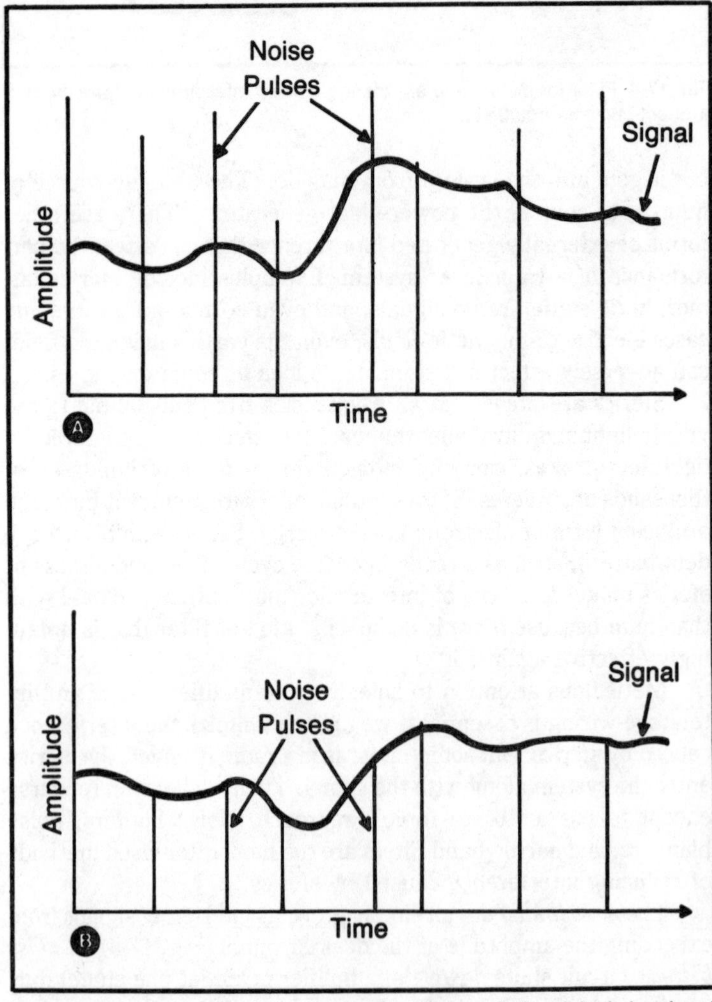

Fig. 17-5. Action of noise limiter and noise blanker. At A, original signal and interference; at B, action of noise limiter.

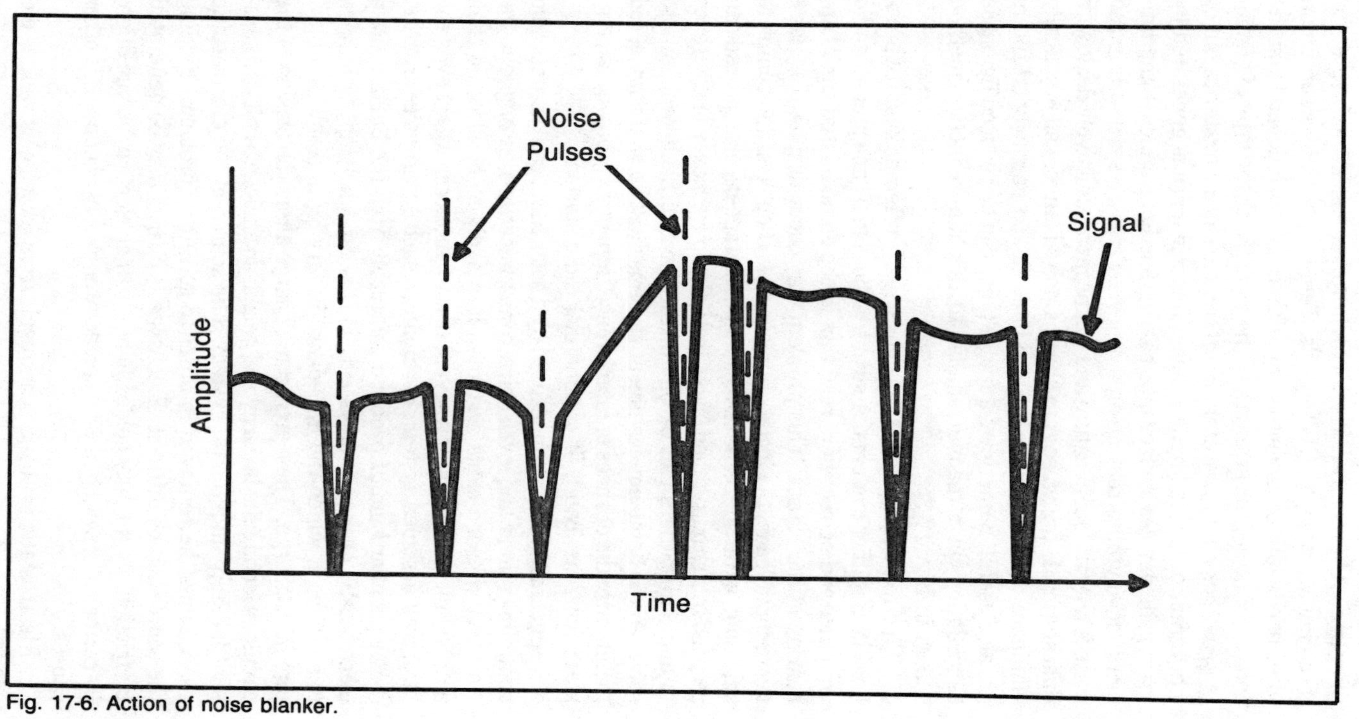

Fig. 17-6. Action of noise blanker.

filter in an AM receiver. We could, however, use a 1-kHz filter for receiving Morse code (CW) signals. There is generally an optimum filter bandwidth in any situation. This bandwidth depends on the amount of noise, the strength of the signal, and the type of signal.

Another kind of external interference, similar to sferics but of a different origin, is impulse noise. Impulse noise is generated by man-made devices such as internal combustion engines, light dimmers, or anything that creates repeated arcs or sparks. Impulse noise is not random in nature and can be more effectively dealt with than sferics. If the pulses are short, a noise blanker can practically eliminate the interference. Noise limiters and narrowband filters are also useful. Sometimes impulse noise can be reduced by phase cancellation, in a manner similar to the hum-reduction method shown in Fig. 17-3.

Electromagnetic fields can cause many different kinds of problems in amplifier systems. A strong radio signal can enter an audio amplifier and be rectified, resulting in the demodulated signal appearing at the output. This problem may occur despite rf bypassing of input leads, the installation of series rf chokes, and complete shielding; a transducer such as a magnetic tape-recording head can respond to a strong rf field all by itself. Radio-frequency interference to audio amplifiers can be one of the most difficult problems to solve. Audio home-entertainment systems must be well engineered to minimize the chances of interference from nearby broadcast, amateur, citizens' band, or commercial radio stations.

A geomagnetic storm, caused by a solar flare, can disrupt radio communications and even wire communications throughout the world. A shielded-cable system is less likely to be affected than a radio link, obviously, but severe disturbances can affect even the most well-designed system. Fortunately, such occurrences are rare.

A peculiar kind of problem can result from the buildup of an electrostatic charge in a system. This is especially apt to happen in a circuit in which the impedance is very high, resulting in low rate of discharge. The electrostatic charge alters the biasing of the circuit, affecting the linearity and gain in an analog circuit and causing "latchup" in a digital system. The source of the electrostatic charge must be removed in order to solve this problem. Finding the source can be difficult. Electrostatic charge may be generated by friction, by a faulty solder joint or other electrical connection, by corrosion of switch or plug contacts, or by any of numerous other causes.

Electrostatic problems can result from internal as well as external causes. In low-level circuits the type of solder used may affect the performance because of thermally generated direct-current voltages. Some kinds of solder are more susceptible to thermal voltages than others; special solders are available for use in devices operating at microvolt or nanovolt levels.

Heat itself can be a source of interference. The random motion of atoms in any substance is a source of broadband noise. In general, the amount of noise generated in this way is directly proportional to the absolute temperature. There is very little that we can do to cool down the external environment, but internally, it is often quite practical to lower the temperature. This can be done by means of liquid gases, such as helium, which have absolute temperatures just a few degrees above zero Kelvin. In extremely low-level radio-frequency devices, such as radio telescopes, significant improvement can be had by the use of temperature-lowering apparatus.

INTERNAL INTERFERENCE

Direct-current and thermal interference can be generated either outside a system or within, as we have just seen. Certain types of interference are caused only internally, however.

Unwanted feedback can produce disastrous malfunction of an amplifier system. In the worst case it results in oscillation, sometimes at more than one frequency, in conjunction with degradation of gain and linearity. Unwanted feedback can even cause physical damage to circuit components; an example is thermal runaway in a power amplifier as a result of parasitic oscillation.

Proper circuit layout and design is the best protection against unwanted oscillation in an amplifier system. In audio-frequency circuits, shielding of interstage wiring is often sufficient. In radio-frequency power amplifiers, neutralization is usually necessary. Two methods of neutralizing an rf amplifier are shown in Fig. 17-7.

Negative feedback can also occur, causing nonlinearity and degradation of circuit gain, although this type of feedback will not result in oscillation. It should be noted that negative feedback is often employed deliberately in an amplifier, and this of course is not interference.

Internal interference can take the form of unwanted coupling between or among different parts of a circuit. For example, in a stereo audio amplifier we might have poor isolation between the channels as a result of improper design or layout. Unwanted cou-

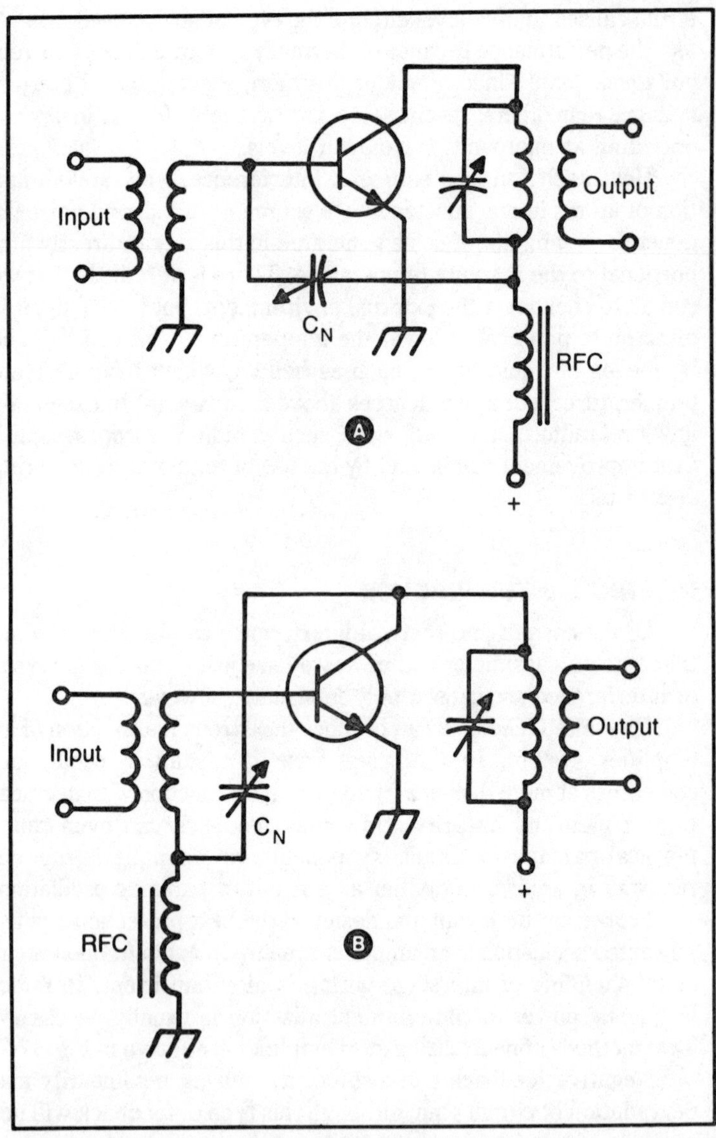

Fig. 17-7. Two methods of amplifier neutralization. At A, output-circuit neutralization; at B, input-circuit neutralization.

pling can be practically eliminated by individually shielding different parts of a system. In a stereo amplifier, for example, we might put the left-hand and right-hand channel circuits in entirely separate metal enclosures.

Grounding schemes can literally "make or break" any amplifier system. The commonly accepted method of grounding is illustrated in Fig. 17-8A. Each grounded (or common) point is connected to a single bus. A variation of this method, in which several buses are tied together at the common point, is shown in B. (Actually, the use of the word "ground" is not completely accurate in most cases. We do not connect all common circuit points to the earth itself but to a bus or chassis that may or may not ultimately have an electrical connection with the earth.)

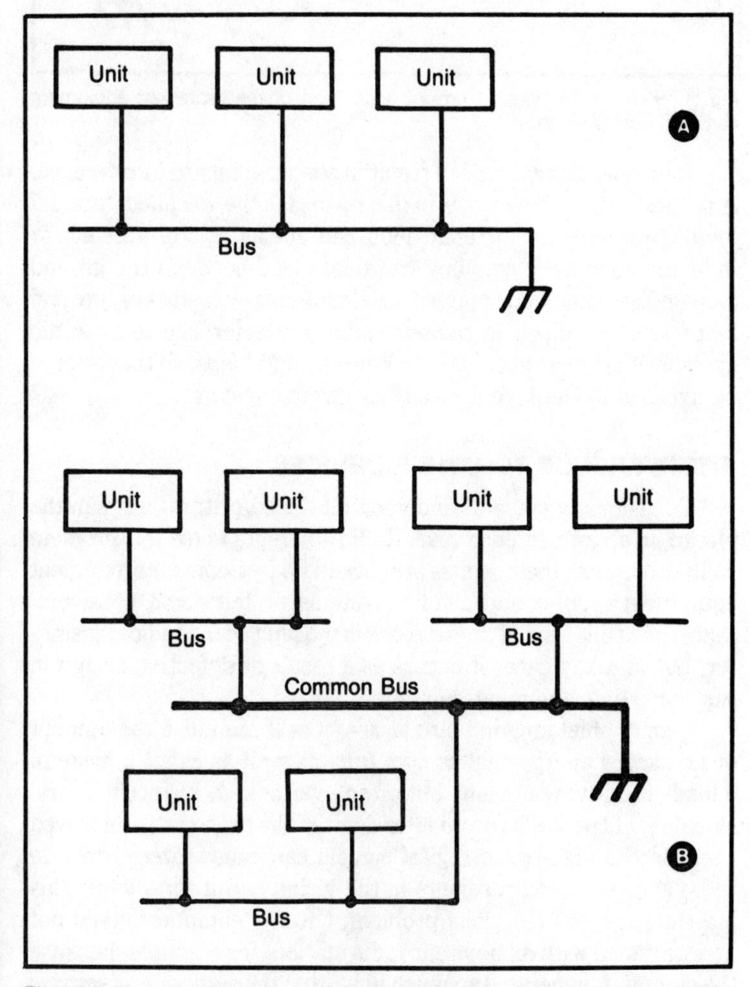

Fig. 17-8. Two acceptable grounding schemes. At A, single bus; at B, multiple bus.

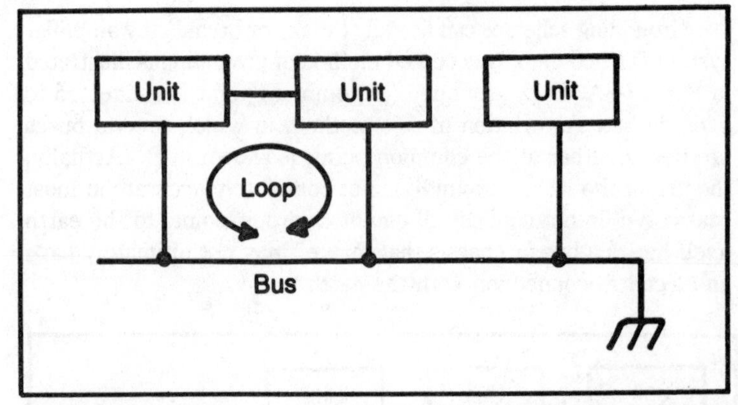

Fig. 17-9. An example of a ground loop. The boxes represent equipment common connections.

Improper grounding can result in susceptibility to interference. A notorious problem-causer in this respect is the so-called "ground loop" (Fig. 17-9). A ground loop can act as a loop antenna or inductor—magnetic coupling can take place between the ground loop and strong electromagnetic fields. In many instances, ground loops are the culprit in radio-frequency interference to an audio system. It is never necessary to have ground loops, so they should be avoided in the layout of any electronic system.

INTERFERENCE TO OTHER DEVICES

Occasionally you will find yourself the culprit, rather than the victim, in an interference case. Radio amateurs know this problem well; sometimes their signals are picked up by home-entertainment equipment such as high-fidelity systems or television receivers. Sometimes this kind of interference is the fault of the radio transmitter, but in many cases it occurs as a result of defective design in the home-entertainment equipment.

Proper shielding and circuit layout will minimize the amount of unwanted energy that *escapes from* as well as *enters* a system. This is intuitively obvious but often overlooked. In recent years, shielding of home microcomputer devices has been vastly improved because the high-speed digital signals can cause interference to radio and television receivers in the vicinity. But for a while this interference was a serious problem. Often a computer could not be interfaced with an amateur radio station, for example, because the computer generated so much noise that the station receiver was rendered useless.

It is sometimes possible to compensate for generated interference, even though it is the fault of one particular device, by modifying the "victim" devices. For example, in the case of microcomputer interference to a radio receiver, we might restrict receiver operation to those frequencies on which the interference is minimal; or we might use a noise blanker or a narrowband filter and put up with what interference remains. This is not by any means an optimal solution to the problem, but it may be all that can be done. Ideally, the problem should be corrected by attacking it at the source.

If the "victim" device is at fault, then there is nothing that can be done to the generating device to get rid of the interference. Consider the case of a defective television set that experiences interference from a properly operating radio transmitter nearby. Although the radio transmitter might be operated at a different frequency, or the power output reduced (or the transmitter shut off entirely), and the interference thereby eliminated, the optimum solution is to correct the defect in the receiver. This can be not only hard to do but also difficult diplomatically—again using the example of a radio-amateur station—nobody wants to be told that he or she is the owner of a defective piece of equipment.

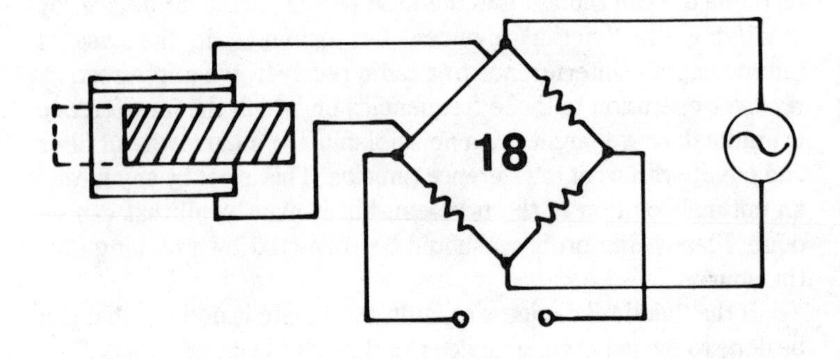

Pressure Transducers

Pressure is not, as sometimes mistakenly thought, a force but a force per unit area (pressure = force/area). This is a fine distinction as far as a pressure transducer is concerned, for any device capable of converting applied mechanical force into some other form of energy is also, in fact, a pressure transducer converting a force per unit area into another form of energy. Thus, force transducers can also be used for pressure measurement, provided the *area* over which the force is applied is restricted and known. Also, because the application of pressure is commonly associated with resulting mechanical movement or displacement of something to which it is applied, a displacement transducer is also capable of pressure measurement.

The basis for all pressure transducers is a force-collecting member, which is some form of elastic element. The force resulting from applied pressure is then converted into some other form of energy, usually electrical energy. The main exception to this rule is the Baudon tube, or metallic bellows, which converts pressure energy directly into mechanical movement. These types of mechanical transducers are widely used as pressure gauges, less so as specified pressure transducers.

As an illustration of this distinction, a Baudon tube pressure gauge was once the common form of oil pressure gauge used with automobile engines. To work as a gauge it needed to be piped directly to the oil supply. The usual modern alternative is the *pressure transducer* mounted directly on the engine block at a suitable point exposed to oil pressure. The converts the oil pressure level into a proportional electrical signal, which is taken to an electric meter (millimeter) by wiring to indicate the pressure. Such a system has greater integrity because it is "fail safe." In the event of a broken lead (or broken instrument), there is no possible loss of oil, only a loss of instrument reading. Also it is more readily adaptable to electrical warning signaling (for example, tripping an electric current to operate a low-level warning light when the indicating instrument itself can be dispensed with).

However, a movable or "elastic" element remains a necessary feature of a pressure transducer; the usual form is a plain diaphragm, often of metal but sometimes of other materials, such as silicon. This forms the force-collecting member, the mechanical movement of which is then convertible into an electrical signal by a true transducer, such as a strain gauge, piezoelectric crystal, variable capacitance, or variable reluctance. The earlier forms of microphones using packed carbon particles behind a diaphragm were another (movable resistance) type of pressure transducer for sound-pressure-level conversion to electrical signals; these have been replaced with condenser and piezoelectric types.

Before we dismiss the Baudon tube entirely as a pressure transducer, though, it should be said that it does have certain useful applications. It is particularly suitable for accurate high-pressure measurement, for example, up to about 100,000 lb/in^2. Also, its mechanical movement can be connected to another type of transducer, such as a potentiometric device, strain gauge, or linear variable-differential transformer, to give an electrical signal output. This is also the case with metallic bellows, which are far more sensitive than the Baudon tube for sensing low pressure (think of the aneroid capsules used in barometers, barographs, and altimeters). The primary disadvantages of the Baudon tube and metallic bellows as elastic elements in a pressure transducer are the relatively large volumes involved and their very limited frequency response because of their bulk and inertia. Thus, simple diaphragms are normally preferred.

DIAPHRAGM PRESSURE TRANSDUCERS

The most widely used type of elastic element in a pressure transducer is the flat circular diaphragm clamped all around its edge. Under applied pressure this produces an immediate deflec-

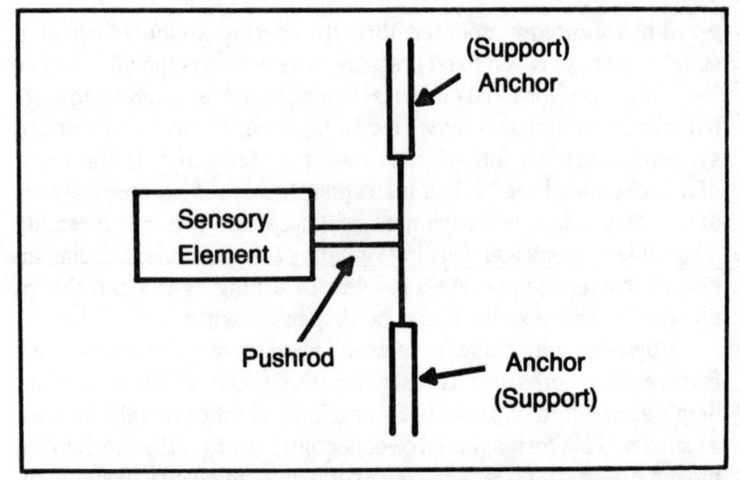

Fig. 18-1. Edge-on diagram of the basic configuration for a pressure transducer.

tion of the diaphragm, which can be sensed in a number of ways, such as with the center of the diaphragm mechanically coupled to the sensory element by a short pushrod (Fig. 18-1), or deflection of the diaphragm monitored directly. The four main types of sensory elements used are resistance strain gauges, piezoelectric crystal, capacitance, and inductance or reluctance.

STRAIN-GAUGE PRESSURE TRANSDUCERS

These are the most widely used types, with strain gauges bonded to, or into, the diaphragm. The trend is towards using semiconductor strain gauges, which provide high sensitivities in smaller sizes with lower input voltages and higher frequency capabilities. Sometimes the actual diaphragm is a silicon chip into which strain gauges are inorganically bonded and automatically diffused.

This solid-state integrated-circuit technique permits formation of microminiature strain gauges within the silicon diaphragm, enabling transducer diaphragms as small as 0.050 in. (1.27 mm) with active diameters as small as 0.028 in. (0.71 mm). This is not achievable with stainless steel diaphragms and precludes their use in applications requiring transducers smaller than 0.125 in. (3.18 mm) diameters. Other secondary advantages of silicon diaphragms are that they tend toward better stability of zero offset and long-term drift.

The major disadvantages of silicon diaphragms are their difficulties of providing water and chemical media protection and their tendency to shatter under particle impingement. Silicon is a brittle material, crystalline in structure, and can crack or shatter on impact. Protective screens are available to minimize this property, but they cannot always be used conveniently.

Stainless steel diaphragms have the obvious advantages of ruggedness and the ability to maintain pressure seals even after electrical failure due to accidental overload. They are easy to waterproof and can be beam welded to resist a variety of corrosive media. However, they are not available in diameters less than 0.125 in. (3.18 mm) with suitable performance specifications.

Semiconductor transducers provide measurements from as low as 110 dB up to 50,000 psi and natural frequencies up to 1 MHz and higher. They are offered in gauge, sealed gauge, and absolute and differential modes and have the advantage of simplicity of application. All that is required is a stable power supply, and their inherently high output is sufficient to drive a variety of indicating and recording devices.

A pressure transducer may incorporate a reference pressure supply. This can have four options:

☐ Gauge-psig: Transducer is referenced to ambient pressure through an open reference tube.

☐ Sealed-psis: Transducer is referenced to 1 atmosphere pressure sealed within the transducer.

☐ Absolute-psis: Transducer is referenced to absolute zero pressure either by sealing a vacuum within the transducer cavity (true absolute) or electrically referencing to absolute zero within the compensation module.

☐ Differential-psid: Transducer is referenced to second pressure source through the reference tube. Differential units must use a nonconductive noncorrosive media that will not affect the construction; for example, water and media containing water may not be permissible. The reference port is the low-pressure side in all differential measurements.

CIRCUITRY

Strain-gauge pressure transducers may have a half-active bridge of two strain gauges or a fully active bridge with four strain gauges. In the former case the bridge circuit needs to be completed externally with two additional resistors or resistor decades. A typical half-bridge circuit is shown in Fig. 18-2, R1 and R2 being the external resistors. Resistors R_C are calibration adjustment resistors, normally an integral part of the transducer.

Color coding is not always the same with different makes of transducers, although green is normally + out and white − out. Excitation input is red and black, which may be + and − or − and +, respectively.

FREQUENCY RESPONSE

Standard pressure transducers are normally single-degree-of freedom systems. The useful frequency range is linear within ±5-20 percent of the resonant frequency. From 20 percent of resonance to resonance the pressure transducer increases sensitivity to a maximum. Above resonance the sensitivity decreases to a point at which the pressure transducer no longer responds to a pressure input (Fig. 18-3). Most semiconductor pressure transducers operate in a dc or static mode (0 Hz). This allows the pressure transducer to read static base pressure as well as the dynamic pressure changes. Exciting the transducer above its linear range (20 percent of resonant frequency) can be dangerous and should be done with extreme care.

INPUT AND OUTPUT IMPEDANCES

In the case of Entran pressure transducers, the nominal resis-

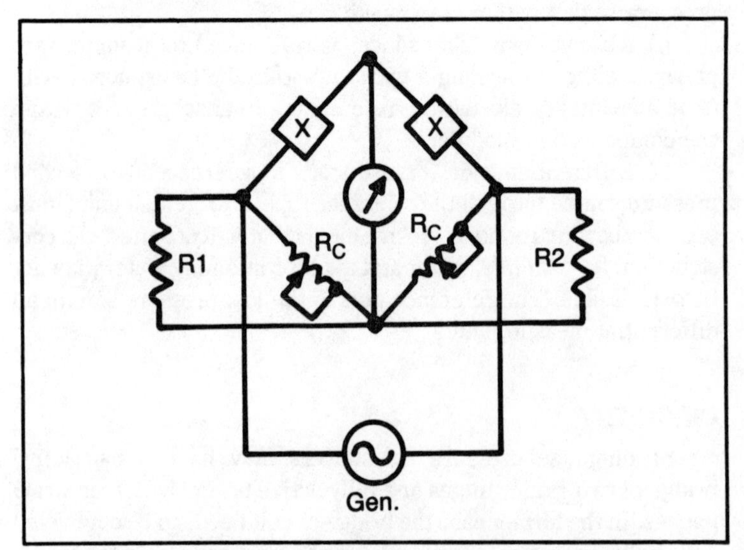

Fig. 18-2. A half-bridge circuit for use with pressure transducers (X).

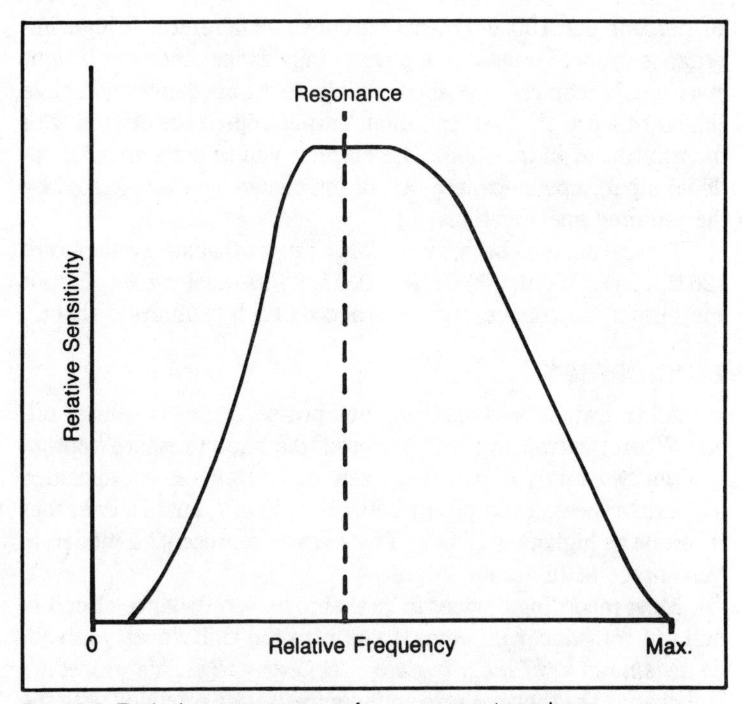

Fig. 18-3. Typical resonance curve for a pressure transducer.

tance of the bridge (and its individual strain gauges) is the output impedance. All units that are compensated for thermal sensitivity shift have a voltage-dropping resistor on the bridge input side (usually in the + In Voltage wire, but it can be − In Voltage wire or both). The input impedance is the nominal value of the dropping resistor (in series with the bridge) plus the bridge impedance. For best results the readout instrumentation input impedance should be at least 1 M. For an impedance of 20 times the output impedance of the bridge, the sensitivity is reduced by 5 percent. With an input impedance of 50 times, the reduction is 2 percent; an input impedance of 100 times yields a 1-percent reduction. For application requiring the most stable zero offset drift, the high value of Entran's custom option bridge impedance should be chosen. Zero offset drifting is a result of selfheating effects of the gauges. As the gauge impedance increases, the power dissipated for a given excitation voltage decreases.

When selecting optional transducer impedances, note that the actual resistance of semiconductor strain-gauge bridges varies greatly with temperature and, in some cases, may be as high as

20 percent per 100 degrees Fahrenheit. Therefore, bridge impedance cannot be used as a form of impedance matching if tight matching is required. Optional input/output impedances will have similar ratios as the standard input/output impedance offered with the transducer in question. Therefore, if you select a specific optional input impedance, the output impedance will be dictated by the required specifications.

Typical nominal bridge impedance ranges that are available are 120 Ω, 250 Ω, 350 Ω, 500 Ω, 1000 Ω, 1500 Ω. Actual values of input and output impedances are recorded on each calibration sheet.

ZERO OFFSET

All transducers have some "null pressure" or "baseline" offset. When the transducer is powered, the "null pressure" output will not be exactly 0 mV. If the zero offset value is not specified in the data sheet, it is typically between ±10 mV, but in some cases it can be as high as ±15 mV. This usually represents a moderate percentage of full-scale output.

Most recording devices have their own zero-balance circuit to null out transducer off sets. If you zero the transducer yourself, *do not shunt one of the transducers legs externally.* This procedure will change the thermal zero shift compensation provided with the unit. *Do not shunt the pressure transducer to reduce zero offset. This will alter the thermal zero shift performance.*

Adjustable zero offsets are also available from Entran as an option. A blackbox module can be provided with an adjustable zero-offset trim pot, or a no-charge five-wire output is available for you to install your own zero-offset trim pot. A tight zero trimming with fixed resistors can be expensive; the five-wire output is an ideal way of achieving zero offset control at no additional cost. On five wire versions, connecting two wires will revert to a standard four-wire system when adjustability is not needed.

Zero offset can also be affected adversely by transducer mounting. Any stresses placed on or near the diaphragm will automatically result in changes in the zero offset. These changes can also be thermally sensitive. Over-torquing of threaded transducers can have the same effect. For threaded devices a recommended installation torque is indicated on the calibration sheet. Zero offsets are trimmed at Entran with the indicated torque applied.

The zero offset will move to its final value while the pressure transducer is being "warmed up." Typical warm-up times can vary from five minutes to several hours, depending upon the transducer and desired level of stability. For critical dc measurements, where ultimate stability is required, a 4-hour warm-up may be advisable. Once the zero reaches equilibrium, it will then exhibit a small drift with time. When minimal drifting is required, the following four alternatives yield better results:

☐ Choose high-impedance options. Because drifting is a function of power dissipated, the higher gauge resistance is more stable for a given input voltage.

☐ Operate the transducer with a lower input voltage. Power dissipated is proportional to the square of the input voltage. Using one-half the typical input voltage dissipates one-fourth the power.

☐ Specify that your pressure transducer is to be temperature compensated in the medium in which the device will be used.

☐ When making dynamic measurements, the output of the pressure transducer can be ac coupled. This completely eliminates the zero offset and all its effects may be ignored.

EXCITATION VOLTAGE

For a standard (Entran) unit the sensitivity is directly proportional to the input voltage. By lowering the voltage, with no other changes in circuitry, the sensitivity decreases, but all other specifications remain the same. If it is necessary to lower the input voltage with either no reduction or a partial reduction in sensitivity, the value of the thermal sensitivity resistor (R_s) can be reduced. This allows more of the applied voltage to be directed toward the bridge but, in turn, reduces the effectiveness of the thermal sensitivity compensation. The compensating resistor (R_s) is expressed as the difference between the input impedance (R_j) and the output impedance (R_0). With no resistor a typical thermal sensitivity shift (tss) might be approximately −15 percent per 100 degrees Fahrenheit.

The following relationships are approximate and indicate the effect of sensitivity change versus input voltage and thermal sensitivity shift for one particular series of transducers. Check the Entran directly for the particular series in which you desire this adjustment for specific values. By selecting the maximum tss you can accept and the desired input voltage, you can calculate approximate sensitivities.

$$V_0 = 2\left(\frac{V_I}{V_{ids}}\right)\left(\frac{R_o}{R_s + R_o}\right) V_{ods}$$

Example.

$$R_s = R_o - \left(\frac{(tss)\ R_o}{-15\%/100°\ F.}\right) \text{for tss between 0 and } -15\%/100°\ F.$$

For V_{ids} = Input voltage from data sheet
V_i = Desired input voltage
V_{ods} = Sensitivity from data sheet
V_0 = New sensitivity
tss = Thermal sensitivity shift (0 to $-15\%/100°$ F.)
R_o = Output impedance form data sheet

Example. Entran's EPN-300-100 pressure transducers

V_{ids} = 15 V
V_{ods} = 3.5 mV/psi
R_o = 500 Ω

Would like to operate 10 Vdc with maximum thermal sensitivity shift of $-5\%/100°$ F.

$$R_s = 500 - \left(\frac{(-5\%/100°\ F.)\ 500\ \Omega}{-15\%/100°\ F.}\right) = 333\ \Omega$$

$$V_o = 2\left(\frac{10\ V}{15\ V}\right)\left(\frac{500\ \Omega}{333\ \Omega + 500\ \Omega}\right) 3.5\ mV/psi$$

$$V_o = 2.8\ mV/psi$$

For maximum stability of zero offset, lower input voltages are recommended. Sensitivity can also be increased by placing a higher voltage on the pressure transducer.

PIEZOELECTRIC PRESSURE TRANSDUCERS

A cross section of a basic piezoelectric pressure transducer is shown in Fig. 18-4. The quartz crystal is contained in a subassembly called the element. The element typically contains at least one but usually several quartz crystals, an insulator, a metal end piece, an electrode to collect the charge, and a fine wire that conducts the

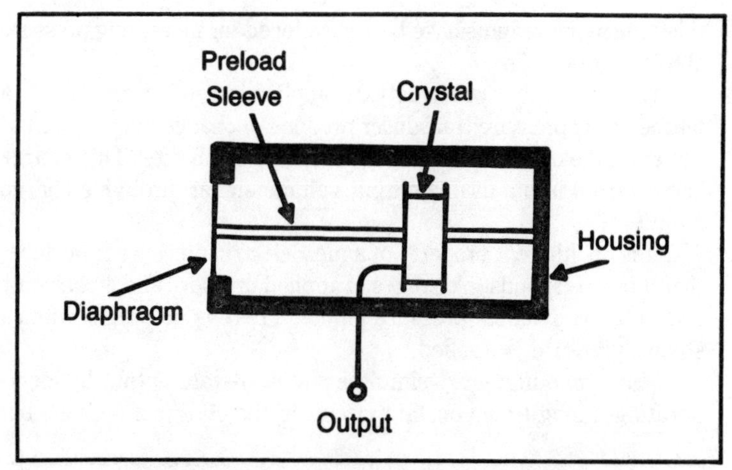

Fig. 18-4. Construction of a piezoelectric pressure transducer.

charge to the connector pin. The element is held together by a preload sleeve, which is a very thin metal cylinder. During assembly the sleeve is stretched slightly, then welded in place in the elongated condition. This preloads or clamps the components of the element tightly together, ensuring the most rigid structure by removing all relative motion between these components.

The element is contained in a steel outer housing or case, which provides for the mounting and sealing of the transducer, contains the electrical connector assembly, and most importantly, supports the outer edge of the diaphragm.

The diaphragm is a thin metal disc that has two main functions. It seals and protects the quartz element and, as previously mentioned, converts the pressure to a compressive force that deforms the element, producing the charge output.

In a typical quartz pressure transducer the composite modulus of elasticity of the element is slightly less than that of the steel housing, because a small portion of its length is made up of quartz, whose modulus is one-third that of steel. This means that the element will deflect slightly more than the outer housing when subjected to pressure.

This presents no real problem for low-pressure transducers, and thin relatively flexible diaphragms can be used because deflections are very slight.

When it comes to higher-pressure measurement, however, more rigid construction may be necessary. In fact, spherical constructions based on a piston movement instead of a diaphragm, or thick flush-diaphragms have been developed for measuring pressure above 50,000 lb/in^2.

Whatever the construction, application of pressure to a piezoelectric pressure transducer produces a change in stress across the crystal element, resulting in a change of charge. This charge is converted into a usable output voltage signal through a charge amplifier.

It is an inherent property of a piezoelectric pressure transducer that it only responds to changes in applied pressure and is thus only accurate for dynamic pressure inputs. There is no output when a steady pressure is applied.

Some manufacturers simulate a steady-state output by incorporating a long time constant circuit in the charge amplifier, but

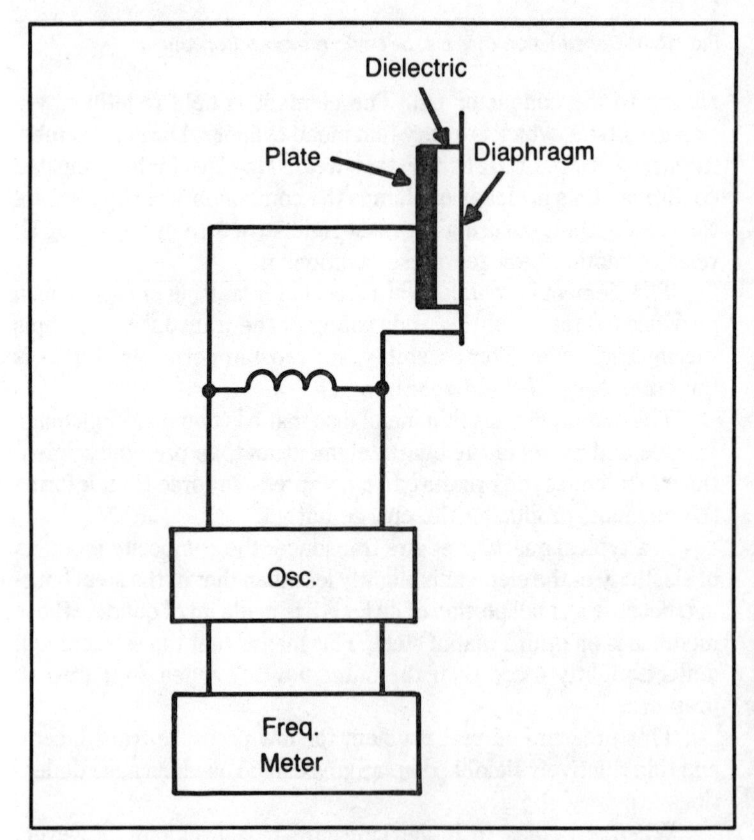

Fig. 18-5. A capacitive pressure transducer can be connected into the resonant circuit of an oscillator, resulting in an output frequency that varies with the applied pressure.

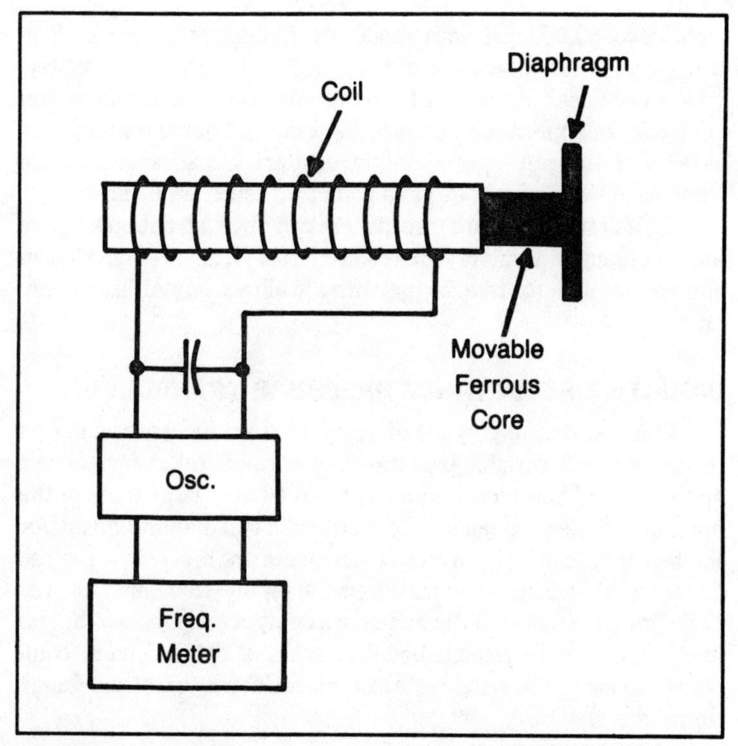

Fig. 18-6. An inductive pressure transducer can be connected into the resonant circuit of an oscillator, resulting in an output frequency that varies with the applied pressure.

this is subject to low-frequency rolloff. The major advantages of the piezoelectric device are the relatively high inherent output and the very-high-frequency response that can be achieved. However, piezoelectric pressure transducers are not easy to install and use because of the charge amplifier, which also makes their initial cost high.

CAPACITANCE PRESSURE TRANSDUCERS

Capacitance-type transducers employ a diaphragm and a fixed plate (Fig. 18-5). Application of pressure to the diaphragm changes the dielectric strength between the plates and, hence, the capacitance of the device. This capacitance is normally used as part of a tuned circuit, where pressure applied will result in a change in frequency of output.

This type of transducer can be employed for pressure ranges

from 5 to 50,000 psi with moderate frequency response. The capacitive device always requires associated electronic circuitry, which is a disadvantage in terms of cost per measurement and simplicity of application but does allow control of the output, such as 0-5 V or 4-20 mA for pressure transmitters. Capacitance pressure transducers are widely used as sound pressure transducers.

Capacitance pressure transducers can offer advantages where high accuracy and stability are required from small-size transducers and are suitable for interfacing with a wide variety of instrumentation.

INDUCTIVE/RELUCTANCE PRESSURE TRANSDUCERS

There are various types of pressure transducers working on the principle of variable inductance or variable reluctance. Their application is, however, limited by the relatively large mass of the moving iron element (an iron core attached to a diaphragm) and/or the fact that relatively large displacements are involved (as in the linear variable-differential transformer). Figure 18-6 illustrates the basic principle. Some of the more compact types use the diaphragm itself as the iron element, but they are still bulkier, more complicated, and less sensitive than piezoelectric or strain-gauge transducers.

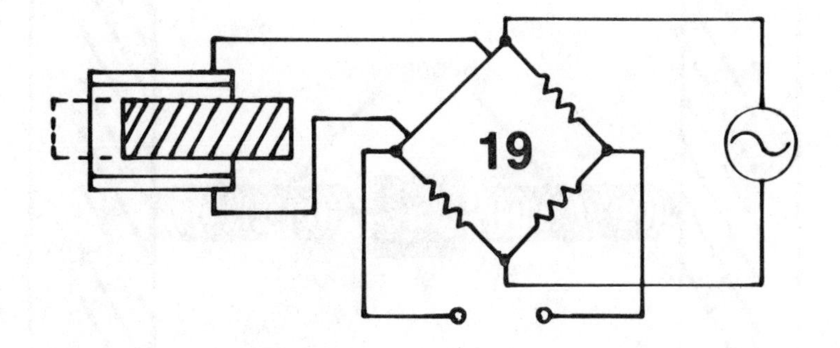

Accelerometers

An accelerometer is a transducer that responds to acceleration in one or more axes. The sensing element consists of a spring-mass system that deflects when subject to acceleration in the direction of its sensitive axis (Fig. 19-1). This mass is normally known as the *seismic element*, and the spring-mass combination as the *seismic system*.

The seismic mass responds to acceleration by producing a force proportional to applied acceleration. The spring deflects until an equal reaction force is developed. This deflection is a linear function of applied acceleration within the constraints imposed by the natural frequency and damping ratio of the seismic system. The actual "spring" involved does not have to be a mechanical one. It can just as well be electrical, and the seismic mass can be part of the transducer itself (for example, the core of an LVDT). Alternatively, a seismic system can be coupled to a separate displacement-sensitive transducer element to produce a transducer whose output is proportional to the acceleration applied to the seismic system. The transducer thus formed is referred to as an open-loop accelerometer because the measurement does not involve output-to-input feedback.

In a simple spring-mass system the mass position is directly proportional to acceleration only as long as the acceleration is considerably less than the undamped frequency of the system. This affects the choice of transducer normally employed. At low frequen-

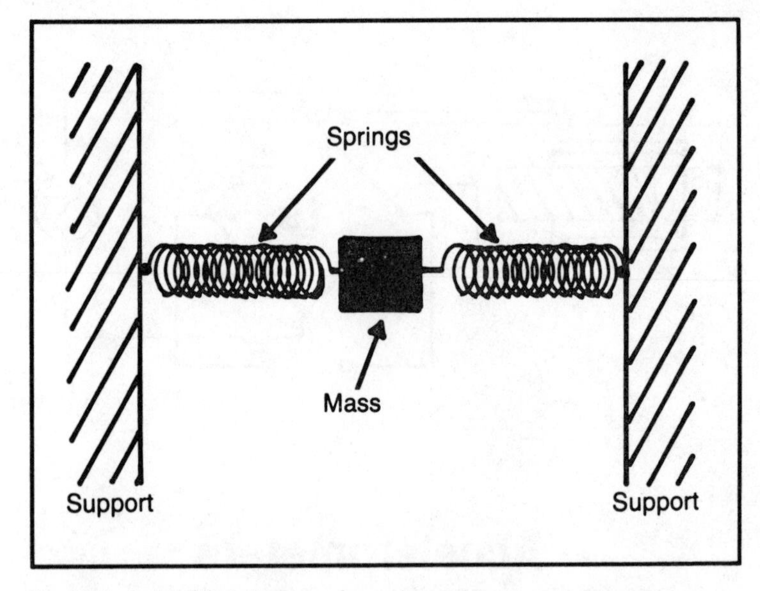

Fig. 19-1. A simplified diagram of a spring-mass type accelerometer.

cies the position of the mass can be detected with a displacement sensor. At higher frequencies a strain-gauge transducer or piezoelectric transducer becomes a more realistic choice. Even more demanding requirements are set by accelerometers designed for measuring shock and vibration.

OPEN-LOOP ACCELEROMETERS

Two important parameters of any open-loop accelerometer are natural frequency and damping ratio. The natural frequency of the spring-mass system is the frequency at which the seismic mass will vibrate with no damping. It is a measure of the speed with which the system can move in response to an impulse. The natural frequency of an undamped accelerometer must be several times the highest frequency of interest in the measurement if the output of the instrument is to be a correct representation of the applied acceleration. However, high natural frequency implies a relatively stiff, and therefore relatively insensitive, measuring system. The sensitivity of an undamped accelerometer is inversely proportional to the square of the natural frequency (Fig. 19-2). Thus, a high natural frequency permits acceleration measurement over a wider range of frequencies, whereas a low natural frequency gives greater sensitivity.

DAMPING

This relationship can be modified by adding damping to the system. Unless an open-loop accelerometer is damped, the seismic system will continue to vibrate and give readings long after the applied acceleration has disappeared. Further, a very small acceleration at frequencies near the natural frequency of an undamped accelerometer will cause an unduly large response from the instrument. Damping ratios of approximately 0.6-0.7 of critical will extend the range of usefulness of an accelerometer to nearly three-fourths of its natural frequency.

In reproducing a complex wave pattern, phase distortion generally results from the lack of linearity between the phase lag of the responding element (seismic system) and the frequency components of the applied wave pattern. A damping ratio in the range of 0.6 to 0.7 produces an almost linear phase lag relationship for frequencies up to the natural frequency of the instrument, and thus this range of damping ratios eliminates phase distortion within the operating range of the instrument.

Open-loop accelerometers use either fluid (viscous) or magnetic (eddy-current) damping. Silicon fluids having a low temperature coefficient of viscosity must be used for viscous damping to make any change in damping ratio with temperature relatively small, if not negligible. Magnetic damping is commonly used in applications

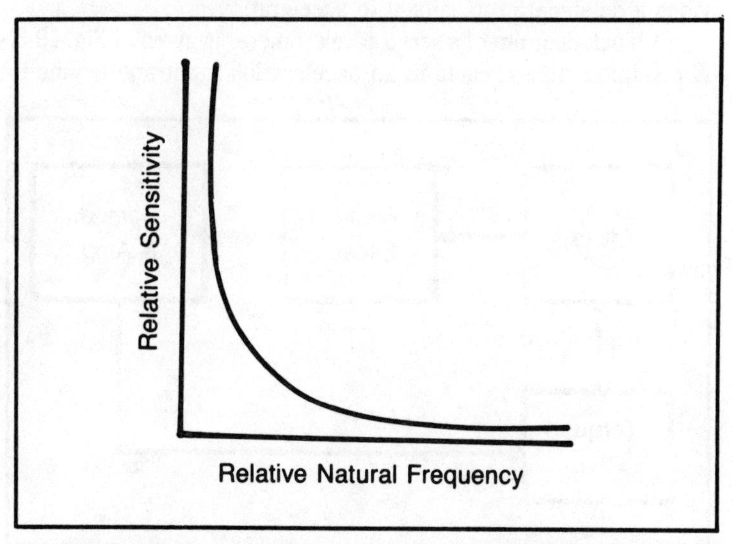

Fig. 19-2. Sensitivity-vs.-damping curve for a spring-mass accelerometer. As the natural frequency increases, the sensitivity drops off rapidly.

requiring a constant damping ratio, because magnetic damping is essentially unaffected by temperature.

Specifically, the output of an accelerometer or spring-mass system is linearly proportional to input acceleration only for frequencies up to about 40 percent of its damped natural frequency. However, when the frequency components of the motion under study are higher than about 2.5 times the natural frequency of the spring-mass system, the relative motion of the seismic system is proportional to the applied displacement.

The response of a spring-mass system is thus proportional to the applied acceleration or to the applied displacement, depending on whether the frequencies involved are lower of higher, respectively, than the natural frequency of the spring-mass system. For example, an accelerometer having a natural frequency of 20 Hz and designed to measure low-frequency accelerations up to about 10 Hz may also be used to measure displacements involving frequencies beyond about 50 Hz.

SERVO ACCELEROMETERS

Servo accelerometers are closed-loop, force-balance transducers with much greater accuracy and stability than that obtainable with open-loop accelerometers. They normally feature a rugged acceleration sensor integrated with dc-operated solid-state circuitry that provides a dc signal proportional to acceleration.

A block diagram of a servo accelerometer is given in Fig. 19-3. A pendulous mass reacts to an acceleration input and begins to

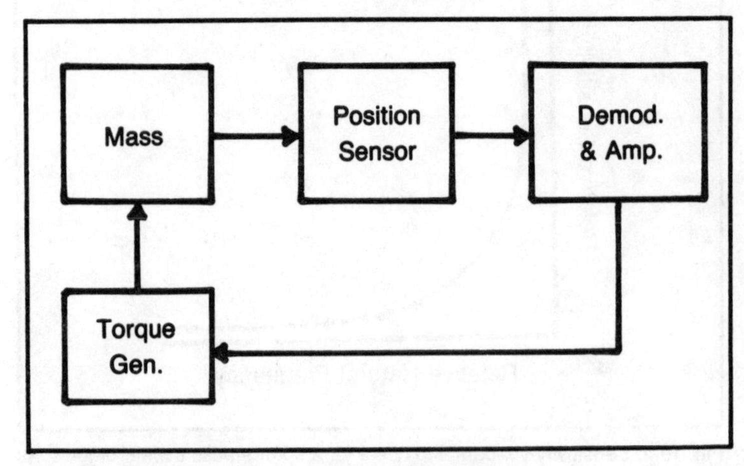

Fig. 19-3. Block diagram of a servo accelerometer.

move. A position sensor detects this minute motion and develops an output signal. This signal is demodulated, amplified, and applied as negative feedback to an electrical torque generator (or torquer) coupled to the mass. The torquer develops a torque proportional to the current applied to it. The magnitude and direction of this torque just balance out the torque attempting to move the pendulous mass as a result of the acceleration input, preventing further movement of the mass.

Because both torques are equal, and because the torque generator output is proportional to its input current, the input current is proportional to the torque attempting to move the pendulous mass. This torque is equal to the product of moment of inertia (a constant) and acceleration. Therefore, the torque generator current is proportional to applied acceleration. If this current is passed through a stable resistor, the voltage developed is proportional to applied acceleration.

The dynamic ratings of the servo accelerometer are unusual in that neither the natural frequency nor the damping ratio is directly and unavoidably related to the accelerometer range, as in open-loop types. The damping in the sensitive axis is accomplished electrically by a passive network.

The natural frequency of a servo accelerometer is a function of loop gain multiplied by the moment of inertia of the pendulous mass. Thus, the natural frequency can also be determined electrically, within reasonable limits. It is possible therefore to provide high natural frequency for low-range accelerometers without sacrificing sensitivity or affecting scale factor. However, small accelerations are generally related to the relatively slow movements of large bodies, and large accelerations are generally related to the relatively fast movements of small bodies. So, as a practical matter, high response is not generally required in low-range accelerometers.

PRACTICAL ACCELEROMETERS

The majority of accelerometers in use today are of the piezoelectric or semiconductor strain-gauge type. Piezoelectric accelerometers fall into two distinct categories: piezoresistive (pr) and piezoelectric (pe). Piezoresistive (pr) accelerometers have the advantage of offering dc response with a sensitivity usually sufficiently high that no preamplification of output is necessary. Also, because they use an external source of energy, they have an inherently low output impedance. They are, however, largely limited to low-frequency measurement yielding signal pulses of relatively long duration.

Note. The piezoresistor (pr) transducer is a *strain gauge*, although not all strain gauges are of pr type (see later description of strain-gauge accelerometers and also Chapter 18 on pressure transducers).

Piezoelectric (pe) accelerometers are self-generating as regards electrical output signal and so do not require any external power supply. Artificially polarized ceramic crystals, having a much greater a shape configuration chosen to further enhance specific characteristics:

☐ *Single-ended compression* for high sensitivity and high resonant frequency. A suitable choice for low-level measurement and for general-purpose use.

☐ *Shear* for miniaturized accelerometers with low mass. By locating the sensitive element isolated from the base, shear accelerometers provide the best protection against pickup from base bending and acoustic noise. Also, shear excitation reduces crystal sensitivity to temperature transients. Shear crystals are, however, less sensitive than compression types.

With pe accelerometers, low-frequency amplifier cutoff is usually 2-5 Hz to reject piezoelectric output produced by many pe transducers. Special isolated designs can be used at much lower frequencies. Piezoelectric accelerometers tend to have very predictable nonlinearity, which can be expressed as a percentage increase in sensitivity with applied acceleration (typically on the order of 1 percent per 500 g). The upper limit of g is normally determined by peak crystal stress and/or maximum nonlinearity.

Note. Careless handling of pe transducers can produce very high g levels (several thousand if dropped from a height of 4-5 feet onto a hard floor).

THERMAL EFFECTS

The sensitivity of piezoelectric acceleration also varies with temperature. A temperature-range rating gives the operating range over which this effect can be regarded as negligible for most practical purposes. Typically, this is from 0 to 180°-260° C. for pe types, and − 50° to + 120° C. for pr types.

A sudden *change* in temperature may also affect the output, giving a momentary spurious signal generated by pyroelectric effect.

EXAMPLES OF DESIGN

Modern pe accelerometers are highly developed designs, largely individual to specific manufacturers and aimed at overcoming the limitation of conventional designs. With conventional compression designs, nonvibrating forces can be transmitted through the mass or crystal, producing a spurious noise signal. With conventional shear design, extraneous forces transmitted through the mass or base again produce noise, but not to the same extent.

The following are examples of optimized designs.

Isobase

Isobase accelerometers (Fig. 19-4) are a premium compression design. They use a specially contoured internal base to make the crystals less sensitive to stress at the mounting point. This gives a significant improvement in isolation from base bending and thermal transients. *Isobase* accelerometers also operate in high g levels and at high temperatures.

Shear Designs

Shear designs also give improved isolation. Electrical output

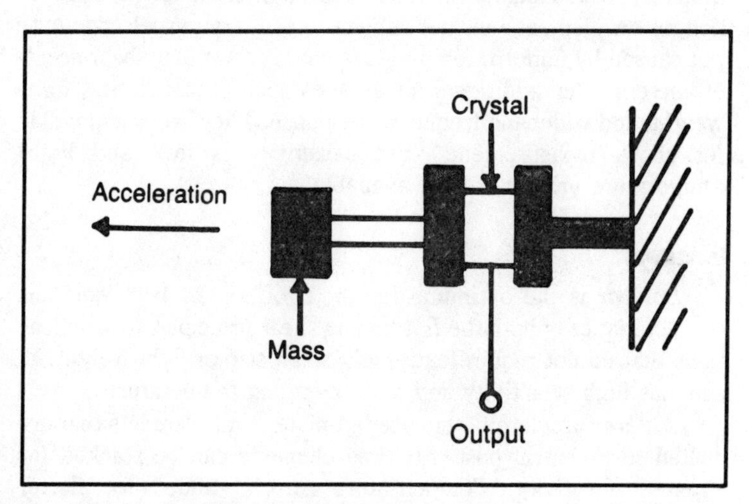

Fig. 19-4. A compression (isobase) accelerometer.

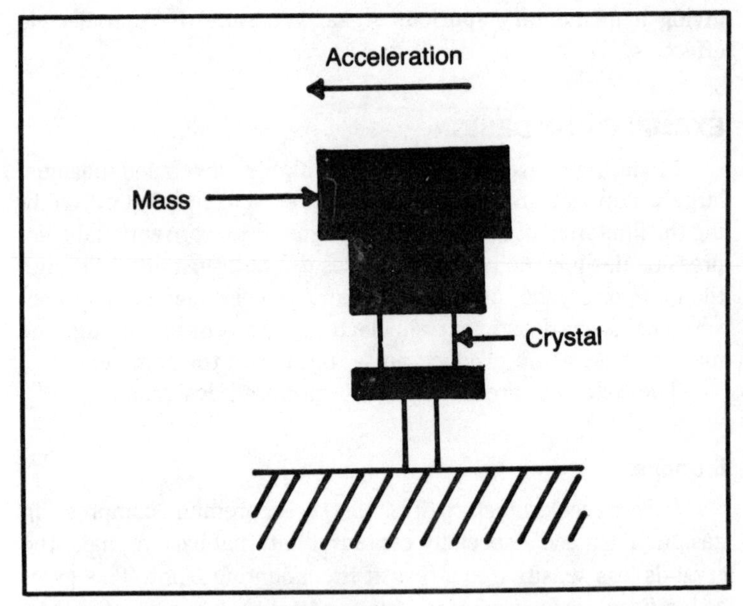

Fig. 19-5. A shear-type accelerometer.

is derived from shearing forces caused by the mass which is mechanically attached or bonded to the periphery of the pe element (Fig. 19-5). This provides excellent base strain protection because base bending produces negligible shearing forces on the element. Similarly, with acoustic excitation and thermal transients. In addition, shear elements are not subject to primary pyroelectric output caused by uniform temperature change. Annular shear accelerometers offer additional advantages such as small size, light weight, and wideband frequency response. They are also popular for shock measurements. But sensitivity is low, and high-temperature models are not available.

Isoshear

Isoshear is the optimum design. It offers the best isolation characteristics of both the *Isobase* and shear principles for applications that do not require extremely small size or light weight. It also has high sensitivity and high operating temperatures.

Isoshear accelerometers use flat-plate shear elements that are bolted to a central post. Multiple elements can be stacked for greater sensitivity. Compensators can be added for flatter temperature response. And electrical insulators can be added to isolate the signal from the case. *Isoshear* offers such excellent insulation from nonvibration environments that measurements can readily be made down to 0.1 Hz in environments of severe thermal, acoustic, or base-bending conditions.

SIGNAL CONDITIONING

A signal conditioner is normally used to interface a pe accelerometer to a readout instrument or recorder. In the case of a pr-type circuit, requirements can be minimal. A pair of transducer elements and a pair of fixed resistors are connected in a Wheatstone bridge configuration, with the external power supply connected as in Fig. 19-6. Output is then taken from the bridge; it may or may not need amplification before being fed to a dc-indicating readout. For increased sensitivity all four arms of the bridge may be active (transducer) elements.

Piezoelectric type accelerometers may be treated either as voltage or charge generators requiring no external supply. The voltage-generator mode is the simplest to condition by using a voltage amplifier (Fig,. 19-7). However, the disadvantage of this system is that sensitivity is affected by cable length.

For that reason the charge-generator mode, which is indepen-

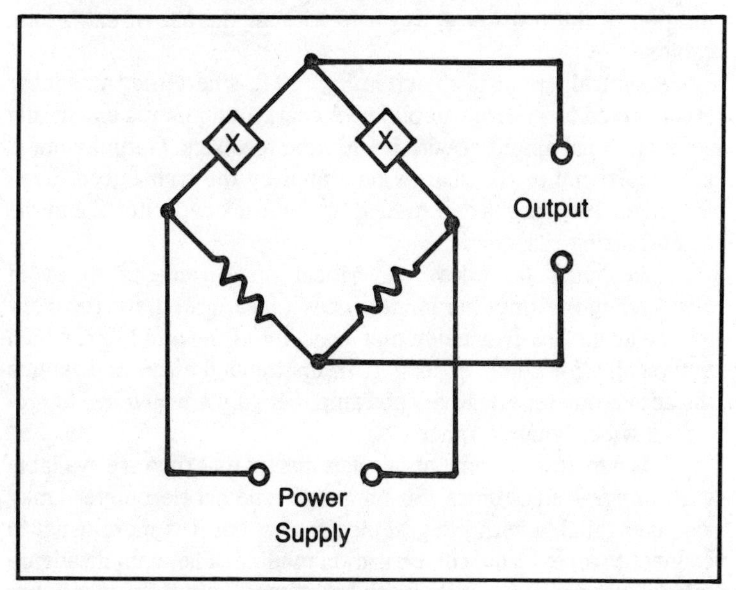

Fig. 19-6. A method of signal conditioning for use with accelerometer transducers (X).

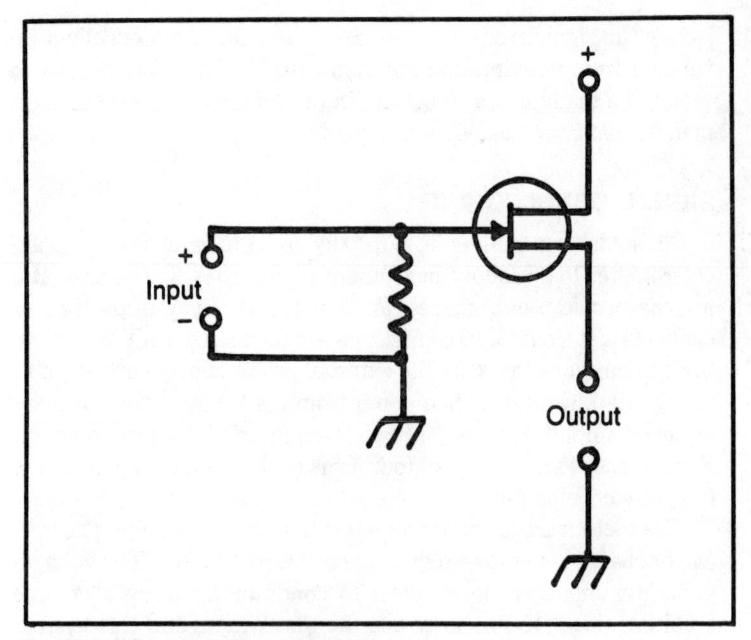

Fig. 19-7. A simple voltage amplifier.

dent of cable length, is normally preferred, and thus the initial calibration of the transducer is made without the use of extra long cables.

A typical circuit is shown in Fig. 19-8, where the transducer is connected to a charge amplifier. A charge amplifier is essentially an operational amplifier with integrating feedback. Output voltage is proportional to the charge generated by the transducer. Low-frequency response is determined by feedback capacitor C and the dc stabilizing resistor R.

Care must be taken on critical measurements to avoid overloading the amplifier input because, if saturated, the recovery will be at the low-frequency time constant of the amplifier, which is typically 2 seconds or longer. In unattended airborne systems the accelerometer sensitivity and amplifier gain are matched to provide a wide dynamic range.

It is a feature of some pe accelerometers that they are available with integral electronics. Strain-gauge-type accelerometers may also be available with integral electronics, but it is more usual to connect to an external compensation module. The main disadvantage of integral electronics (with both types) is that the maximum series temperature of the accelerometer is that for the electronics, which may be considerably lower than that of the accelerometer itself.

STRAIN-GAUGE ACCELEROMETER

An example of modern strain-gauge accelerometer construction is given in Fig. 19-9. The moving mass is mounted on a cantilever beam to which strain gauges are bonded top and bottom. Under acceleration forces, the g effect on the mass causes the beam to bend; the resulting bending movement produces a strain proportional to acceleration and detected by the strain gauges. The strain gauges are connected to a bridge circuit with either two or four "active" arms (two or four strain gauges on the beam, respectively). With a voltage applied to the bridge, bridge unbalance results in a circuit charge proportional to acceleration.

Accelerometers of this type can have an extremely linear response, such as a design full-scale range well within a 1-percent error band for nonlinearity, or overall nonlinearity of better than 99 percent. If it is used for higher g measurement than the design range, however, nonlinearity will become increasingly marked.

Accelerometers of this type are also designed with different natural frequencies as well as different g ranges—the higher the

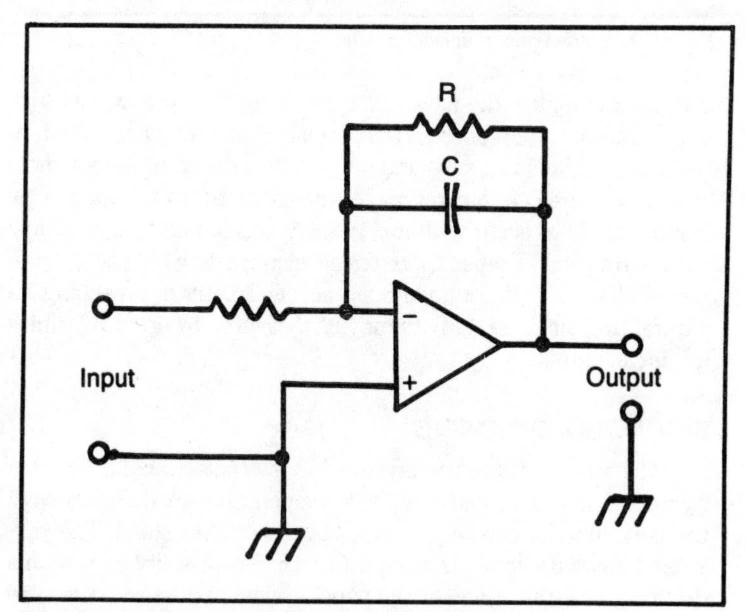

Fig. 19-8. A charge amplifier using an operational amplifier.

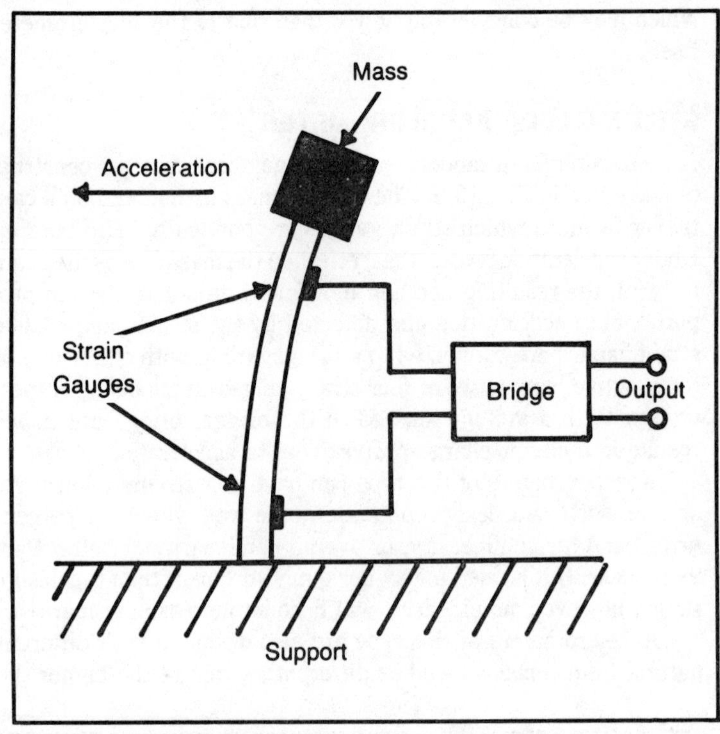

Fig. 19-9. A strain-gauge accelerometer.

g range the higher the natural frequency of the system, and vice versa. The main point here is that the maximum frequency of vibration applied should be substantially less than the natural frequency; otherwise the accelerometer may be damaged by overranging. This is most likely to occur with low-g-range accelerometers (with low natural frequency), when a *damped* accelerometer would be the preferred choice. With an undamped accelerometer approaching its natural frequency, sensitivity may be measured by up to 100 times its linear output.

SIGNAL CONDITIONING

The strain gauges are arranged in a Wheatstone bridge configuration, normally connected to a compensation module (although the same function can be performed by discrete resistors). This provides sensitivity compensation for thermal sensitivity shift with a dropping resistor in one of the supply lines, zero offset trim, and zero offset drift.

It is important to use the specified excitation voltage for the specific model of accelerometer used, because higher or lower voltages can affect both the sensitivity and the amount of compensation required (resistor values). However, it is easy to adjust the circuit, if you know what you are doing.

In the case of a typical Entran accelerometer, for example, the sensitivity is directly proportional to the input voltage, and all compensation and trimming has been done in the compensation module for the design voltage. If the exciting voltage is lowered, with no other changes in circuitry, the sensitivity decreases, but all other specifications remain the same. If it is necessary to lower the input voltage with either no reduction or a partial reduction in sensitivity, the value of the thermal sensitivity resistor (R_s) can be reduced. This allows more of the applied voltage to be directed toward the bridge but, in turn, reduces the effectiveness of the thermal sensitivity compensation. The compensating resistor (R_s) is expressed as the difference between the input impedance (R_i) and the output impedance (R_o). With no resistor at all, the thermal sensitivity shift (tss) is approximately −15 percent per 100 degrees Fahrenheit for most of Entran's standard devices.

The following relationships are approximate and indicate the effect of sensitivity change versus input voltage and thermal sensitivity shift. By selecting the maximum tss you can accept and the desired input voltage, you can calculate approximate sensitivities.

$$V_0 = 2 \left(\frac{V_i}{V_{ids}}\right) \left(\frac{R_o}{R_s + R_o}\right) V_{ods}$$

where $R_s = R_o - \left(\frac{(\text{tss})\, R_o}{-15\%/100°\,\text{F.}}\right)$ for tss between 0 and −15%/100° F.

For

V_{ids} = Input voltage from data sheet
V_i = Desired Input voltage
V_{ods} = Sensitivity from data sheet
V_0 = New sensitivity
tss = Thermal sensitivity shift (0 to −15%/100° F.)
R_o = Output impedance from data sheet

Example. Entran's EGA-125-250 accelerometer

$$V_{ids} = 15\text{ V}$$
$$V_{ods} = 1\text{ mV/g}$$
$$R_0 = 500\ \Omega$$

would like to operate 10 Vdc with maximum thermal sensitivity shift of −15%/100° F.

$$R_s = 500 - \frac{(-5\%/100°\text{ F.})\ 500}{-15\%/100°\text{ F.}} = 333\ \Omega$$

$$V_o = 2\left(\frac{10\text{ V}}{15\text{ V}}\right)\left(\frac{500\ \Omega}{333\ \Omega + 500\ \Omega}\right) 1\text{ mV/g}$$

$$V_o = 0.80\text{ mV/g}$$

For maximum stability of zero offset, lower input voltages are recommended. Sensitivity can also be increased by placing a higher voltage on the accelerometer. This should be discussed with Entran before purchase.

The following notes also apply to Entran accelerometers.

Zero G Offset

All transducers have some no load or zero g offset. When the transducer is powered, the no-load output will not be exactly 0 mV. If the zero offset value is not specified in the data sheet, it is typically between ±10 mV, but in some cases it can be as high as ±15 mV. This usually represents a small percentage of full-scale output. Offset values closer to 0 mV can be provided on special request. Most recording devices have their own zero balance circuit to null out transducer offsets. If you zero the transducer yourself, *do not shunt one of the transducer legs externally*. This procedure will change the thermal zero shift compensation provided with the unit. Do not shunt the accelerometer to reduce zero offset. This will alter the thermal zero shift performance. Adjustable zero offsets are also available from Entran as an option.

The actual value of the zero offset will drift to its final value while the accelerometer is being "warmed up." Typical warm-up times can vary from 5 minutes to 2 hours, depending upon the transducer and desired level of stability. For critical dc measurements, where ultimate stability is required, a 4-hour warm-up may be advisable. Once the zero reaches equilibrium, it will then exhibit a small drift over time. When minimal drifting is required, the following four alternatives yield better results.

For Best Zero Stability

☐ Choose Entran's high-impedance option. Because drifting is due to power dissipated, the higher gauge resistance is stabler for a given voltage.

☐ Operate the transducer with a lower input voltage. Power dissipated is proportional to the square of the voltage. Using one-half the typical input voltage dissipates one-fourth the power.

☐ Select a damped accelerometer. Oil damping creates a better heat distribution internal to the accelerometer and is inherently stabler.

☐ Provide better heat sinking for the accelerometer when mounted.

When making dynamic measurements, the output of the accelerometer can be ac coupled. This completely eliminates the zero offset, and all its effects may be ignored.

Zero Offset Shift with Temperature (Thermal Zero Shift)

The change of zero offset with temperature is minimized in Entran's thermal compensation process. Standard units are provided fully compensated and the actual maximum shift is described in the accelerometer bulletins. Typical compensations are in the range of ±1 to ±2 percent of full scale per 100 degrees Fahrenheit. This means that the stabilized zero offset will change a maximum of 1 to 2 percent of the full-scale output per 100 degrees Fahrenheit. For example, for a transducer with 1-percent compensation and a 250-mV full-scale output, if the stable zero is +3 mV, it will shift to a maximum of either +0.5 mV or +5.5 mV over a 100° F. change. These values apply within the compensated temperature range of the transducer.

Sensitivity Shift with Temperature (Thermal Span Shift)

The change of sensitivity with temperature is minimized in Entran's thermal compensation process. Standard units are provided fully compensated (except for low-cost units), and the actual maximum shift is described in their accelerometer bulletins. Typical compensations are in the range of ±1 to ±2.5 percent per 100 degrees Fahrenheit. This means that the sensitivity of the accelerometer will only change a maximum of 1 to 2.5 percent of the calibrated value per 100° F. For example; a transducer with 1-percent compensation and a 1.00 mV/g sensitivity, will shift to a maximum of either 0.99 or 1.01 mV/g over a 100° F. change. These values apply within the compensated temperature range of the transducer.

Compensated Temperature Range

The *compensated temperature range* is the range in which the accelerometer will meet the specifications for zero and span shift as posted in the data sheets. Above and below this range, the transducer will continue to operate, but the specification will gradually increase from the data sheet values. The transducer is compensated for equilibrium values of temperature, not for fast temperature changes, pulses, or excursions. If the accelerometer is compensated from 80° F. to 180° F. and the actual 100° F. differential occurs in a rapid excursion, the accelerometer must be allowed to come to an equilibrium temperature before it will meet the listed specifications. *Compensation is only valid for equilibrium or slow changes in temperature, not for thermal shocks.* In cases where fast temperature changes are required, we suggest using a damped accelerometer or a unit with a larger housing to act as a better heat sink.

All standard accelerometers are compensated from 80° F. to 180° F. unless otherwise specified. However, Entran will compensate your accelerometer over any 100° F. band within the operating temperature without additional charge. For intervals of greater than 100° F., compensation can be achieved by special request. For large differentials, such as a 200° F. or 300° F. bandwidth, the tightest compensation can only be achieved over a 100°-150° F. section, but you may specify this section. For example, an accelerometer can be compensated from 0° to 300° F. with any 100° F. section at a compensation of percent per 100 degrees Fahrenheit while the other intervals are at a compensation of percent per 100 degrees Fahrenheit. Most of Entran's standard units are available for compensated ranges from −40° to +250° F.; however, ranges from −100° to +500° F. are available on special order.

See also, Chapter 28 on vibration monitoring.

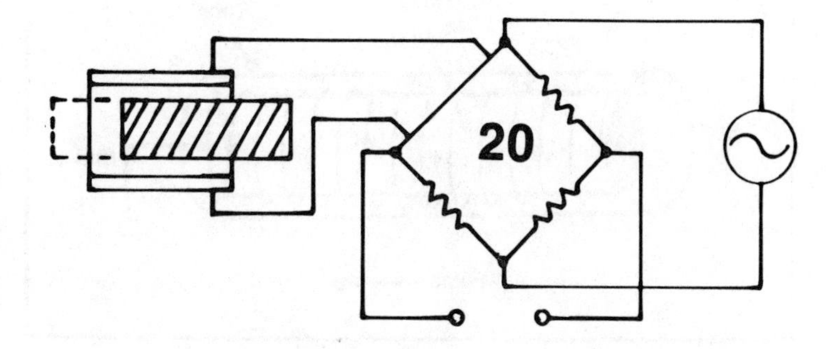

Measurement of Force and Torque

One of the most effective devices for measuring force is a coil spring. Strictly speaking, this is not a transducer as such because it converts mechanical force into proportional mechanical movement; thus, it is an all-mechanical system rather than an energy converter. It works on the same principle as a spring balance.

You can easily make such an instrument by mounting a spring in a tube, together with a rod passing through the center. One end of the spring is fixed to the tube, the other end to the rod (see Fig. 20-1). End A of the rod can be bent at right angles to lock onto something to measure pull or *tensile* force. End B is plain and pressed against something to measure *compressive* force. Thus, regardless of where tensile force or compressive force is being measured the spring is extended in the same way.

The amount of spring extension is a measure of the force involved. This can be indicated directly by extending the tube length to accommodate the full length of the spring when fully extended and adding a pointer attached to the rod at the point at which the spring is attached to it. The tube is slotted to accommodate movement of the pointer, and a graduated scale can be added to indicate actual pointer movement (Fig. 20-2).

It now remains to design a suitable spring size for the range of force to be measured. The basic formula involved is:

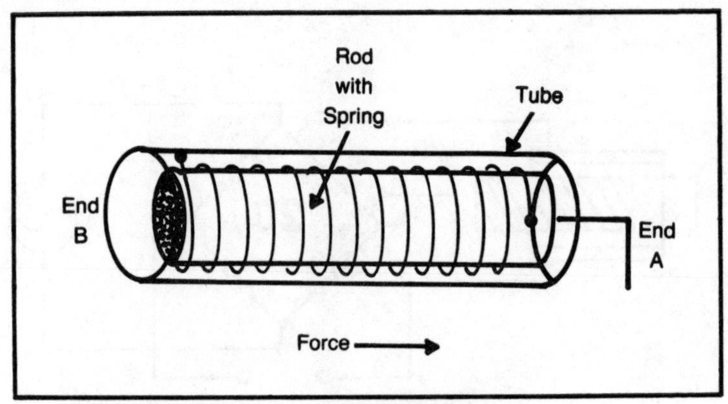

Fig. 20-1. A simple spring device for measuring force.

$$\text{spring deflection} = \frac{8FD^3}{Gd^4} \times N$$

where F is the force

D is the mean diameter of the spring

d is the diameter of the wire

G is the modulus of rigidity of the spring material

N is the number of active coils in the spring

The limiting force the spring can accept is related to the tor-

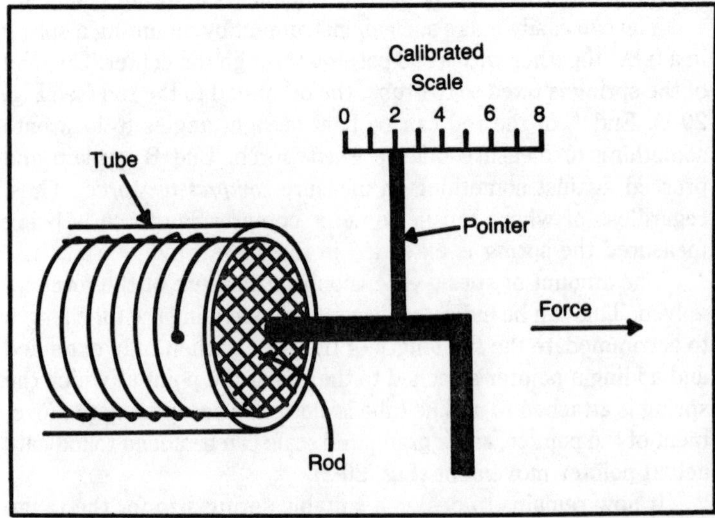

Fig. 20-2. A calibrated scale indicates the relative force in the device of Fig. 20-1.

sional stress developed in the spring when extended, so another formula is involved here:

$$\text{torsional stress (S)} = \frac{8FD}{\pi d^3}$$

As a starting point, let's look at spring *materials* first. Music wire is a logical choice (hardened spring steel wire), followed by oil-tempered steel wire or hard-drawn steel wire. These together with stainless steel wire are the strongest spring materials. For completeness we could also consider materials such as phosphor bronze and hard brass wire. The material properties we are interested in are shown in Table 20-1.

Once the material is chosen, the geometry of the spring needs to be worked out: we need values for D, d, and N. Points to remember are that the stiffness of a spring (resistance to deflection) is proportional to the *fourth* power of the wire diameter d and varies inversely as the *cube* of the mean diameter D.

Both d and D thus have a marked effect on spring performance. Using a wire size only one gauge up can appreciably reduce the deflection, and vice versa. Similarly, only a small increase in spring diameter D can considerably increase the deflection; or a small decrease in D can make the spring much stiffer.

The latter effect, particularly, should be borne in mind when making a helical spring by wrapping around a mandrel. There will be an inevitable "spring back" resulting in a spring inner diameter size greater than that of the mandrel. An undersize mandrel is thus required to form a spring of required diameter. The degree of undersize can only be estimated from experience because it will vary with

Table 20-1. Properties of Spring Materials.

Material	Maximum safe torsional stress (S)	Modulus of elasticity
music wire	180,000	12,000,000
oil-tempered steel wire	150,000	11,500,000
hard-drawn steel wire	150,000	11,500,000
18/8 stainless wire	100,000	9,700,000
phosphor bronze	90,000	6,300,000
hard brass	50,000	5,500,000

the quality of the spring material used and with the coiling technique.

The number of active coils is those actually "working" as a spring. The usual practice is to allow three fourths of a turn (or one complete turn) at each end in the case of a plain compression spring to produce parallel ends. Thus, geometrically the spring has a total number of coils equal to N + 1 1/2 (or N + 2), the number of active coils being calculated for the required deflection performance. Extension springs, on the other hand, commonly have all the coils "active," the ends being made off at right angles to the main coil.

Another important parameter is the *spring rate* (or load rate), which is simply the load divided by the deflection.

$$\text{spring rate} = \frac{P}{\text{deflection}}$$

$$= \frac{Gd^4}{8ND^3}$$

When the spring is of constant diameter and the coils are evenly pitched, the spring rate is constant.

Basically, spring design involves calculating the spring diameter and wire size required to give a safe material stress for the load to be carried. It is then simply a matter of deciding how many coils are required (how many active turns) to give the necessary spring rate or "stiffness" in pounds per inch of movement.

Although the working formulas are straightforward, spring design is complicated by the fact that three variables are involved in the spring geometry: diameter (D), wire diameter (d), and number of active coils (N). However, only D and d appear in the stress formula, which is the one to start with. So here it is a case of "guesstimating" one figure and calculating the other on that basis.

DESIGNING THE SPRING GEOMETRY

The most direct way of deciding the spring geometry is to fix a value for wire diameter d (that is, guesstimate a likely standard gauge size) and calculate the required value of D from

$$D = \frac{\pi S d^3}{8F}$$

Here F is the maximum force to be measured, and S is determined from the wire material value for maximum safe torsional stress.

Alternatively, you may need to start with a fixed value for the spring according to diameter D to fit inside a suitable tube. In this case you calculate the wire diameter size required from

$$d = 3\sqrt{\frac{8FD}{\pi S}}$$

The snag here is that the calculated value for d is unlikely to correspond exactly to an available gauge size. That means you will have to select the nearest available gauge size and recalculate the *actual* value of D required from the first formula.

Having finally arrived at suitable values for D and d, the number of active turns required is

$$N = \frac{Gd^4 \times \text{deflection}}{8FD^3}$$

Deflection in this case refers to the amount of spring movement you want to accommodate, that is, the length of the slot in the tube.

TORSION SPRINGS

Another way of measuring force is with a *torsion spring*. Here one end of the spring is fixed, and the other terminates in an extended arm (Fig. 20-3). A force applied to the end of this arm will produce a deflection related to the amount of applied force and the length of the spring arm. In this case the stiffness of the spring (its resistance to deflection) will be directly proportional to the fourth power of the wire diameter (d) and usually proportional to the mean spring diameter (D). The design formulas involved are

$$\text{stress (S)} = \frac{32\ PR}{\pi\ d^3}$$

$$\text{deflection (degrees)} = \frac{3665\ FRD}{Ed^4} \times N$$

where E is the elastic modulus of the spring material,

$$\text{all spring steel wire E} = 30{,}000{,}000$$

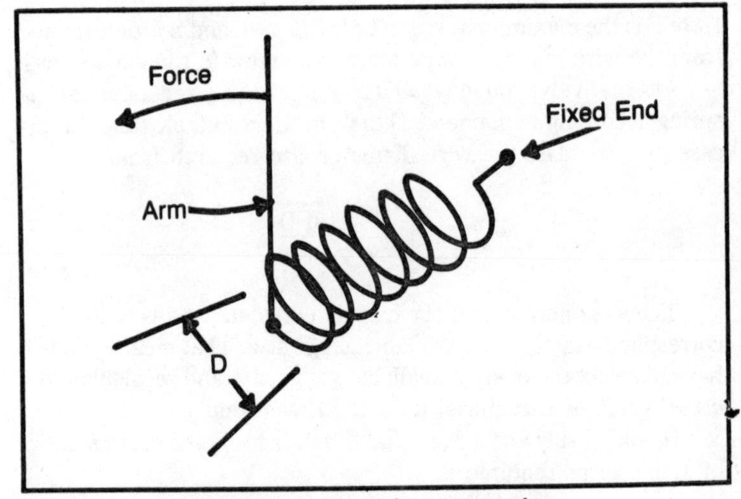

Fig. 20-3. A torsion spring may be used to measure force.

stainless steel wire E	=	28,000,000
phosphor bronze E	=	15,000,000
hard brass E	=	9,000,000

Design calculations are again based on working the spring material within acceptable limits of stress. The force (P) acting on the spring is applied over a radius (R) equal to the effective length of the free arm of the spring. Design calculations can proceed as follows:

☐ Knowing the force to be accommodated and the spring arm leverage required (R), use the stress formula to calculate a suitable wire size:

$$d^3 = \frac{32PR}{\pi S} = \frac{10.18PR}{S}$$

where S is the maximum permissible material stress.

☐ Adjust to a standard wire gauge as necessary.

☐ Calculate the angular deflection of such a spring, using a specified value of diameter D from the deflection formula and ignoring the factor N. This will give the deflection per coil. Then simply find out how many coils are needed to produce the required deflection.

This stage may, of course, be varied. The load moment FR may be the critical factor: the spring is required to exert (or resist) a certain force (F) at a radius R with a specific deflection. In this case, having adopted a specific value for D, the deflection formula can be used to find a solution for the number of turns required. If the spring is to be fitted over a shaft or spindle, you should check that in its tightened position it does not bind on the shaft. Incidentally all springs when deflected have a store of energy (mechanical force has been connected into mechanical energy).

MEASURING TORQUE

Springs can also be used to measure torque. In this case they take the form of a flat helix or clock spring, best wound from flat-strip spring material (Fig. 20-4). Any load applied in such a way as to wind the spring up (clockwork fashion) generates a torque in the spring proportional to the applied turning movement and the (unwound) spring diameter.

This principle is utilized in simple, mechanical, static, torque-measuring instruments incorporating a calibrated spring and attached pointer movement, the spring being rotated by a protruding shaft or check or similar device for attachment to a workpiece. Rotating the workpiece or instrument then gives a torque indication. And, allied to a slipping brake mechanism, measurement of dynamic torque is possible.

A slipping brake system is, in fact, the basis of mechanical dynamometers. The brake applies a load to the machine on test, the corresponding load force being determined over a specified mo-

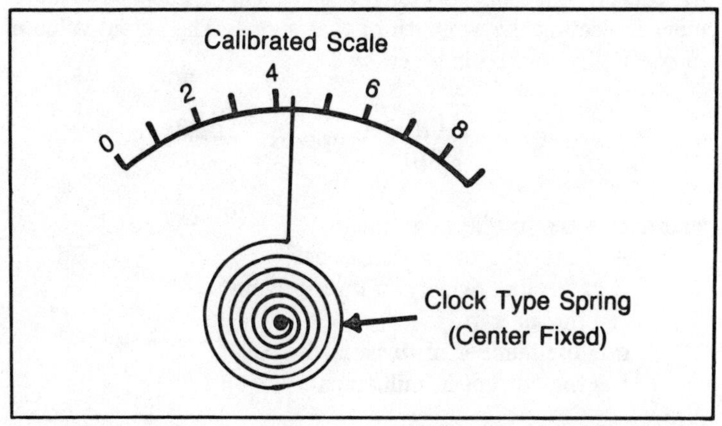

Fig. 20-4. A spiral-wound, clock-type, spring device for measuring torque.

ment arm; the product of brake load and moment arm is the torque developed.

TRUE TRANSDUCERS

Devices that are true transducers are also widely used for measuring force and torque, normally generating an electrical signal that is suitable for interfacing with instrumentation. Load cells, for example, are widely used for measuring tensile and compressive loads, not just for weighing applications. Force measurement is, after all, only another form of weighing. Strain gauges, too, are widely used for both static and dynamic torque measurements.

ELECTRIC CURRENT INTO TORQUE

Just as specific types of transducers convert mechanical movement into an electrical signal, the same process can be used in reverse. A simple example is the movement of readout instruments (ammeters and voltmeters). The signal circuit they receive is passed through a pivoted coil to which a needle is attached. The coil faces a fixed permanent magnet. Current passing through the coil generates an electromagnetic field in opposition to that of the magnet, causing the coil to rotate away from its initial position by an amount proportional to the electromagnetic field strength. This in turn is directly proportional to the signal current flowing through the coil. A compass can be surrounded by a coil, producing deflection of the needle in a similar way (Fig. 20-5).

A moving-coil instrument movement thus works as a transducer, connecting an electrical signal into a proportional movement indicating the strength of that signal. The actual value of torque (Q) generated is given by

$$Q = \frac{ABnI}{9.810} \text{ or approx. } \frac{ABnI}{10}$$

where Q is the torque in gram-cm
A is the area of the coil in cm^2
B is the flux density in lines per cm^2
or the air gap
n is the number of turns in the coil
I is the current in milliamps

With typical small meter movements the torque generated is

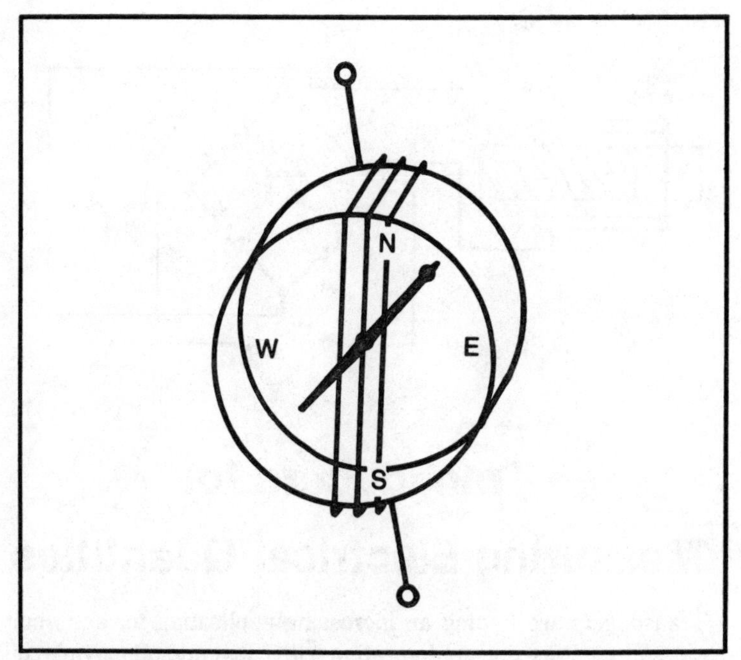

Fig. 20-5. Electric current is converted into torque by means of a compass-galvanometer device.

of the order of 0.01 gram-cm per milliamp. Thus, a 0-5-mA milliammeter movement develops a torque of about 5 × 0.01 = 0.05 gram-cm. A far more robust coil would be used, developing a torque of about 5000 × 0.01 = 50 gram-cm torque. At the other end of the scale, lighter movements with very low friction bearings are needed in microammeters, with rather higher pro rata torque developed by using more turns in the coil.

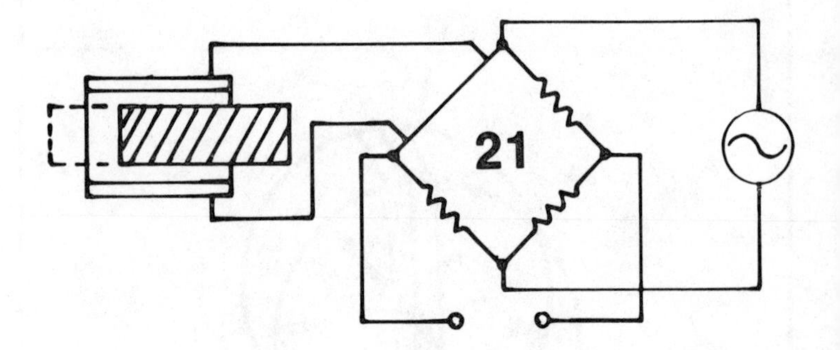

Transducers for Measuring Electrical Quantities

Transducers are finding an increasing application for ac circuit measurement used in configuration with conventional moving-coil or digital instruments or recorders. Basically these transducers are complete, special circuits designed for the following measurements: current (sensing mean or rms current), voltage (sensing mean or rms voltage), power (watts and VAR), phase angle, frequency, suppressed zero voltage, and linear inverse voltage.

Such transducers can be provided with a current output or a voltage output. A true current output is normally preferred because this compensates automatically for variations in total loop resistance, such as temperature changes affecting the resistance of pilot wires, or changes in receiving equipment. A voltage-output transducer is normally only used when the indicating instrument or receiving device requires a true voltage input and, in consequence , current is consumed.

Transducers can be connected into the circuit to be measured in a similar manner to that of any other measuring instrument, either directly or coupled in by a transformer, depending on the magnitude of the quantity being measured. They then provide a dc analog output signal that is proportional to the input parameter being measured (see Fig. 21-1). Alternatively, for working a digital instrument, the output signal can be rendered in digital form.

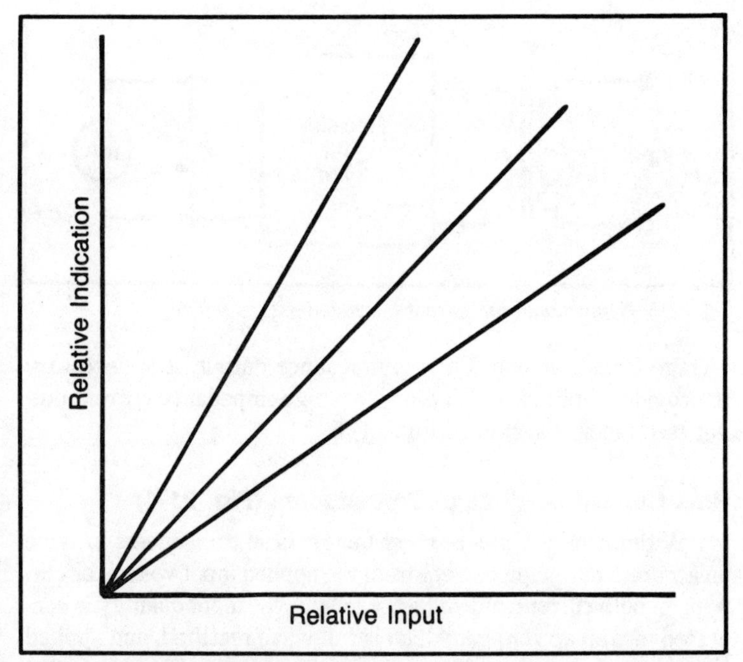

Fig. 21-1. Ideally, an electrical transducer produces an output that is directly (linearly) proportional to the input. This is graphically shown by straight-line output-vs.-input functions.

ANALOG TRANSDUCERS

The design and construction of (ac) electrical measuring transducers is specific to individual manufacturers. The following descriptions are based on the GEC "1 state" range, representing typical state-of-the art development in static circuit design.

Current Transducer

A block diagram of a measure-sensing current transducer is shown in Fig. 21-2. It is self-powered (requires no auxiliary power supply), with input isolated by a small internal transformer. Output from the transformer is rectified and smoothed, with voltage limitation and protection against transients provided by zener diodes.

Voltage Transducer

Here the circuitry is basically the same; however, because a

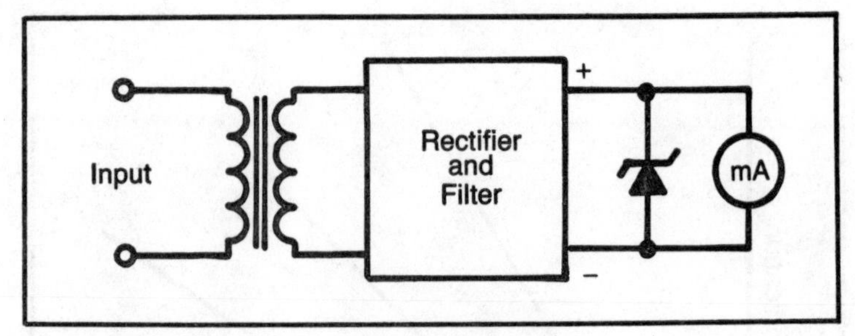

Fig. 21-2. A transducer for current measurement.

voltage transformer has a low-impedance output, it is necessary to provide amplification to produce a high-impedance current output (see block diagram of Fig. 21-3).

Rms Current or Voltage Transducers (Fig. 21-4)

Within stated limits of crest factor these transducers provide an accurate dc analog current from the applied input voltage or current. In both current and voltage versions, the input quantity is converted into an ac voltage, which is full-wave rectified, and applied to a square-law circuit. Although this circuit is nonlinear in operation, it provides a dc voltage output that is a linear function of the rms value of the applied input. The output from the square-law circuit is converted to a current signal in the amplifier, which in this case is powered from a separate supply.

Rms voltage transducers also require a separate power supply. Alternatively, power can be taken from the measured voltage source when the rated measurement is within ±20 percent.

Power Transducer

A block diagram of a power transducer is shown in Fig. 21-5.

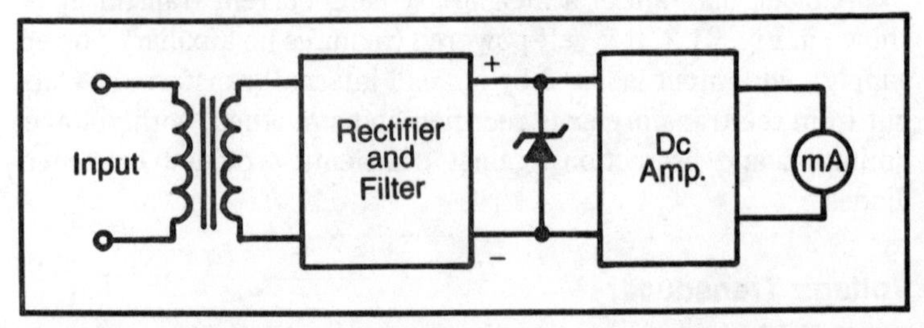

Fig. 21-3. A transducer for voltage measurement. An amplifier is incorporated, providing a low-impedance output.

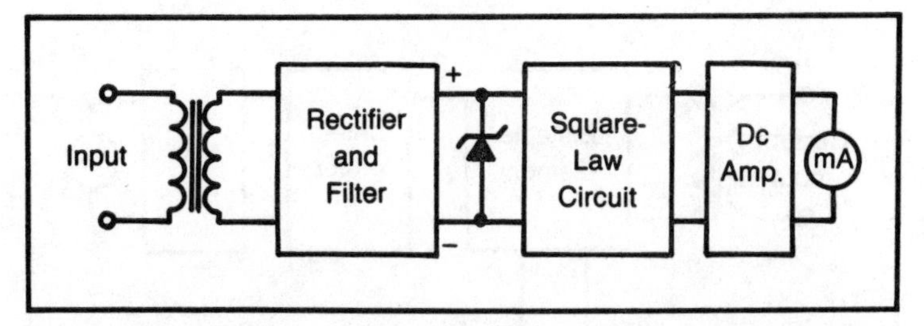

Fig. 21-4. A transducer for measuring root-mean-square (rms) values.

The circuit comprises an oscillator, a modulator, an amplifier, and an integrator. The current and voltage components of the input power are used to produce a train of rectangular pulses; each has height proportional to the instantaneous voltage and width proportional to the instantaneous current. The integral of this signal is proportional to the level of the power being measured. Protection against transient and overload conditions is provided. Links can be included for coarse adjustment and potentiometers for fine adjustment of the conversion ratio and calibration to cover a range 70 percent to 200 percent of the nominal input.

Power transducers require an auxiliary power supply, but because the burden is relatively low, this supply can be taken from the measuring voltage transformer if necessary. A single circuit can be used for measuring single phase or for balanced-load three-phase power applications. Two circuits mounted in one housing can be used for unbalanced-load three-phase power applications.

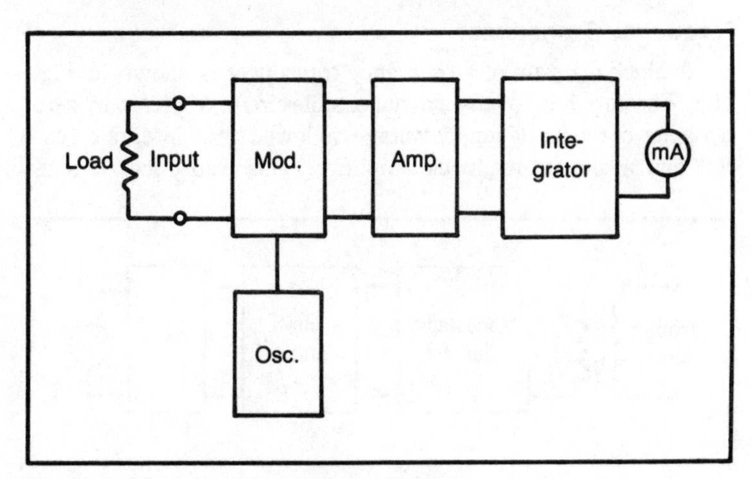

Fig. 21-5. A transducer for measuring power.

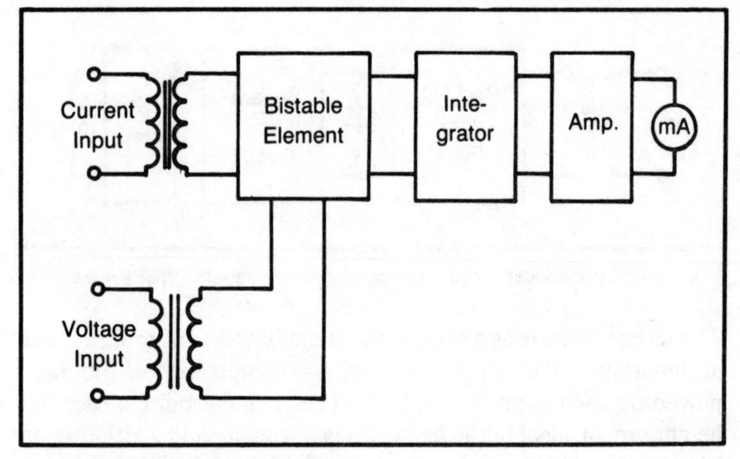

Fig. 21-6. A phase-angle transducer.

Phase-Angle Transducers

A block diagram of a phase-angle transducer is shown in Fig. 21-6. The circuit has internal transformers that feed the current and voltage inputs into a bistable element. Consequently, the output changes state when the inputs pass through zero. The signal from the bistable element is integrated, and the resultant dc voltage is fed to the output amplifier. Transducers for monitoring phase angles in the range 0°-±60° and 0°-±180° are available.

Phase-angle transducers require an auxiliary power supply.

Frequency Transducer

A block diagram of a frequency transducer is shown in Fig. 21-7. The circuit is based on a monostable circuit triggered by zero crossings of the input supply voltage, followed by an integrated circuit and a current-feedback amplifier. This transducer is self-

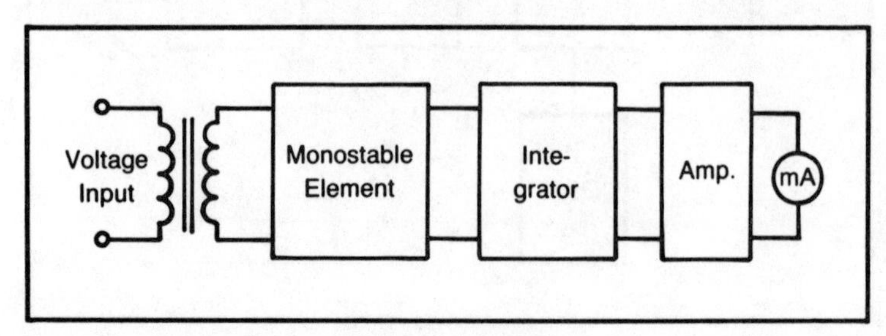

Fig. 21-7. A transducer for measuring frequency.

powered in the sense that it measures the frequency of the input to the power supply.

Suppressed-Zero Voltage Transducer

A block diagram of a suppressed-zero voltage transducer is shown in Fig. 21-8. Used with a negatively biased amplifier, the transducers give a range of suppression rates and a powerful high-impedance output. The accuracy is expressed as a percentage of the output span.

After rectification and smoothing, the measured voltage input is held in a negative state by the bias stage until it reaches a value equal to the bias level. Further increases of measured input result in a positive input to the amplifier. This provides a true current output that is proportional to the measured input voltage.

Linear Inverse-Voltage Transducer

A linear inverse-voltage transducer provides maximum output at zero input and zero output at full scale input. It is used where the frequency and voltage of a generator output must be synchronized with the supply levels already in operation before final connection is made. The transducer provides 100 percent output only when the running and incoming voltages are equal both in magnitude and in phase. The inverse output is a safety feature that prevents wrong synchronization if a fault develops in the measuring equipment.

A power supply stage fed from the running input energizes the bias stage, and the output is achieved by the bias stage when the running input is zero. A difference in level between the running and incoming voltages is rectified and presented as a voltage of opposite polarity to the bias-stage voltage. Consequently, as the input voltage increases, the output current decreases.

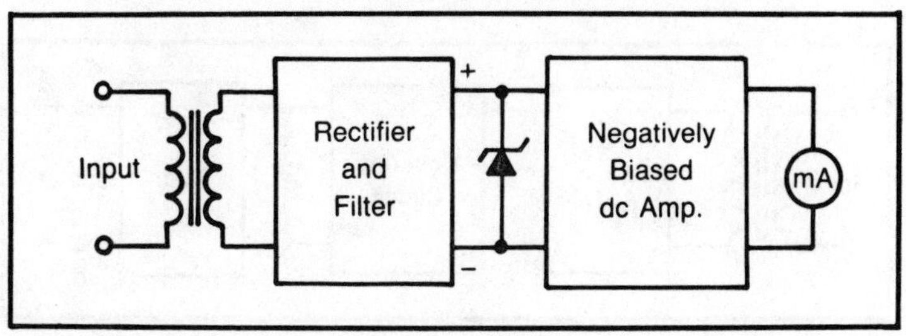

Fig. 21-8. A suppressed-zero voltage transducer.

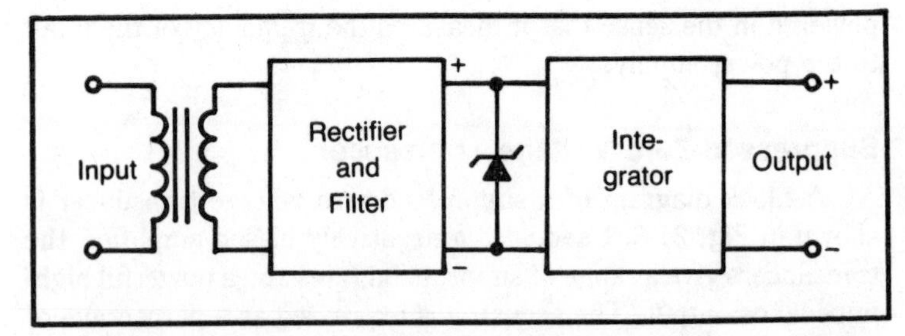

Fig. 21-9. A transducer/converter for measurement of current or voltage.

TRANSDUCERS FOR DIGITAL INSTRUMENTS

A large number of digital instruments are limited to the measurement of dc current and voltage. It is necessary to use special transducers with these instruments to ensure that a smooth dc output is connected to the instrument to mean ac current, voltage, power, phase angle, or frequency. These are sometimes known as transducer/converters.

Current and Voltage Transducer/Converters

The circuit is similar for both current and voltage transducers, the difference being in the transformer and tapping arrangements. Input is an ac voltage that is rectified, smoothed, and integrated to give a smooth dc output (Fig. 21-9).

Rms Current and Voltage Transducers/Converters

Again current and voltage transducer circuits are identical, except for the transformer and tappings. The output from the transformer is full-wave rectified and passed to a squaring circuit. This produces a signal that is proportional to the rms value of the input. The signal is then smoothed, tapped down, and linearized before being fed to the output terminals (Fig. 21-10).

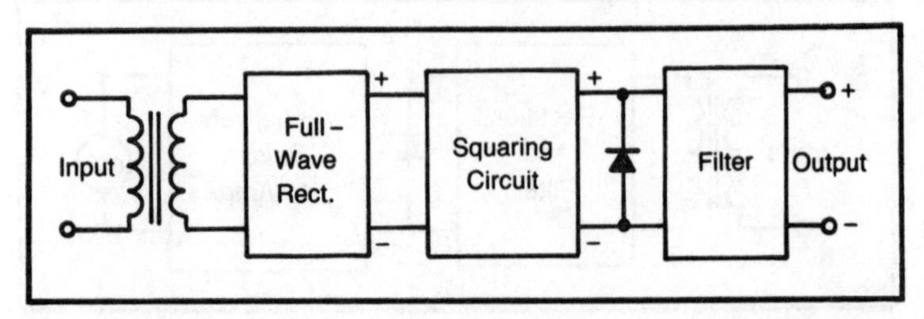

Fig. 21-10. A transducer/converter for determination of rms current or voltage.

Power Transducer/Converter

The circuit is calibrated for two voltage tappings. It requires an auxiliary supply and has a zero adjuster and a phase-angle adjuster. The current and voltage inputs are used to produce a train of rectangular pulses; each has height proportional to the instantaneous voltage and width proportional to the instantaneous current. The area under the pulses is then $\int vi\, dt$, which is a true measurement of watts.

Phase-Angle Transducer/Converter

The transducer consists of two switching circuits, each fed from a resistor connected across the secondary of the input transformer. These circuits are adjusted to switch at near-zero input voltage. Thus, the output changes state each time the input quantity passes through zero. The resulting pulses are differentiated and switch a pair of gates connected as a bistable element.

Both outputs from the bistable element are integrated, and the resultant dc voltage is fed to the output terminals in such a manner that it is a function of the difference between the integrated voltages.

Frequency Transducer/Converter

The circuit consists of a transistor pump circuit that delivers a train of pulses to the output. Integration of the signal produces a dc voltage proportional to the frequency of the supply.

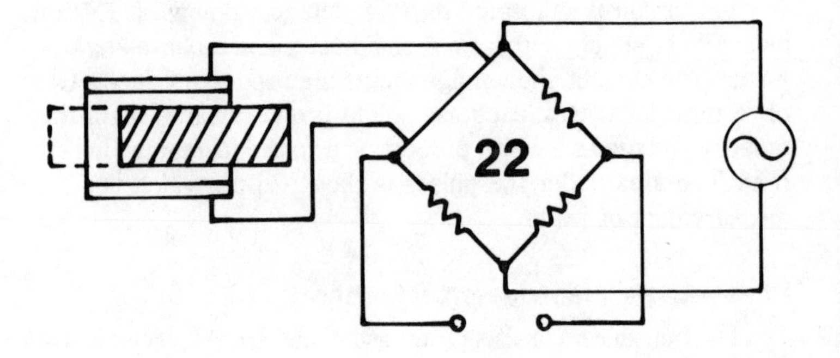

Measurement of Noise

Sound and noise are simply *pressure waves* traveling through the air at the speed of sound (1122 ft/sec or 765 mph) under standard atmosphere conditions. (Sound waves can also travel through other media at higher velocities, but that is outside the subject of transducers.) It logically follows that to measure sound it is only necessary to measure the *pressure* of the wave(s) caused, using a suitable transducer.

The actual physical *pressures* involved are quite small but extend over a considerable range. Sound pressure levels are expressed in decibels (dB), the significance of which will be explained later. The actual pressure produced by any sound pressure level is then given by

$$\text{pressure (lb/in}^2) = 29 \times 10^{dB/20} \times 10^{-10}$$

For example, 80 dB is a fairly loud sound. Calculating the equivalent pressure, we get

$$\begin{aligned}\text{pressure} &= 29 \times 10^{80/20} \times 10^{-10}\\ &= 29 \times 10^{4} \times 10^{-10}\\ &= 0.000029 \text{ lb/in}^2\end{aligned}$$

We need a very sensitive transducer to respond to such low pressure levels and connect it into a measurable/indicatable quantity (for example, a meter reading). Fortunately, we have one in the *microphone.* Microphones are, in fact, the standard form of transducer used for measurement of sound.

Now to return to the subject of decibels. A human being (and presumably also animals, birds, and other creatures) detect sound as an energy *intensity* level. The lowest sound intensity level that can be detected by the average individual is 10^{-12} W/m^2. This level is described as the *threshold of hearing.*

With increasing intensity levels, sound appears louder and louder, until eventually the sensation received changes from hearing to feeling. This occurs at a sound intensity level of about 1 W/m^2 and represents the *threshold of feeling.*

The intensity or sound energy range covered between just hearing and the threshold of feeling is thus from 10^{-12} to 1, or a range of 1 million million. Such a simple linear scale is far too large to work with, so we use one related to the *ratio* of intensity levels, called *bels*, and expressed by the formula

$$\text{bels} = \log_{10}\left(\frac{I_h}{I_0}\right)$$

where I_h is the sound intensity level of the sound heard
I_0 is the sound intensity level at the threshold of hearing

Even this proved too broad a scale, so a smaller unit of one-tenth of a bel, called a *decibel*, is used: thus,

$$\text{decibels (dB)} = 10 \log_{10} \frac{I_h}{I_0}$$

This gives a range of 120 dB between the threshold of audibility and the threshold of feeling. It also establishes a useful figure to remember: a doubling in sound *intensity* corresponds (almost exactly) to a rise of 3 dB.

This can be proved as follows. If I_1 represents the original sound intensity, and I_2 the doubled sound intensity, the dB relationship is

$$\text{dB} = 10 \log_{10}\left(\frac{I_2}{I_1}\right)$$

$$= 10 \log_{10} \frac{2I_1}{I_1}$$

$$= 10 \log_{10} \frac{2}{1}$$

$$= 3.01$$

This is not the complete story, especially concerning sound transducers. They detect sound *pressure* level. Now sound intensity is proportional to the *square* of the corresponding sound pressure level, so the dB relationship for sound pressure levels becomes

$$\text{dB} = 20 \log_{10} \left(\frac{P_2}{P_1}\right)$$

In other words, doubling the sound pressure level is equivalent to a rise of 6 dB.

Thus because sound is *measured* in terms of sound pressure level, a change of 6 dB represents a doubling or halving of sound *intensity* as distinguished by the human ear.

BASIC SOUND MEASUREMENT TECHNIQUE

A microphone converts sound pressure impinging on it into an electrical signal that can be amplified as necessary to provide a readout. However, the transducer signal output will normally be a ragged ac-type signal alternating rapidly from positive to negative levels, which is impossible to display except on an oscilloscope. To obtain a meaningful readout for instrumentation, therefore, you must interface the transducer with circuitry that modifies the output signal so that both positive and negative parts are rendered as a positive signal, although varying in amplitude at audio frequencies. For a steady "average" reading the signal is put through a *time constant* circuit; the square root of the average signal is then extracted to give a root-mean-square (rms) average reading.

In the practical sound-level meter the time constant circuit normally provides two averaging time constants: one *slow* (about 0.9 sec), and one *fast* (about 0.125 sec). Slow speed is selected for measuring fluctuating sounds; fast speed is selected for measuring continuous, reasonably steady noise. If in doubt as to the character of the noise, you can measure it on both settings and compare them. If the fluctuations in meter reading on fast-setting range are less than 6 dB, then the average reading should agree with the reading attained on slow setting.

In addition, the sound-level meter incorporates a calibrated alternator network, enabling reading range to be switched to various levels (usually in 10-dB steps). This basically is equivalent to extending the length of scale of the meter reading by the number of alternative steps provided.

WEIGHTING NETWORKS

Averaged sound pressure levels read as a single number indicated on a meter still have one basic limitation. They are a purely *quantitive* measurement of sound, which differs from subjective response to sound level at different frequencies (see Fig. 22-1). Here the curves correspond to different "loudness levels," as heard, showing how apparent loudness changes with frequency. It will be

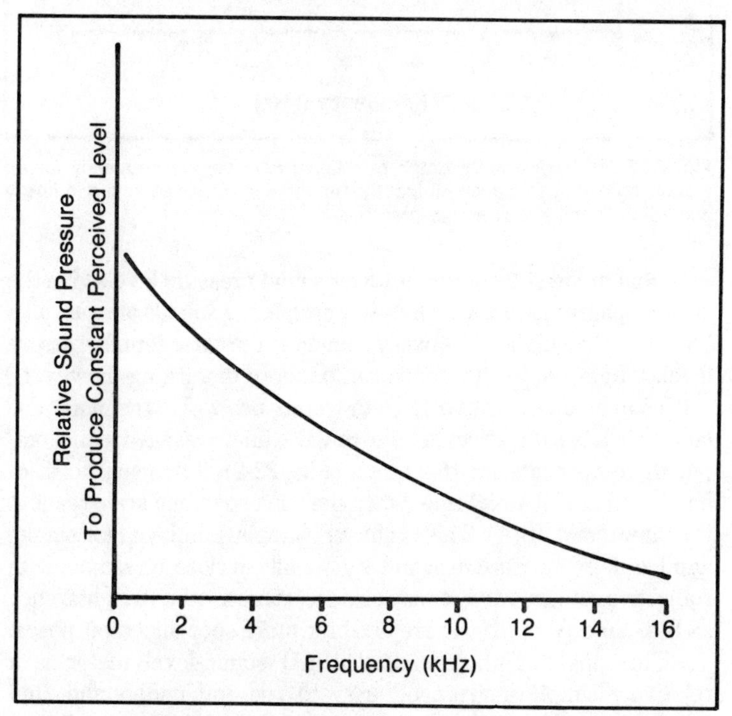

Fig. 22-1. Relative sound pressure required to produce a given perceived level of sound as a function of frequency.

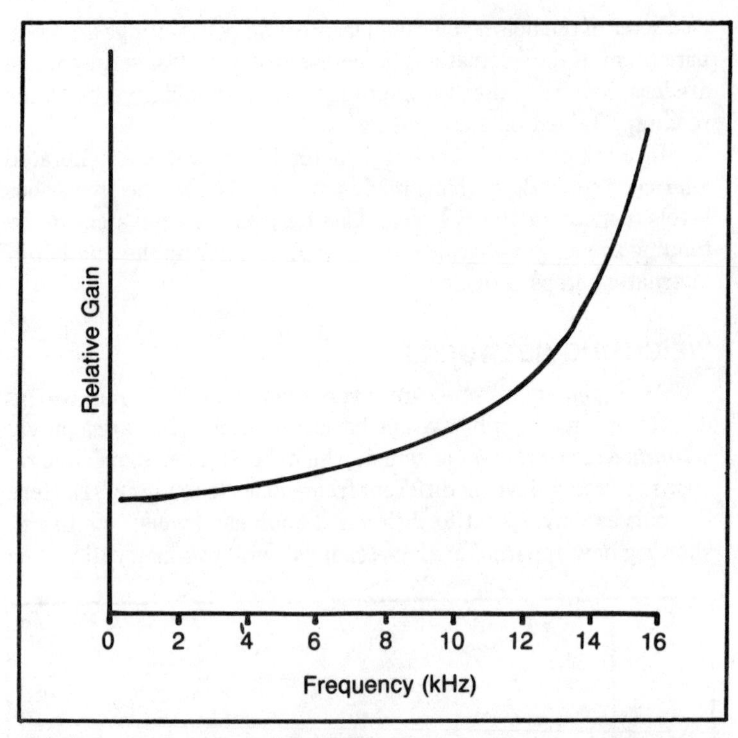

Fig. 22-2. An approximate graph of relative gain versus frequency for an equalizing circuit, designed so that the human ear perceives sound intensity according to the actual sound pressure.

seen that at lower frequencies lower sound pressure levels give the same apparent loudness. Thus, low-frequency sounds appear to be "louder" than higher-frequency sounds at the *same* sound pressure level. (High-frequency sounds can be more disturbing, however.)

To overcome this, we add a *weighting network* to the meter circuit, which has the effect of adjusting the final averaged signal output to compensate for this effect (Fig. 22-2). The usual form of employment is A weighting, when the corresponding scale readout is calibrated in dB(A). This is now universally employed for single-number noise measurement and is generally in close agreement with subjective ratings. Meters may also incorporate other weights, such as B, C, and D, but those are used for more specialized purposes.

The single-number reading dB(A) sound-level meter is a relatively simple instrument, easy to use and understand, and generally suitable for comparative purposes. It still has many limitations, however. It is measuring only a (weighted) *average* sound pressure level that gives no indication of the active *frequency content* of the sound. More complex instruments employ additional filters and circuitry to measure sound pressure levels at different frequencies, which enables the actual distribution of sound content to be analyzed; from this a frequency spectrum can be plotted or presented on a display. This is concerned entirely with electronic circuitry and is outside the subject of transducers. The transducer (microphone) provides the original signal information only. To what degree it is amplified, averaged, weighted, alternated, and analyzed depends on the following electronic circuitry and readout system adopted.

MICROPHONES

Microphones are the only transducer involved in noise-measuring instruments. The two main types are *piezoelectric* and *condenser*. Piezoelectric microphones employ a crystal of piezoelectric material as a transducer. They may also be known as *crystal* or *ceramic* microphones. *Condenser microphones,* also known as *electrostatic* or *capacitor* microphones, work on the principle of variations in electrical capacitance to provide an output signal. They are generally more prone to self-noise and require more extensive instrumentation than ceramic types, but they can offer advantages for specific applications—for example, when good high-frequency characteristics are required.

Other types of microphones that may be used include moving-coil or *dynamic* types and special directional microphones, which may incorporate any one of the three types of elements. Dynamic microphones are, however, relatively limited in application and have the basic disadvantage that comparatively large sizes are necessary to avoid distortion at low frequencies.

Hydrophones are a further type of microphone, designed specifically for underwater sound measurement. They normally employ piezoelectric elements.

Table 22-1 gives a general comparison of different types.

The choice of microphone depends on a number of factors. Low self-noise is obviously desirable, for example; for a microphone used to measure low sound levels, this is essential. The ceramic type is the preferred choice in this case, because even under the best conditions the inherent self-noise of condenser-type microphones can seldom be reduced below about 20 dB. A further requirement for the measurement of low sound levels is that the output voltage

Table 22-1. Types of Microphones.

(i) **Piezoelectric microphones**	
(a) quartz crystal	response, linear sensitivity, very good stability, very good remarks—very rugged type of transducer
(b) ADP	response, linear sensitivity, very good stability, very good
(c) barium titanate	response, linear sensitivity, fair to good stability, good remarks—can be affected by water
(d) lead zirconate	response, linear sensitivity, good stability, very good remarks—very rugged type of transducer
(e) ceramic	response, linear sensitivity, good to very good stability, good to excellent remarks—specially treated ceramic piezoelectric crystals are now usual choice with good, stable performance at economic price
(ii) **Condenser microphones**	response, linear sensitivity, good stability, very good to excellent remarks—clamped or stretched diaphragm construction, electrostatic or capacitor modes. The latest types are electric foil microphones (capacitor type) permanently polarized with an electrostatic charge during manufacture. Condenser microphones are more prone to self-noise.
(iii) **Moving coil**	response, falling at low and high frequencies sensitivity, good stability, very good remarks—response tends to be poor at low frequencies.
(iv) **Hydrophones**	ceramic crystal type, used for underwater sound measurement.

generated by the microphone must override the circuit noise of the amplifier in the sound-level meter.

The measurement of high sound levels requires that the system be free from microphonics or spurious signals generated by vibration. This is generated in the circuit components rather than the microphone, but mechanical vibration can also affect the microphone itself. A relatively soft type of mounting and special types of low-sensitivity microphones for measurement of very high sound levels may be required.

Microphonics are not normally experienced, except where sound pressure levels in excess of 100 dB are being measured, unless the microphone or instrument is mounted in such a way that mechanical vibrations are transmitted directly to it.

Both the piezoelectric and condenser microphones are generally well suited for the measurement of low-frequency noise, although moving-coil types can have distinct limitations imposed by their physical size. Accurate measurement of high-frequency sounds calls for the use of a small-size microphone. Regarding frequency response, the condenser type offers a narrower deviation range and the possibility of extending accurate measurements up to frequencies of 20,000 Hz. Both types can, of course, be calibrated for correction. Special types of microphones are required for accurate measurement of frequencies in excess of 20,000 Hz, or above the audio-frequency range.

The final choice of type can also be affected by humidity and temperature. A number of piezoelectric crystals, for example, are sensitive to humidity and can even be damaged if humidity is too high. In unfavorable ambients the piezoelectric element must therefore be rated as suitable for working at the prevailing humidity and/or temperature.

Condenser microphones are also sensitive to humidity, although they are not necessarily damaged by high humidity. Sensitivity, in this case, is caused by electrical leakage across the microphone, which tends to increase with increasing humidity. Exposed insulating surfaces are generally treated to maintain low leakage with high humidity, but it is not necessarily positive protection under all conditions. Some authorities recommend that the microphone be kept at a higher-than-ambient temperature to reduce leakage.

Other microphones have much higher maximum service temperatures. Condenser microphones normally have a maximum service temperature of the order of 100° C. In the case of portable instruments, temperature limitations are normally restricted to battery performance. Battery life will tend to be shortened at temperatures in excess of about 50° C., and battery output may be insufficient at near-zero or subzero temperatures, depending on battery type. Special batteries are required for satisfactory operation at subzero temperatures.

All microphones are temperature sensitive. Calibration is carried out at a normal ambient temperature, and sensitivity will vary at different temperatures. This is generally negligible in the case of piezoelectric microphones, where the temperature coefficient is only of the order of −0.01 dB per degree Celsius. With a typical condenser microphone the temperature coefficient can be as much as four times greater.

MICROPHONE CALIBRATION

Microphones for sound-level meters are calibrated by the manufacturers with microphone reciprocity calibrators, whereas sound-level calibrators are used for overall calibration of sound-level meters. Acoustic calibrators can provide an overall calibration check in field use, but not necessarily with original accuracy. Many meters have built-in electrical calibration and internal calibration controls that enable *electrical* calibration to be carried out, but this is only a check on the stability of the electrical system.

The following comments largely cover the question of calibration as far as the average user of sound-level meters is concerned:

☐ Initial (manufacturer's) calibration of the instrument can usually be regarded as adequate for about a year, after which a calibration check should be made. Such calibration is normally based on free-field response to noise of random incidence. The accuracy given by the calibrated meter then depends on the stability of the microphone and stability of the electrical system; the validity of the measurement is dependent on the characteristics of the sound field present when the measurement is taken.

☐ Electrical calibration is a useful check on the stability of the electrical system of the calibrator and can be carried out quite simply at regular intervals, such as monthly (especially if meters have built-in electrical calibration and internal calibration controls).

☐ The use of an additional (calibrated) microphone can provide a simple comparative check as to whether the original microphone has changed in characteristics and thus needs calibration. This method, used with electrical calibration, provides the simplest overall calibration check. Note, however, that this depends on the second microphone being used only for checking and being carefully stored.

□ Acoustic calibrators can provide an overall calibration check under "field" conditions, but the accuracy of such a calibration check depends both on the suitability of the instrument and its proper method of employment (particularly regarding sealing of the microphone in the sound chamber).

□ An exact calibration check can only be carried out under laboratory conditions. It is generally recommended that instruments be returned to the manufacturer(s) at suitable intervals (about once a year) for such a check.

□ Calibration should always be checked if the instrument is subject to shock, such as being dropped or receiving rough handling.

USING SOUND-LEVEL METERS

Most people using sound-level meters adopt a "point-and-read" technique. This is generally (but not completely) satisfactory when sound levels are being measured in a large open area (for example, outdoors), or under *free-field* conditions. This broadly implies that there are no surfaces present to produce sound reflection at the point where the measurement is being taken. Most meter microphones are omnidirectional and thus pick up sounds coming from all directions.

Wind blowing directly onto the microphone can produce false readings, although this effect is minimized when the microphone itself is fitted with a windshield. Note also that wind can affect the strength of noise measured from a distant source, depending on wind strength and direction and the distance from the source. The same noise source will give a higher reading downwind than upwind at the same distance from the source, for example.

Although most microphones used for sound measurement are omnidirectional, this may only apply as far as low-frequency sounds are carried. At higher frequencies, where the *wavelength* of the sound may be comparable to the size of the microphone itself, considerable directional effect may be present. In this case the measured response will vary with the direction in which the microphone is pointed.

As a general rule, for accurate measurement a sound-level meter should be held at arm's length sideways with the microphone pointed away from the noise source, that is, with the sound impinging on the microphone at grazing incidence (90°). However, this will depend on the type of microphone. Some may need to be pointed directly at the noise source for free-field measurement. Requirements in this respect will be specified in the manufacturer's instructions. Errors as large as ±6 dB may occur through bad positioning of a hand-held sound-level meter, due mainly to the presence of the operator. For most accurate results the meter (or separate microphone) should be clamped in position, and operator and any other instrumentation moved a short distance away.

MEASURING INDOOR NOISE

Accurate measurement of noise levels in rooms or buildings is complicated by the fact that sound reflections are inevitably present, the strength of which will vary at different points. Thus, the actual position at which the reading is taken is significant. Also if the noise generated by a particular source is being measured (for example, a machine), the meter records not only machine noise but background noise as well.

To complicate the question still further, the behavior of noise in the near field differs from that in the far field; it is also affected by the *shape* of the noise field.

Here "field" refers to the hemispherical distance of the point of measurement from the source—call this R—or the radius of the hemisphere, which is simply the linear distance between noise source and the point considered.

Normally sound level decreases according to distance squared, that is, by $1/R^2$. In the near field, or for small distances (small values of R), sound intensity will decrease at a *lower* rate than given by this law. At some value of R, propagation of sound-level intensity then changes to follow the inverse-square law, or the condition changes from near field to far field.

Measurements made in the near field would, therefore, give a false value if extrapolated on the noise-law basis to estimate such level at a greater distance. Similarly, measurements taken in the far field would give a false (low) rate if used to estimate noise level at a short distance from the noise source in the near field.

The point at which the near field changes to the far field depends both on the bulk and shape of the noise source and on whether noise is emitted uniformly from that shape. If we assume that the latter applies, then the extent of the far field can be estimated as being at least three times, and no more than five times, the linear dimension of the noise source "facing" the point considered. Taking measurements at a greater distance than this should ensure a true far field reading when the noise law applies for any subsequent recalculations or estimates required.

Note, however, that most practical noise sources have both bulk and nonuniform radiation of sound. In other words, they are *directional* noise sources, radiating sound more in some directions than others. This can only be discovered by taking measurements at various positions around the source and, if necessary, plotting equal sound-level *contours*.

In general, sources of low-frequency noise tend to be nondirectional (uniformly radiated), particularly when the source itself is small in comparison with the wavelength of the sound being generated. Directional effects tend to become more apparent as the sound frequency increases, particularly when the source of the sound becomes large in comparison with the wavelength of the sound.

REVERBERANT FIELD

In a room the walls, ceiling, and floor all form reflecting surfaces. If we assume a point source of sound, the sound field within the room has two components:

☐ The direct sound S_D between the source (S) and receiver (R) (Fig. 22-3A). This is the same as free-field conditions and is not modified by the presence of the enclosure.

☐ The reflected sound S_R, which reaches the receiver after one or more reflections from the enclosure surfaces (Fig. 22-3B). The value of this is determined by the power of the source and the reflecting properties of the enclosure surfaces. With perfect reflection the value of S_R would eventually rise to extreme values, regardless of the value of the source. In practice, this is impossible, but nevertheless the combined value of S_D and S_R can give rise to very high levels at any point R with highly reflective enclosure surfaces. Marked variations in sound pressure level at different points R will also be experienced due to interference effects between the multidirectional waves. This will be further modified by a nonpoint source.

If the size of the source is small relative to the dimensions of the room, there will be a definite distinction between the effects of the direct sound and the combination of direct and reflected

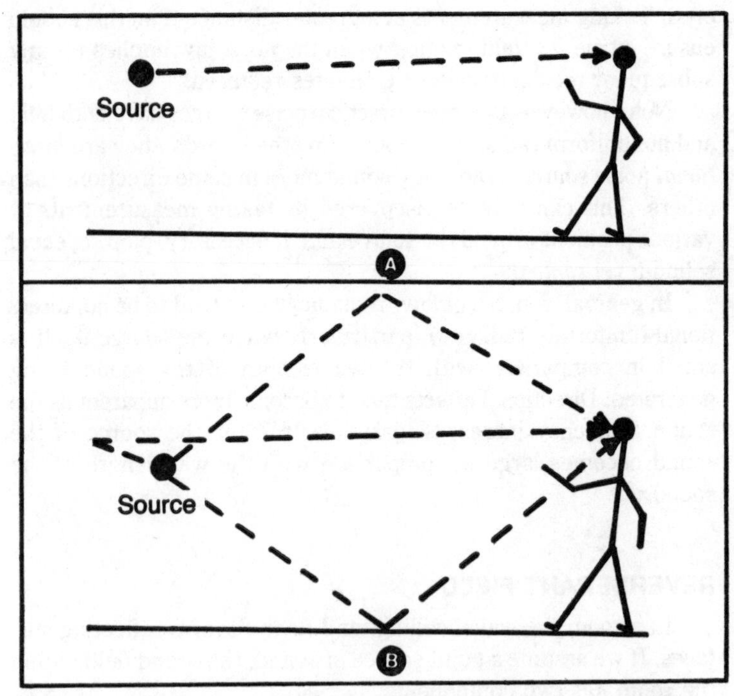

Fig. 22-3. At A, direct propagation of sound; at B, reflected propagation.

sounds. Near to the source, direct sound will predominate, and the effect of reflected sound components will tend to be negligible. At some greater distances, reflected sound will predominate: the region in which this occurs is known as the reverberant field. The inner boundaries of this field will be determined by the size of the room, the absorption characteristics of the reflecting surfaces in the room, and the acoustic power, size, and directivity characteristics of the source.

Fluctuations in sound pressure levels in the reverberant field due to "interference" effects yield patterns that are known as *standing waves*. In general, these are most marked when the source has frequency components corresponding to, or close to, any of the possible resonances of the room. The spacing of such standing waves tends to be slightly greater than one-half of the wavelength concerned.

SEMIREVERBERANT FIELD

In practice, in most rooms in which sound measurements are

semireverberant: the walls, ceiling, and floor are neither completely reflecting nor completely absorbent. If sound measurements are made in such a room in the same way as in a free field, correction is necessary to obtain equivalent free-field sound pressure level readings. This necessitates evaluation of the room characteristics or room constant. Rooms with a large degree of absorption have a large room constant and approach free-field conditions. Rooms with a large degree of reflection and only a small degree of absorption have a low room constant and approach reverberant-field conditions.

The main constant can be expressed in terms of the ratio of sound decay in the room (D). This is normally determined as *reverberation time* or the time in seconds required for the sound pressure level to fall to 60 dB when the source is shut off. The decay rate then follows as

$$D = \frac{60}{\text{reverberation time}}$$

Once the decay rate has been established or estimated, the average sound level in that part of the room where the reverberant field applies can be estimated from the equation

$$SPL = PWL - 10 \log_{10} V - 10 \log_{10} D + 48$$

where PWL is the source power level
V is the volume of the room in cubic feet
D is the decay rate, as defined above

It should be noted that this formula gives approximate answers only. Correction may also be needed to compensate for changes in atmospheric pressure, temperature, and relative humidity.

BACKGROUND NOISE

The question of background noise causing inflated readings when measuring the noise of some individual source is normally not too significant. Take a reading with the noise source switched off. This is the background noise. Now take a reading with the noise source switched on. This is a "combined" reading of the noise source and background noise level. However, if the difference be-

Table 22-2. Combining Near-Equal Noise Levels.

Difference between sound levels	Amount to be added to loudest sound level
0 dB	3 dB
1 dB	2.5 dB
2 dB	2 dB
3 dB	1.75 dB
4 dB	1.5 dB
5 dB	1.25 dB
6 dB	1 dB

tween the two readings you have taken is 6 dB or more, then the effect of the background can be ignored entirely.

Suppose, for example, the background noise level measured was 55 dB. With the noise source switched on the new reading is 65 dB. This is a 10-dB difference, so certainly the combination of the background noise can be ignored. It is only when two nearly equal sound sources are present at the same time that the lower one has any appreciable contribution to noise level measured.

Here, however, we must be careful. The combined sound level of two near-equal noise sources is *not* the arithmetical sum of the two dB values. The amount added to the overall noise level is related to the *difference* between the two. This is somewhat complicated to work out mathematically, but Table 22-2 gives satisfactory answers. For differences greater than 6 dB the amount to be added becomes negligible.

Example 1. Find the combined sound level of two equal sound sources, each of 55 dB.

difference = 0
so amount to be added is 3 dB
Combined sound level = 55 + 3 = 58 dB

Example 2. Sound level A is 68 dB, and sound level B is 71 dB. Find the combined sound level.

difference = 3 dB
so amount to be added to loudest (B) is 1.75 dB.

$$\text{Combined sound level} = 71 + 1.75$$

$$= 72.75 \text{ dB}$$

Note. In practice, sound levels are never quoted in decimal fractions; values are rounded up or down to the nearest whole number. In this case, therefore, the combined sound level could be given as 73 dB.

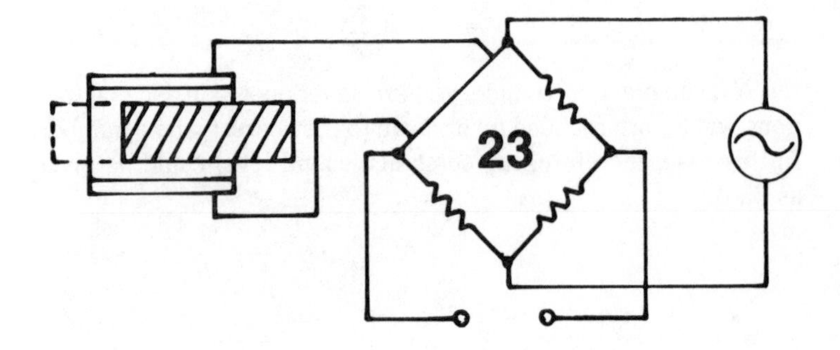

Acoustic Devices

Different kinds of transducers can be used for generating, detecting, or measuring acoustic waves in various media. Some of these, which we have already touched on, are piezoelectric, inductive, or capacitive devices. The constructional details of these devices have been outlined in previous chapters. Here, we will be concerned with applications of acoustic transducers, especially for short-distance communications purposes.

THE DYNAMIC TRANSDUCER

The least expensive, most rugged, and perhaps most versatile acoustic transducer is an ordinary dynamic speaker or microphone. The dynamic device operates by means of electric currents generated by moving magnetic fields (speaker) or by means of mechanical forces caused by varying currents in the presence of a magnetic field (microphone). These principles are illustrated in Figs. 23-1A and 23-1B.

Actually, a speaker can operate as a microphone. You may have tried this by connecting a small dynamic speaker to the input of a tape recorder or public-address system. The use of a speaker as a microphone is somewhat analogous to the operation of a generator as a motor. The converse of this is also often observed—a microphone can be used as a tiny speaker—but this practice is not recommended because a microphone is not normally expected to

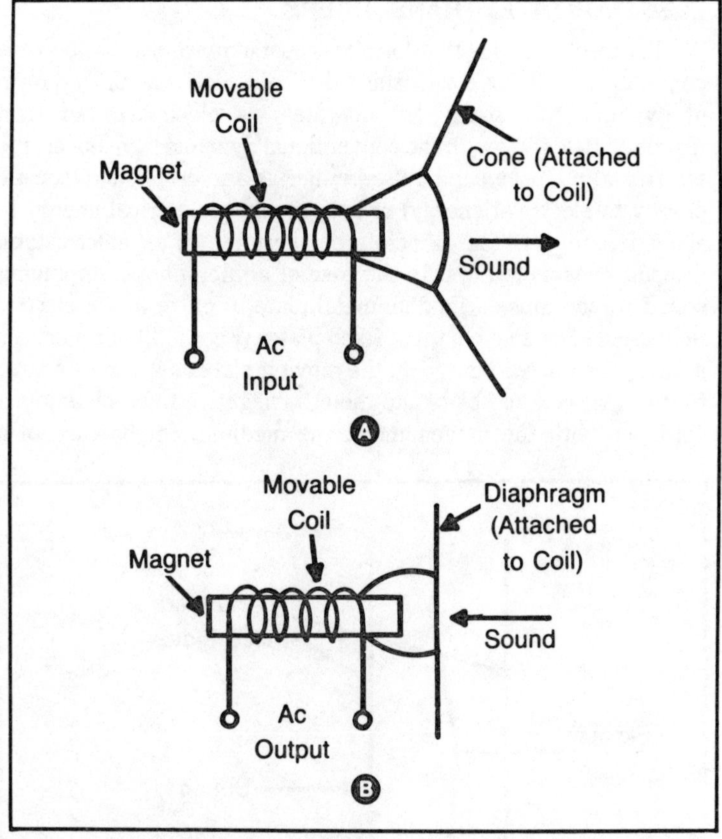

Fig. 23-1. Principle of operation of dynamic speaker (A) and microphone (B). These renditions are greatly simplified.

handle very much current, and damage is likely if a microphone is connected to the output terminals of an audio amplifier!

Dynamic transducers can operate over a wide range of frequencies, from below the human hearing range (less than 16 Hz) to tens or even hundreds of kilohertz. The design of a dynamic transducer is dictated by the range of frequencies over which it is to be used. In general, for a given power requirement the size of a transducer is proportional to the acoustic wavelength over which it is operated. A low-frequency device is therefore larger than a high-frequency device. The hi-fi enthusiast will recognize this by comparing the sizes of a woofer (bass speaker), midrange speaker, and tweeter (treble or high-frequency speaker).

ELECTROSTATIC TRANSDUCERS

In recent years another form of acoustic transducer has become common. This is the electrostatic device, which is actually a form of dynamic transducer that operates via electric rather than magnetic fields. As with the conventional dynamic transducer, the electrostatic device can be used either as a microphone (acoustic energy to electrical energy) or as a speaker (electrical energy to acoustic energy). The principle of operation of an electrostatic transducer is as follows. In the case of a microphone, impinging sound waves cause a flexible metal plate to move in the electric field created by two charged, fixed plates (Fig. 23-2). This causes a fluctuation in the charge on the moving plate as it comes nearer to the positive fixed plate and then the negative fixed plate in accordance with the movement of the medium. In the case of a

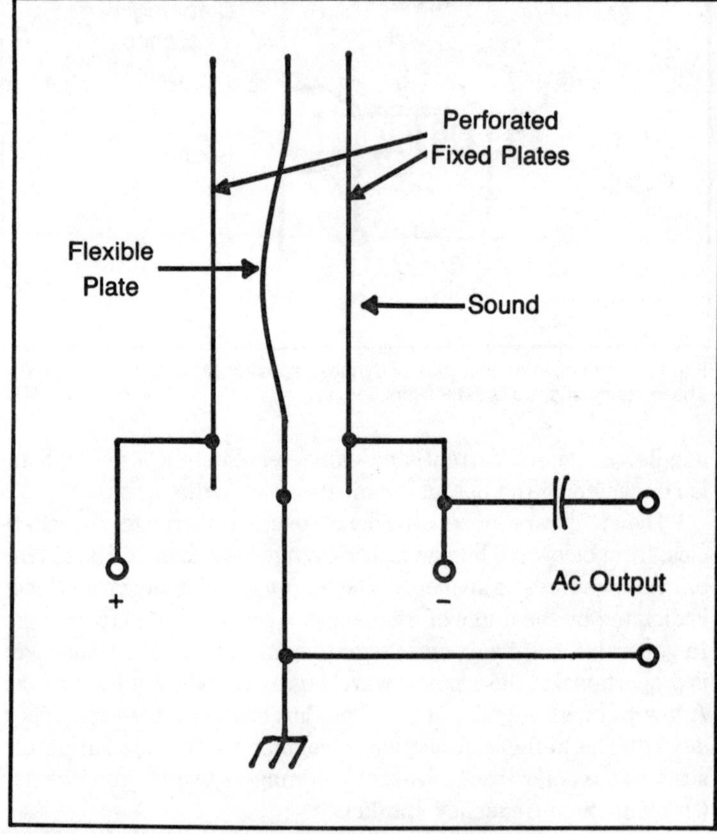

Fig. 23-2. Principle of electrostatic speaker or microphone.

speaker a varying charge on the movable plate causes it to be alternately attracted and repelled by the fixed plates.

Electrostatic transducers are somewhat more expensive than their magnetic counterparts, but they offer certain advantages. The impedance of the electrostatic transducer can be made very high (in fact, practically infinite), so that the sensitivity and power requirements are miniscule. Also, small electrostatic transducers exhibit excellent high-frequency response and can be used with good results at ultrasonic wavelengths. Another advantage of electrostatic transducers is their relatively light weight and small physical depth.

Both the magnetic and electrostatic types of dynamic acoustic transducer are physically rugged.

PEIZOELECTRIC TRANSDUCERS

The piezoelectric effect can be used to advantage in acoustic transducers, particularly those devices that convert sound energy into electrical impulses. The principle of the piezoelectric transducer is discussed in Chapter 6.

The main advantage of the piezoelectric transducer is its exceptional high-frequency response. Crystals can in fact respond to acoustic disturbances at megahertz frequencies. This is the principle of operation of a selective filter, for example, used in a radio receiver. The natural (resonant) frequency of a quartz crystal may be as high as 10 MHz or more.

Piezoelectric transducers are light in weight, small in physical size, and fairly sensitive. Their main drawback is that they are somewhat fragile, both mechanically and electrically. They cannot tolerate excessive input currents or voltages. Thus, they have limited usefulness as "speaker-mode" devices, except at low power levels.

AN ULTRASOUND RECEIVER

With a sensitive acoustic transducer, an audio amplifier, and a signal converter, it is possible to literally listen to ultrasound. The normal upper limit of human hearing is approximately 16 kHz to 20 kHz for a young person and decreases with age until it might be as low as 6 kHz to 8 kHz for an elderly person. With the device described here, it is possible to "hear" sounds up to several tens of kilohertz—frequencies higher than even dogs can hear.

A block diagram of the circuit is shown in Fig. 23-3. The

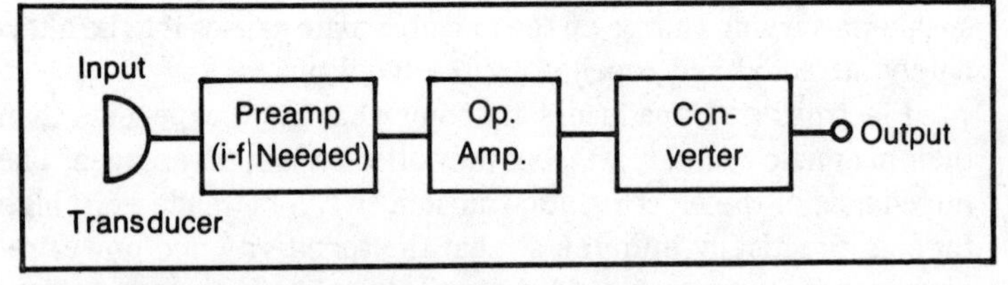

Fig. 23-3. Block diagram of an ultrasound receiver.

transducer may be a tweeter from a hi-fi speaker system, or it may be an electrostatic or piezoelectric device (commonly available in electronics stores). The audio amplifier can be an ordinary module obtainable in an electronics store, but a better design is to use an operational amplifier circuit. The input impedance of the operational amplifier should be high if a piezoelectric or electrostatic transducer is used (Fig. 23-4A), low if a magnetic (tweeter) device is used (Fig. 23-4B). With piezoelectric or electrostatic input transducers, a field-effect-transistor preamplifier stage can be put ahead of the op amp for increased sensitivity.

The converter can be designed in either of two ways. One method is to build a small amplitude-modulated (AM) transmitter consisting of a modulated local oscillator operating at a frequency of a few megahertz. The output of the oscillator can be fed directly to the antenna terminals of a communications receiver. The receiver must have a product detector, so that it can receive continuous-wave signals. The ultrasound noises will appear as sidebands above and below the carrier frequency of the modulated oscillator. If the oscillator operates at 3.500 MHz, for example, then a 30-kHz ultrasonic note will appear as two signals—one at 3.470 MHz, and the other at 3.530 MHz. The receiver dial can be tuned from zero (in this case 3.500 MHz) to several tens of kilohertz (it makes the most sense to go upward in frequency on the receiver dial—3.510, 3.520, 3.530, and so on—so that the signals appear "right side up").

The other method of detecting the ultrasound is to construct a variable-frequency oscillator, tunable from perhaps 10 kHz upward to several tens of kHz, and mix the output of this oscillator with the output of the audio amplifier. This scheme is shown in block form in Fig. 23-5. Another audio amplifier at the output, incorporating a selective filter resonant at about 700 Hz, allows some discrimination between various ultrasonic noises. The main problem with this scheme is that it results in double-signal reception, whereas modern communications receivers allow single-signal reception and also offer variable selectivity.

There is another, quite interesting, way to monitor the ultrasonic output of the circuit of Fig. 23-4, if you happen to have access to a spectrum analyzer. (You might want to borrow the one in the lab where you work—borrow the use of it, that is.) Simply connect the amplifier output to the spectrum analyzer through the appropriate attenuator and set the analyzer for zero-left and about 5 kHz per graticule division. The resolution should be well under 1 kHz, perhaps even less than 100 Hz, although the optimum setting will depend on the ultrasonic noise. With this arrangement you will be able to literally see the spectrum of sound from zero up to several tens of kilohertz. It may not even be necessary to use the preamplifier at all. It will be imperative that all leads be carefully shielded, especially between the transducer and amplifier and be-

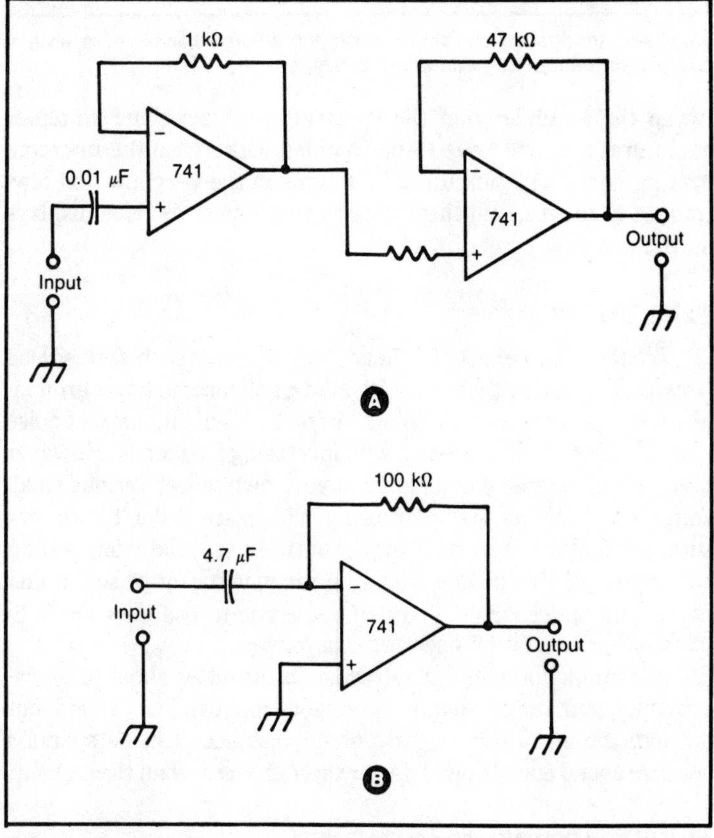

Fig. 23-4. Operational amplifier circuits for high input impedance (A) and low input impedance (B).

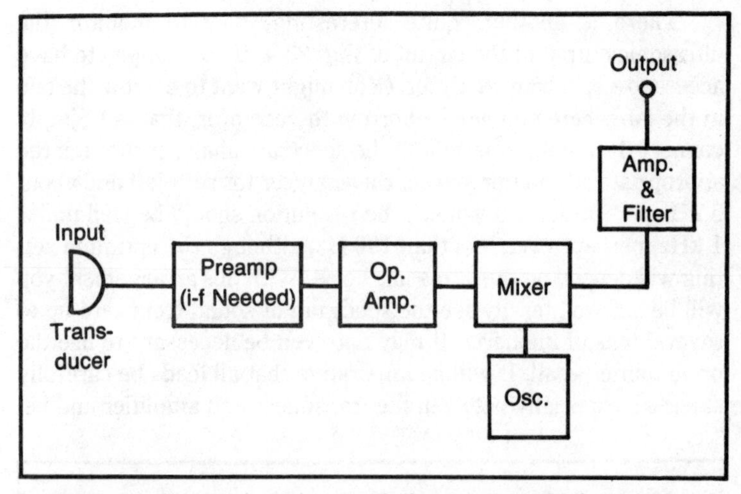

Fig. 23-5. An ultrasound receiver employing a communications receiver in conjunction with an amplifier and a converter.

tween the amplifier and the spectrum analyzer. The amplifier enclosure, too, will have to be shielded. Otherwise the interconnecting wires will pick up radio signals in the very-low and low-frequency ranges, and these signals will appear as false displays on the analyzer screen.

EXPERIMENTS

What can you expect to "hear" with a device such as this? Dog whistles? Actually, there is quite a lot of ultrasonic noise around. Most of it is in the form of "pink" noise, a broadband form of noise that gets less and less intense with increasing frequency. However there are ultrasonic components in your own voice; various small animals and insects may emit ultrasonic noises (I don't know for sure; it might be interesting to find out). An especially interesting application of this device would be for monitoring of sound and ultrasound under water. Some piezoelectric transducers are submersible and could be used for this purpose.

In conjunction with the ultrasonic transmitter about to be described, short-range sound communication can be carried out through the air. Under water, longer distances can be spanned because sound travels much faster through water than through air.

AN ULTRASOUND TRANSMITTER

The ultrasound transmitter is actually simpler to build than the receiver. An oscillator, preferably of variable frequency, is constructed using a tuned circuit as shown in Fig. 23-6. A common 88-mH toroid can be used as the inductor; a 365-pF variable capacitor, available in most electronics stores, can be used at C. The output of the oscillator can be fed to the input of a hi-fi amplifier for use at frequencies up to perhaps 30 to 40 kHz. Alternatively, a broadband power amplifier circuit can be built using a power transistor.

The transducer must be capable of handling the power output from the amplifier. The logical choice is a hi-fi tweeter with a maximum power rating of at least 50 percent more than the amplifier output. Tweeter devices are available that can handle quite large amounts of power. The cost, of course, depends on the power-handling capacity. It is important that a high-pass network be installed in the audio line to the tweeter; a simple means of doing this is to place a 0.1-μF ceramic capacitor in series with the line. This will also prevent the tweeter from being subject to direct currents.

Ultrasound, like ordinary sound, is reflected and diffracted. For this reason a direct line of sight is not necessary for communications. However, diffraction will not take place to as great an extent as with ordinary sound, and most objects do not reflect 100 percent of the sound energy that strikes them. Consequently the best results are obtained with a line-of-sight path. Path loss is af-

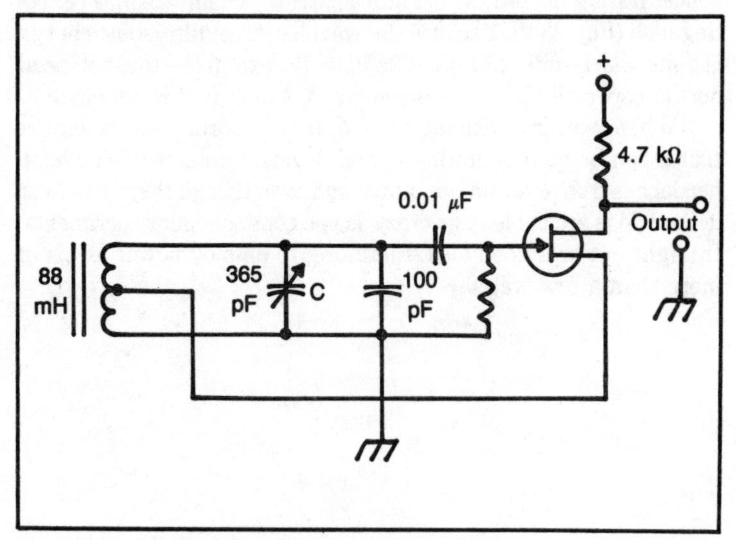

Fig. 23-6. A simple oscillator for producing ultrasound.

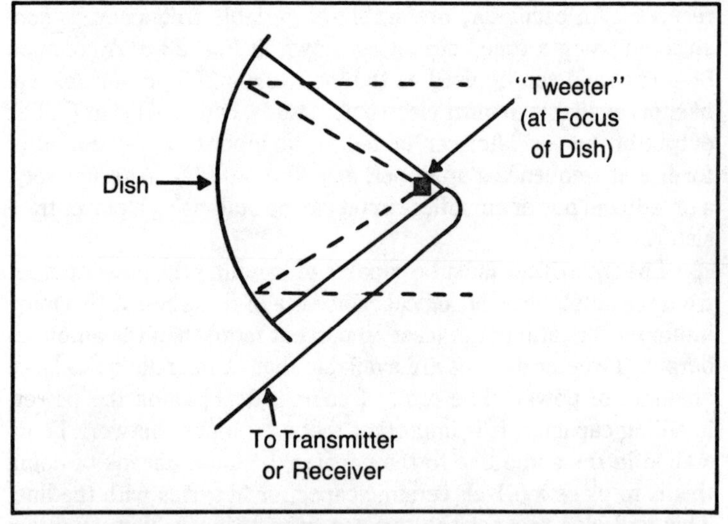

Fig. 23-7. A dish reflector can be used to concentrate the ultrasound emanating from a transducer. The same technique can be used for reception, enhancing the range and reducing interference from stray sources.

fected by rain, fog, or dust in the air.

Using a tweeter with 10 W of audio power and a direct line-of-sight path, signals can be transmitted over fairly long distances (1 or 2 km) under ideal conditions. The range can be increased by concentrating the ultrasound into a narrow "beam" using a reflecting dish (Fig. 23-7). Because the wavelength of ultrasonic energy is quite short—only 13 mm at 26 kHz, for example—the dish need not be especially large. A diameter of 2 or 3 feet is adequate.

When working with ultrasound, it is important to remember that it can be quite irritating at high levels. Some people develop headaches from exposure to ultrasound even though they can't hear it. And dogs are liable to go crazy as you conduct your experiments. It might not be a good idea, therefore, to employ power levels of more than a few watts!

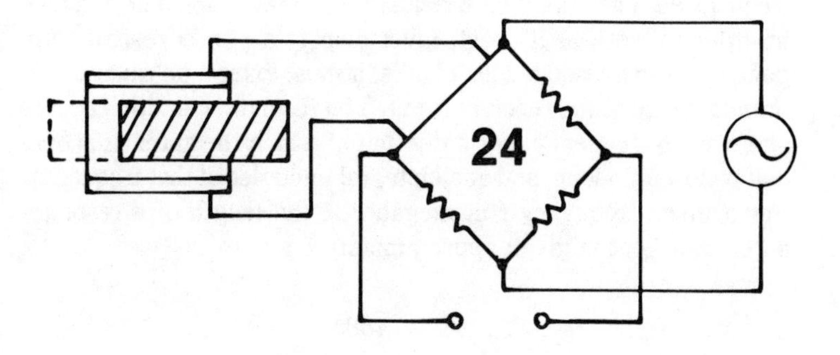

Antennas

One of the most common types of transducers is the radio antenna. An antenna performs either of two functions: conversion of an alternating electric current into an electromagnetic field, or vice versa. For such a simple function, antennas can get awfully complicated. We can only outline some of the most often-used kinds of antennas in this chapter; whole volumes have been devoted to the subject. For more information an antenna text should be consulted.

RECEIVING ANTENNAS

A receiving antenna converts an electromagnetic field into an alternating current having the same frequency characteristics. Note that we say *characteristics*; an electromagnetic field rarely has just one frequency. The simplest form of receiving antenna is a length of conducting material, such as wire or metal tubing. Theoretically, every piece of conducting material (or even semiconducting) has currents flowing in it as the result of electromagnetic energy at radio frequencies. Even your own body is affected to some extent. (In recent years there has arisen some concern that man-made electromagnetic fields might have some adverse health effects because of this fact.)

If you have engaged in shortwave listening, then you know that (in general) the physical size of a receiving antenna makes a difference in how well it will perform. It is not as simple a proposition as "the longer the better," though, except at low and very low frequencies. The important consideration is *resonance*. The antenna, in order to work at its best, must display a purely resistive impedance without reactance. That is, it must exhibit no capacitance or inductance at the receiver input. The resonant condition occurs at a certain frequency, normally that at which the antenna is one-half wavelength long, and at all integral multiples of that frequency. For a given frequency f in megahertz, the length of a resonant antenna in feet is given approximately by

$$L = 468/f$$

where L represents the physical length of a wire antenna. If metal tubing is used, then L is somewhat shorter than would be indicated by the above formula; a typical value might be

$$L = 450/f$$

Radiation Resistance

When an antenna is operating at resonance—the absence of reactance—the impedance $R + jX$ is purely resistive. That is, $X = 0$, and R has a certain finite value. For a half-wave antenna, as given by the above equations, the value of R is about 73 Ω for a thin wire in free space. If tubing is used,or if there are objects in the vicinity of the antenna that might affect the impedance, or if the antenna is not straight, then the value of R is somewhat different from the nominal 73 Ω. The value of R tends to be increased by the presence of objects, especially near the ends of the antenna. The value of R goes down if the apex angle is much less than 180° (Fig. 24-1).

If an antenna is operated at an odd harmonic of the fundamental frequency (that is, a multiple of 3, 5, 7, and so on), then the radiation resistance will be fairly close to the value at the fundamental frequency. But the situation is much different at *even* harmonics (multiples of 2, 4, 6, and so on). Resonance occurs at these frequencies also, but the radiation resistance is many times higher; theoretically, it is infinite, but in practice it can be hundreds or thousands of ohms.

Radiation resistance is important because we must know its value in order to obtain the optimum impedance match between an antenna and the feed line. Ideally, the feed line should have a

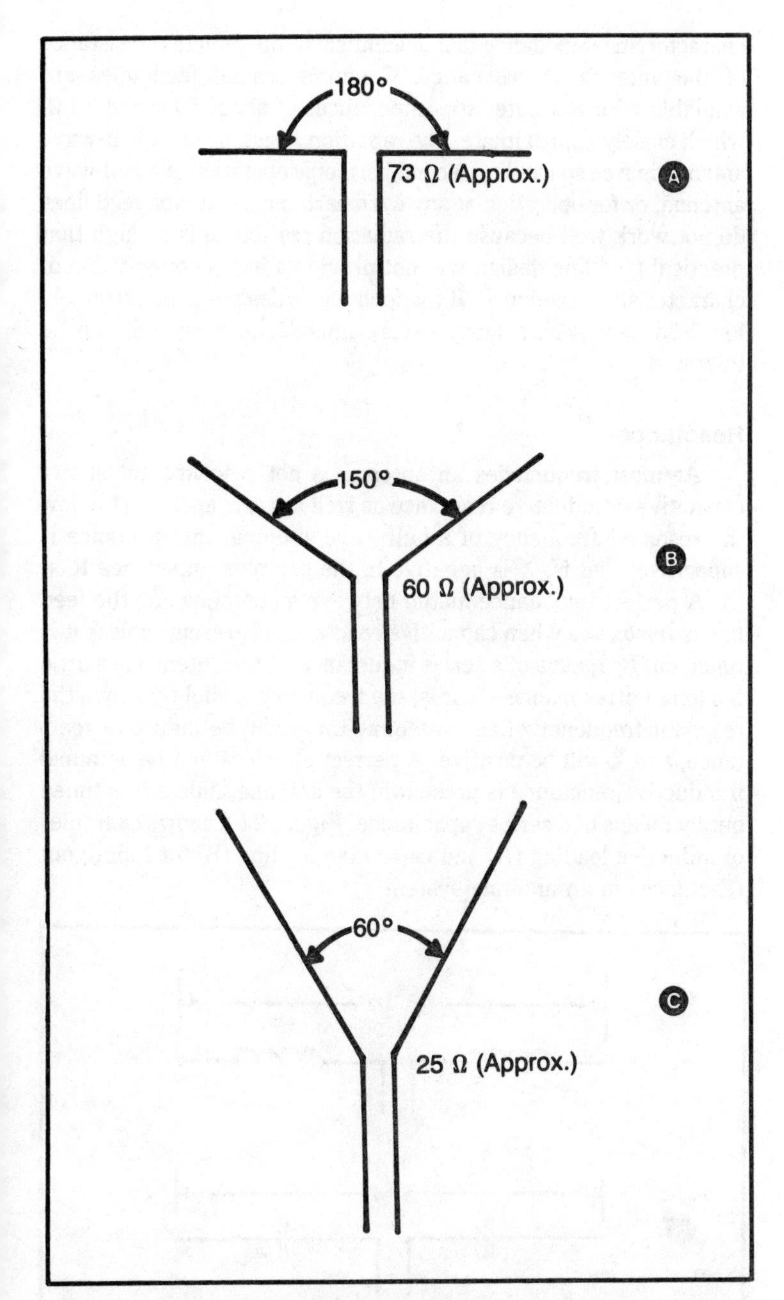

Fig. 24-1. Impedance of dipole feed point as a function of the apex angle. At A, a straight dipole; at B, 150° apex angle; at C, 60° apex angle. These values are approximate.

characteristic impedance that is identical to the radiation resistance of the antenna at resonance. Common coaxial feed lines are available with characteristic impedances of about 50 Ω and 75 Ω, which closely approximates the radiation resistance of a half-wave antenna in free space. For second-harmonic operation of a half-wave antenna, or for operation at any even harmonic, common feed lines do not work well because the radiation resistance is so high that practical feed-line design will not provide a high-enough value of characteristic impedance. If the feed line is designed for extremely low loss, however, a fairly severe impedance mismatch can be tolerated.

Reactance

At most frequencies an antenna is not resonant but shows capacitive or inductive reactance as well as resistance. Just below the resonant frequency of a half-wave antenna, the reactance is capacitive; that is, X is negative in the complex impedance $R + jX$. A perfect impedance match between an antenna and the feed line is *impossible* when capacitive reactance is present, unless it is tuned out by means of a series inductance. If the antenna is a little too long for resonance—that is, the frequency is slightly above the resonant frequency of the antenna—there will be inductive reactance, and X will be positive. A perfect match cannot be obtained if inductive reactance is present in the antenna, unless it is tuned out by means of a series capacitance. Figure 24-2 shows examples of inductive loading (A) and capacitive loading (B) for tuning out reactances in an antenna system.

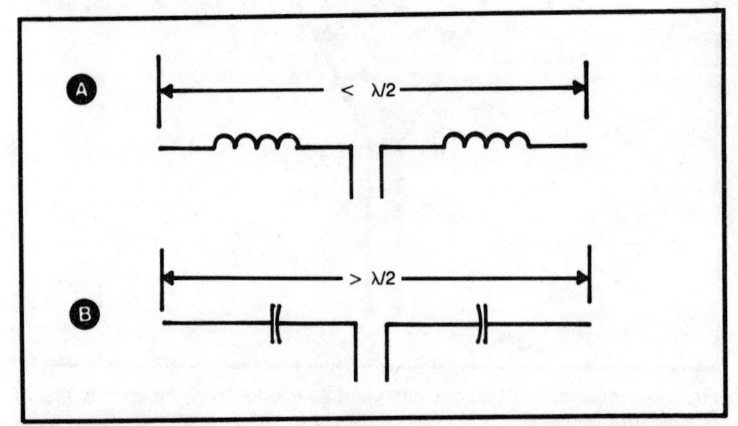

Fig. 24-2. Inductive loading (A) and capacitive loading (B) in a dipole antenna.

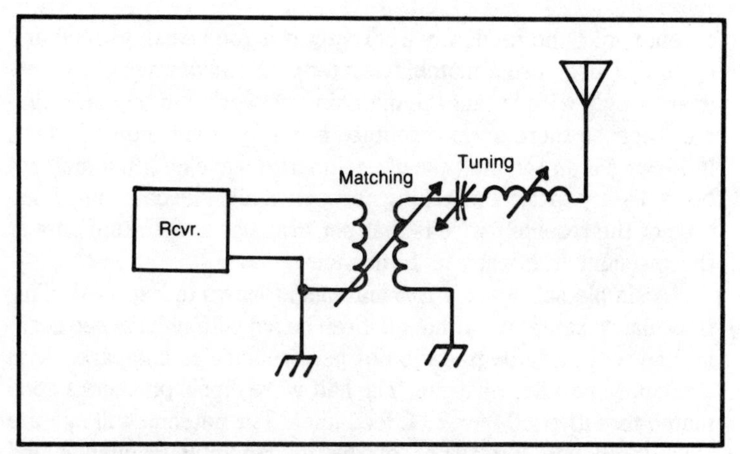

Fig. 24-3. A matching/tuning network for a random-wire antenna.

The above discussion might make it appear as if a perfect impedance match is absolutely necessary between an antenna and the feed line. This is not really true; a little reactance can be tolerated,and it is not essential that the radiation resistance of an antenna exactly match the characteristic impedance of the feed line. A near-perfect match is more important in some situations than in others. Generally, as the frequency of operation increases, the importance of a perfect match becomes greater. At low and very-low frequencies, large mismatches may be of no consequence. But at very-high, ultrahigh, and microwave frequencies, a good match is very important.

TYPES OF RECEIVING ANTENNAS: EXAMPLES

A "random wire" is the simplest form of receiving antenna. In conjuction with a tuning network at the receiver input, such an antenna can provide good reception over a wide range of frequencies. The tuning network eliminates the reactance at the feed point and matches the radiation resistance of the antenna to the characteristic impedance of the receiver input (Fig. 24-3).

A random wire is an unbalanced form of antenna and requires a good earth ground for optimum functioning. In general, the wire should be as high as possible above the ground and should be made as long as possible, especially for reception at low or very-low frequencies.

A specialized form of wire antenna is a single wire having a length that is a multiple of one-quarter wavelength. This kind of antenna, like the random wire, requires a good earth ground and often will also need a matching system. The advantage of the resonant single wire is that the matching network can be made simple, because there is no reactance at the receiver input. In fact, if the wire is an *odd* multiple of one-quarter wavelength, it may not be necessary to use a matching network at all. The chief disadvantage of the resonant wire is that performance is optimum only at the resonant frequency or frequencies.

A simple half-wave *dipole* antenna is shown in Fig. 24-4. This is a balanced antenna, although it can be fed with unbalanced coaxial line with little degradation of performance as compared with a balanced parallel-wire line. The half-wave dipole presents a good match for either 50 Ω or 75 Ω feed lines. The antenna will operate well at all odd harmonics of the fundamental frequency and reasonably well at even harmonics or frequencies not related to the fundamental, provided the feed line is not too long and a matching network is employed at the receiver input.

A quarter-wave *vertical* antenna is shown in Fig. 24-5. This antenna requires a good earth ground for effective performance. It presents a reasonable match for 50 Ω coaxial cable. The ground-mounted vertical relies so heavily on an excellent ground that radial wires are recommended to be installed in a manner similar to that shown. The radials enhance the conductivity of the earth in the vicinity of the antenna, minimizing loss and thereby maximizing efficiency.

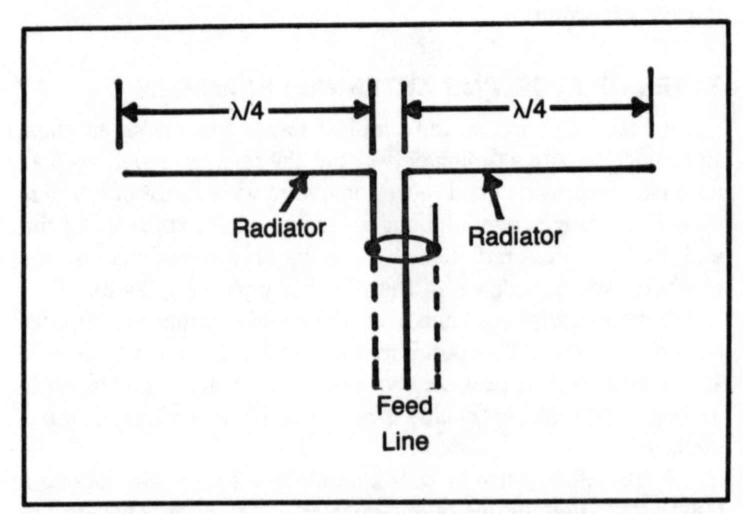

Fig. 24-4. A half-wave dipole can be fed with coaxial line as shown here.

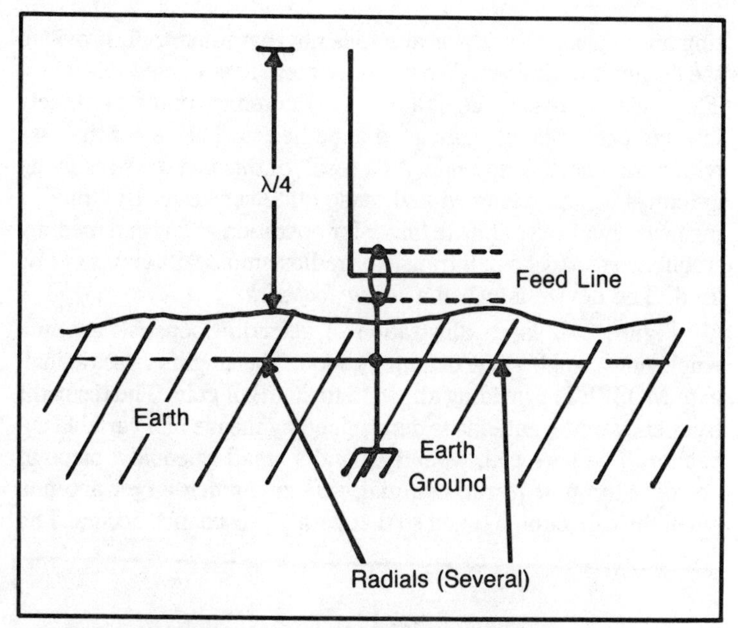

Fig. 24-5. A ground-mounted quarter-wave vertical antenna.

The vertical antenna will function well at odd harmonics of the resonant (quarter-wave) frequency without the use of a matching network. At even harmonics, if the feed line is short, a matching network at the receiver input will provide good performance. If the matching network can tune out reactance, a vertical antenna will work well at any frequency down to about the one-eighth wave value.

A vertical antenna may be elevated above ground and the radial system replaced by three or four horizontal or slightly drooping "spokes" as shown in Fig. 24-6. This kind of antenna is called a ground plane. The fundamental resonant frequency is that frequency at which the vertical element and the radial "spokes" all measure one-quarter electrical wavelength. Performance is similar to that of a ground-mounted vertical antenna with an extensive radial system.

Both of the above-described vertical antennas are unbalanced systems and should, ideally, be used with unbalanced feed lines such as coaxial cable. It is possible to use parallel-wire feed lines with unbalanced antennas, but there will be some sacrifice of feed-line efficiency.

Receiving antennas enjoy a certain advantage over transmitting antennas in that physical size is not that important, provided the design is optimized. (We will have more to say about this later.) The radiation resistance of a receiving antenna can be extremely low and performance quite good nonetheless. This is not the case with a transmitting antenna. As a result of this fact, tiny receiving antennas can be designed and made quite sensitive. By "tiny" I mean as small as your little finger for operation at low and medium frequencies. Most small transistor radios employ antennas of this kind. The device is called a *ferrite loopstick.*

Figure 24-7 is an illustration of a ferrite loopstick antenna employing a single stage of amplification. The amplifier uses a dual-gate MOSFET, providing about 15 to 20 dB of gain. The resonant frequency of the antenna is determined by means of a variable capacitor. The wire coil, wound around a small solenoidal piece of powdered-iron or ferrite material, acts as a much larger antenna when the capacitor is tuned so that parallel resonance occurs. The

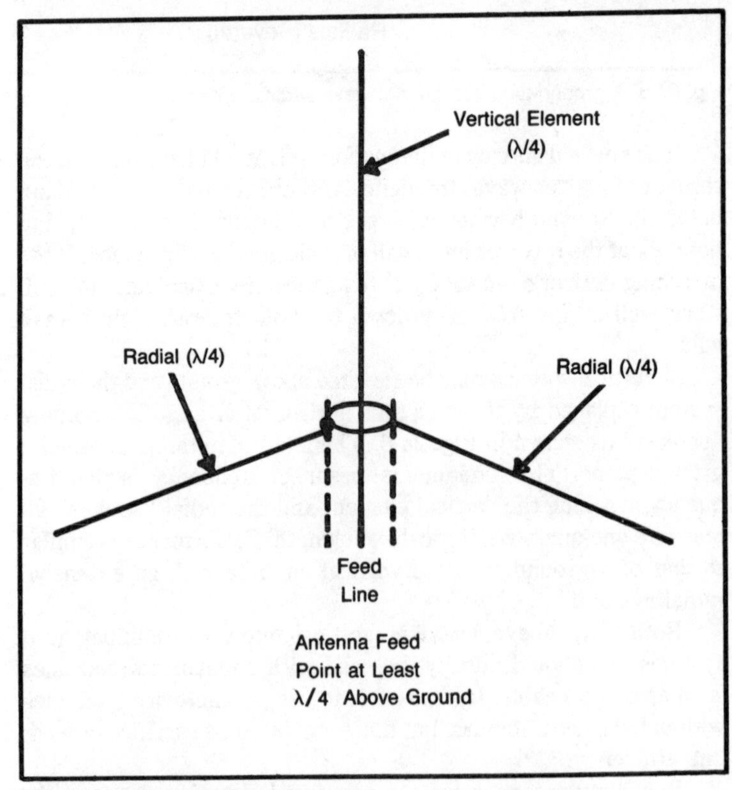

Fig. 24-6. A ground-plane antenna.

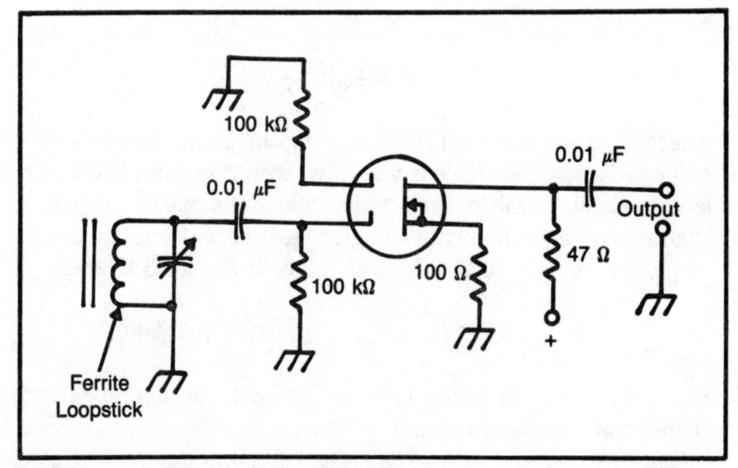

Fig. 24-7. Example of a ferrite-loopstick antenna with one stage of amplification.

impedance then becomes very high—on the order of thousands of ohms—and the tiny coil acts as a much larger resonant system. This kind of antenna works best at frequencies below about 20 MHz.

Another small receiving antenna, also employing an amplifier stage, is shown in Fig. 24-8. This antenna uses a short "whip," perhaps 2 feet in length, and a parallel-resonant tuned circuit. This kind of antenna will function very well into the very-high range, or up to about 300 MHz, provided good circuit design is employed in the amplifier.

TRANSMITTING ANTENNAS

Many types of receiving antennas will also work for transmitting. A transmitting antenna converts an alternating current into an electromagnetic field having the same frequency characteristics. The random-wire, tuned-wire, vertical, ground-plane, and various other types of receiving antennas can be effectively used to transmit signals. But the ferrite loopstick and whip antennas (shown in Figs. 24-7 and 24-8) will not work well for transmitting. Their radiation resistance is very low, and most of the transmitter output power would be wasted as loss if these types of antennas were used for transmitting.

Radiation resistance is a direct function of the physical size (in terms of the wavelength) of an antenna. For receiving, a very low radiation resistance can be tolerated. But for transmitting it cannot. In a transmitting antenna the efficiency E of an antenna is given, in percent, by the equation

$$E = 100\,(R_R/R_R + R_L)$$

where R_R is the radiation resistance and R_L is the loss resistance. If the radiation resistance is very low, then it is more likely to be low in relation to the loss resistance, which is a function of ground conductivity and antenna conductor resistance. Consider the example where $R_L = 10\ \Omega$, a typical value. If $R_R = 73\ \Omega$, then

$$E = 100(73/10 + 73) = 100(73\ \ 83) = 88\%$$

But if $R_R = 1\ \Omega$—a liberal estimate, at best, for a whip antenna at medium frequencies—then

$$E - 100(1/10 + 1) = 100(1/11) = 9\%$$

Very poor. In practice, it might be far less than even this. For a ferrite loopstick antenna, the efficiency might be less than 0.001%. The power not radiated would be dissipated as heat, mostly in the ferrite or powdered iron. It would not take much transmitter power to crack the core in such an instance.

The consideration of radiation resistance is the main difference between a transmitting antenna and a receiving antenna. Transmitting antennas must be physically large—at least a few percent of the free-space wavelength—if they are to function well. In theory,

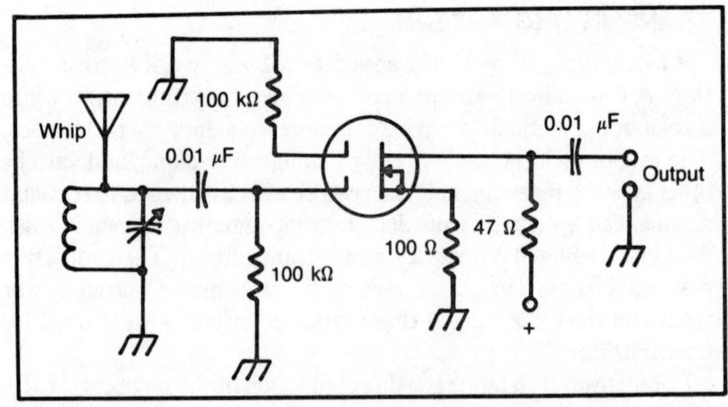

Fig. 24-8. Example of a whip-antenna system employing one stage of amplification.

even a tiny antenna, say a paper clip at 3.5 MHz, can be made efficient for transmitting if the loss resistance can be made small enough, but there is a practical limit to how low the value R_L can be.

Transmitting antennas must not only be at least a certain physical size, but they have to be capable of handling the power dissipated in the conductor(s). A length of AWG No. 40 wire might function well with a 1-W transmitter, but it would melt if subjected to an rf signal of 10 kW.

Outside these fundamental constraints, any kind of antenna can be used equally well for either receiving or transmitting. All gain and loss factors, with the above exceptions, apply to transmitting antennas in the same way as to receiving antennas.

ANTENNA DIRECTIVITY AND GAIN

A truly omnidirectional antenna—a device that radiates and picks up signals equally well in all directions in three dimensions—is a theoretical ideal not achievable in practice. Any antenna has favored directions, in which the radiation and response are maximum, and weak or null directions. The half-wave dipole responds and radiates best in directions perpendicular to the wire, assuming the antenna is perfectly straight. The response and radiation are minimum (theoretically zero) in a line containing the wire. In intermediate directions the response and radiation are variable, generally getting smaller and smaller in directions nearer and nearer to the line containing the wire. The response-radiation pattern for a half-wave dipole antenna is shown in Fig. 24-9.

A vertical (quarter-wave) or ground-plane antenna has similar radiation and response characteristics. The maximum radiation or response is in the horizontal direction; theoretically, zero radiation-response occurs in a vertical direction (straight up). This is shown in Fig. 24-10.

The directional properties of antennas change when they are operated at harmonic frequencies. In the case of a dipole operated at the second harmonic frequency, where it measures 1 wavelength, the lobes are sharper than in the half-wave case, and about 2 dB of gain occurs over the half-wave case (Fig. 24-11A). At the third harmonic, where the antenna measures one and one-half wavelengths, multiple lobes occur (Fig. 24-11B). A similar thing happens with the vertical and ground-plane antennas. If an antenna is operated at a very large harmonic value—say, the 51st

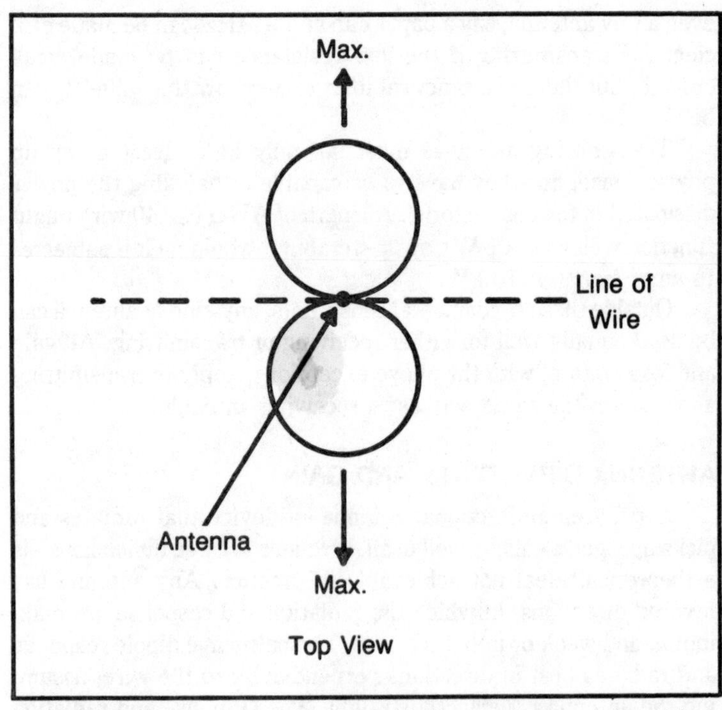

Fig. 24-9. Radiation pattern of horizontal half-wave dipole, as viewed from directly overhead.

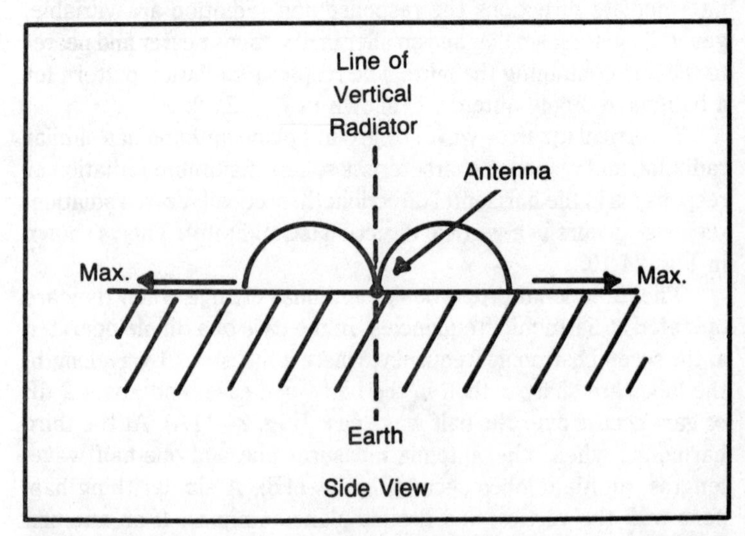

Fig. 24-10. Radiation pattern of vertical quarter-wave antenna, as viewed from the horizontal.

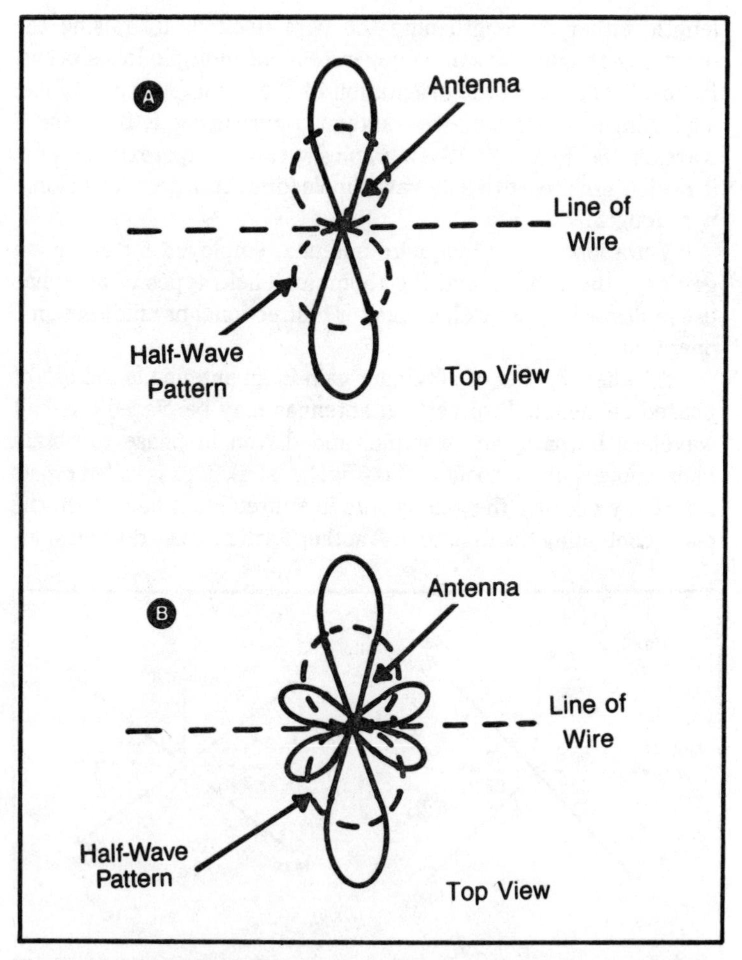

Fig. 24-11. Top view radiation pattern for dipole operated at second harmonic (A) and third harmonic (B).

harmonic—there are many lobes, and there may also be considerable gain in some of them.

A long-wire antenna provides perhaps the most illustrative example of gain resulting from operation at a harmonic frequency. An end-fed long-wire can be operated as a quarter-wavelength radiator against ground,and the radiation pattern will resemble that of a half-wave dipole. The only difference will be a slight loss with respect to the dipole. A half-wave end-fed radiation behaves just like a half-wave dipole, except that the feed-point impedance is very high. But when the radiator is made longer than one-half wavelength, either by lengthening the wire itself or by raising the operating frequency, gain is generated and multiple lobes occur. Figure 24-12 shows the orientation of the main (strongest) lobe, approximately, for long-wire antennas measuring 1, 2, 4, and 8 wavelengths. Figure 24-13 is a graph showing the approximate gain, in decibles relative to a half-wave dipole (dBd), as a function of long-wire length.

Variations of the long-wire antenna, employed for obtaining gain, are the V beam and the rhombic. These types of antennas use multiple long-wire elements for bidirectional or unidirectional operation.

Another method of obtaining gain in an antenna is the use of phased elements. Two vertical antennas may be placed one-half wavelength apart, for example, and driven in phase to obtain bidirectional gain of about 3 dBd (Fig. 24-14A). This is called broadside array because the gain occurs in a direction broadside to the plane containing the radiators. Another form of phased-vertical ar-

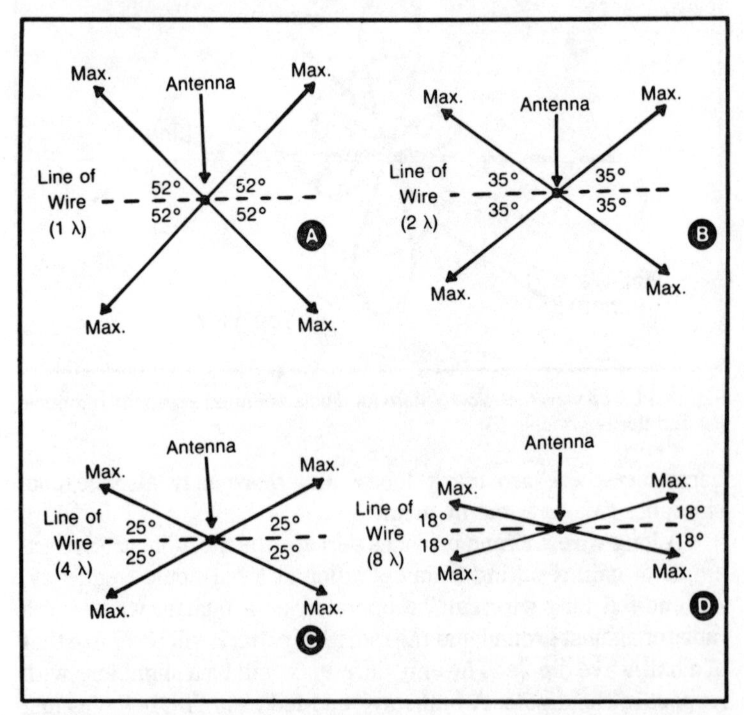

Fig. 24-12. Radiation-pattern maxima for long-wire antenna having lengths of one wavelength (A), two wavelengths (B), four wavelengths (C) and eight wavelengths (D). These are top views.

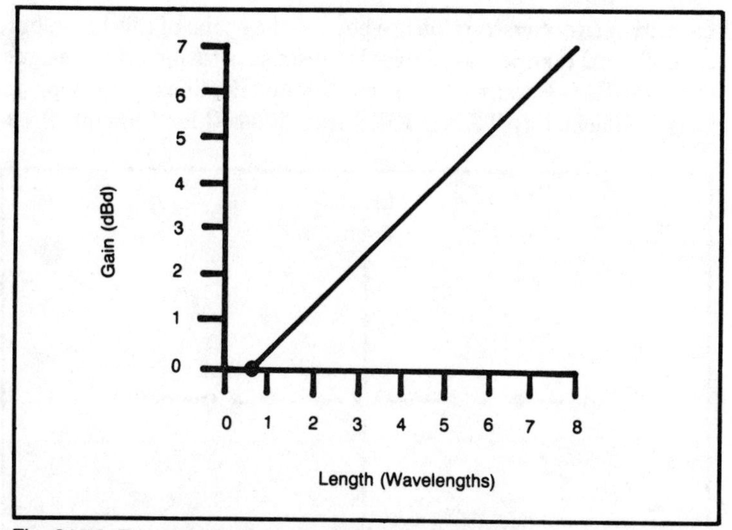

Fig. 24-13. Theoretical gain, over a half-wave dipole, of a long-wire antenna as a function of the length.

ray, known as a collinear antenna, is shown in Fig. 24-14B. The elements are aligned along a common line, and gain occurs in all directions perpendicular to that line (horizontally in the case of the collinear vertical). There are many other methods of obtaining gain and directivity by means of phased elements; Fig. 24-14 is only intended as a sample illustration.

Yet another way to obtain gain and directivity is by the use of parasitic elements. This is a rather interesting phenomenon that involves proximity coupling between an antenna-driven element and a free element nearby. Electromagnetic coupling takes place between an antenna and surrounding conducting objects, and in some cases this coupling can greatly affect the radiation pattern of the antenna. If the nearby conductor is about the same length as the active (driven) element, is close by, and is parallel to the active element, radical changes occur in the directional pattern. The parasitic element may act either as a "reflector" (Fig. 24-15A) or a "director" (Fig. 24-15B). With a single reflector or director, gain of up to about 6 dBd can be realized if the parasitic element is just the right length and is placed at just the right distance from the active element.

Generally, only one reflector is used in a parasitic array. But two or more directors can be employed, along with one reflector, to obtain considerable gain over a dipole. The physical details of parasitic-array construction are beyond the scope of this book, but several good texts are available that discuss parasitic-array design. (See the TAB Handbook of Radio Communications, by Joseph J. Carr, published by TAB BOOKS Inc., Blue Ridge Summit, PA.)

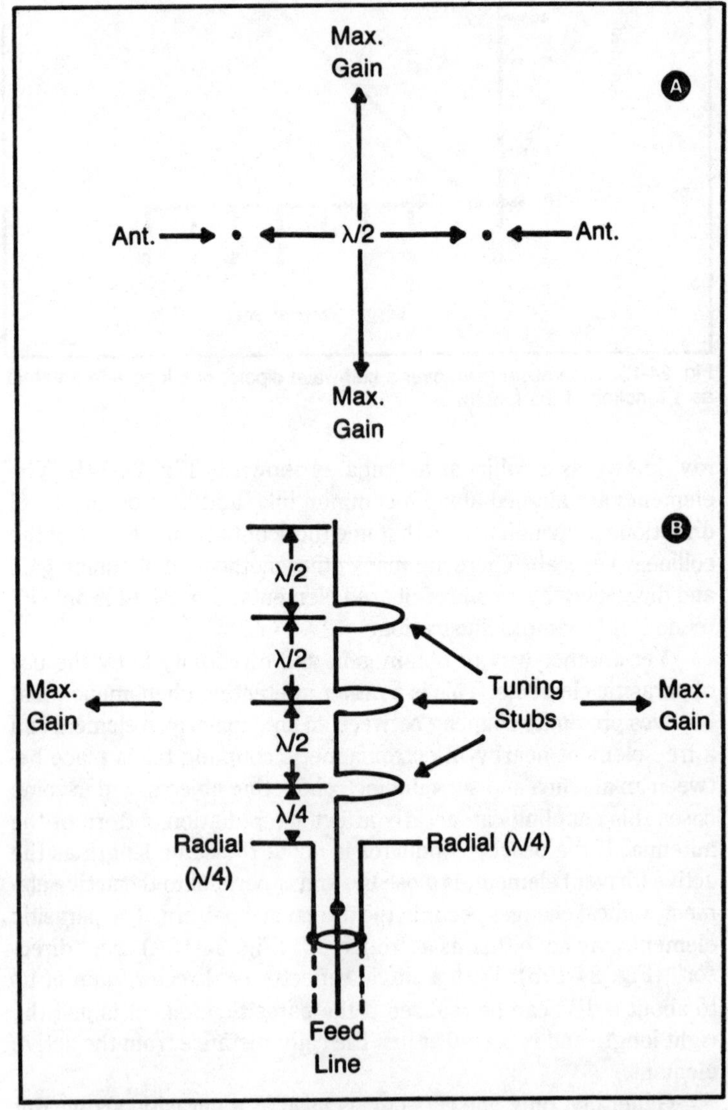

Fig. 24-14. Two methods of obtaining gain by phasing. At A, a broadside phased array, viewed from the top and employing two vertical antennas. At B, a vertical collinear array.

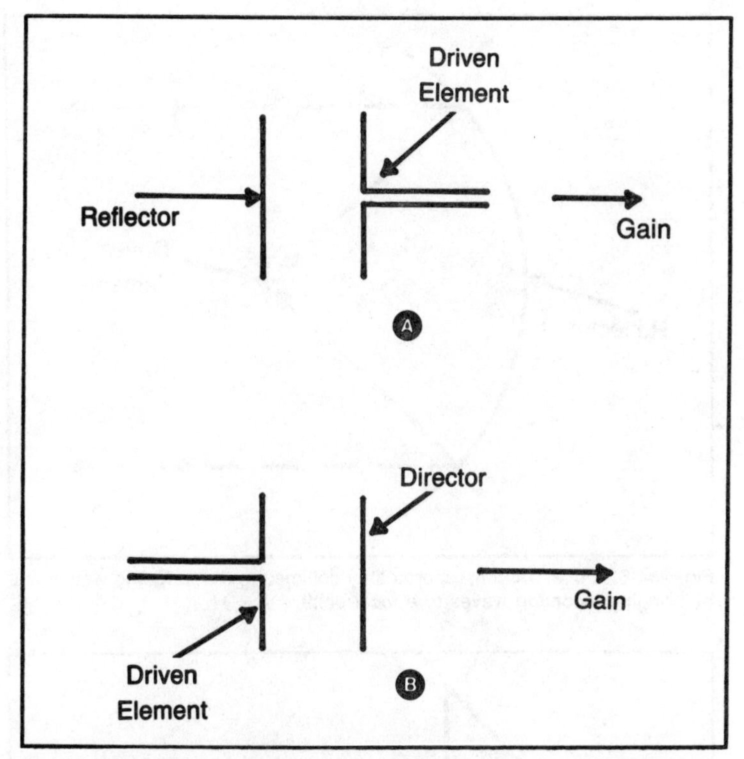

Fig. 24-15. Examples of yagi antennas using a single reflector (A) and a single director (B).

At very-high and ultrahigh frequencies, various high-gain antennas are used. A dish antenna, in which a large spherical or parabolic reflector is used, can give extreme gain—up to 30 dBd or more—if the reflector is large enough. At short wavelengths a truly large reflector is not that unwieldy. The principle of operation is just like that of a visible-light reflecting device (Fig. 24-16). The driven element is placed at the focal point of the reflector, which serves to collimate, or make parallel, the transmitted rays of electromagnetic energy. For receiving, the rays converge at the focal point where they are picked up by the active element.

Another often-used high-gain antenna at very-high and ultrahigh frequencies is the helical antenna. As its name implies, it consists of a helically wound conductor, measuring about one-third wavelength in diameter and having at least several turns (Fig. 24-17). A reflector is generally used to enhance the gain. The helical antenna has circular polarization; that is, the electric lines of flux

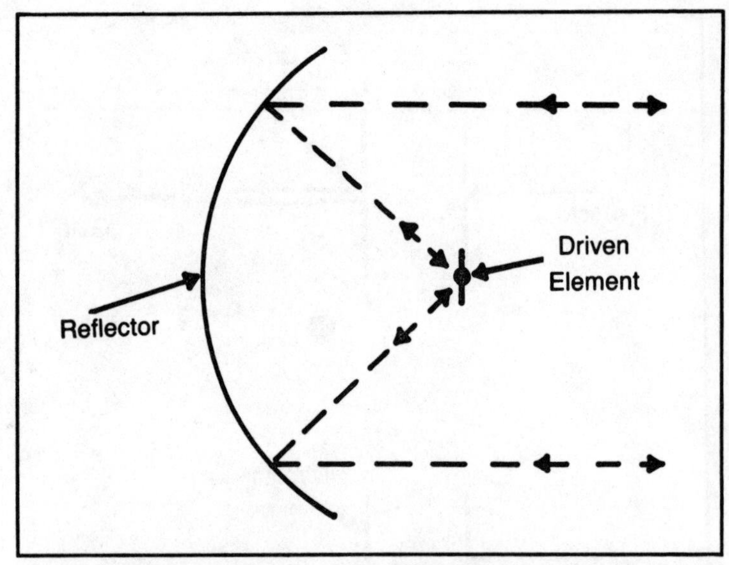

Fig. 24-16. A dish antenna operates by collimating the outgoing waves and by bringing incoming waves to a focal point.

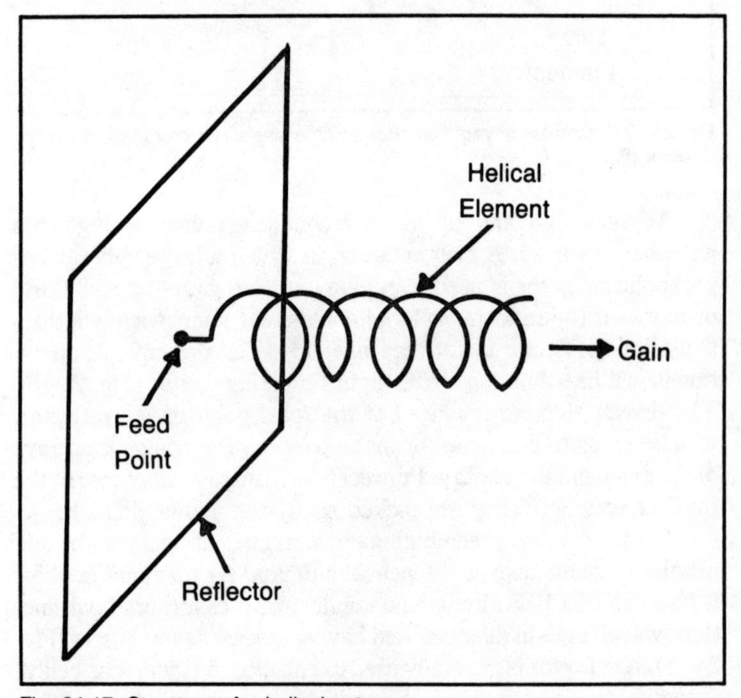

Fig. 24-17. Structure of a helical antenna.

rotate as they are emitted during transmission, and the antenna is not sensitive to the orientation of the received lines of flux. Electromagnetic-field polarization is an important consideration especially at very-high and ultrahigh frequencies; it can also have an effect at lower frequencies.

POLARIZATION

An electromagnetic field can be thought of, on a small scale, as electric and magnetic lines of flux that continually alternate in polarity (direction) and are always mutually perpendicular. An electromagnetic field propagates outward from the source in spherical wave fronts that travel at about 186,282 mi/sec (299,792 km/sec) in free space. If we could see the electric and magnetic lines of flux in an electromagnetic field at a great distance from the source, they would appear as shown in Fig. 24-18.

The *polarization* of an electromagnetic field is generally defined as the orientation of the electric lines of flux. The electric lines of flux are normally parallel to the orientation of the radiating element. A horizontal half-wave dipole antenna, for example, produces electric lines of flux that are horizontal; a vertical ground-plane antenna produces electric lines of flux that are vertical. Certain types of antennas (for example, the helical antenna) produce electric lines of flux that turn around and around as they travel through space; this kind of polarization is said to be circular.

At very-low and low frequencies, vertical polarization is preferable to horizontal polarization. This is because the electromagnetic waves at very-low and low frequencies tend to be propagated in contact with the surface of the earth, and vertical electric lines of flux travel better under such conditions than horizontal electric fields. At medium frequencies, vertical polarization is still desirable for surface-wave propagation, but long-distance skywave propagation is possible with either vertical or horizontal polarization.

At high frequencies the polarization is of little consequence. Good long-distance skywave propagation can be had with either vertical or horizontal polarization. At the extreme low end of the high-frequency spectrum—perhaps from about 3 to 6 MHz—some surface-wave propagation occurs with vertical polarization favored over horizontal, but the effect is not too significant.

At very-high frequencies and also at ultrahigh frequencies (30 to 300 MHz and 300 MHz to 3 GHz, respectively), polarization takes on a different significance. Frequency-modulated commercial and amateur signals are usually vertically polarized, whereas continuous-wave and single-sideband signals are horizontally polarized. Broadcast stations use combined vertical and horizontal polarizations. There is no technological reason for this; it is only a matter of convention.

Infrared, visible-light, and even ultraviolet and X-ray energy can be polarized. Normally it is not; the infrared from hot coals, the visible light from an incadescent lamp, and the ultraviolet rays

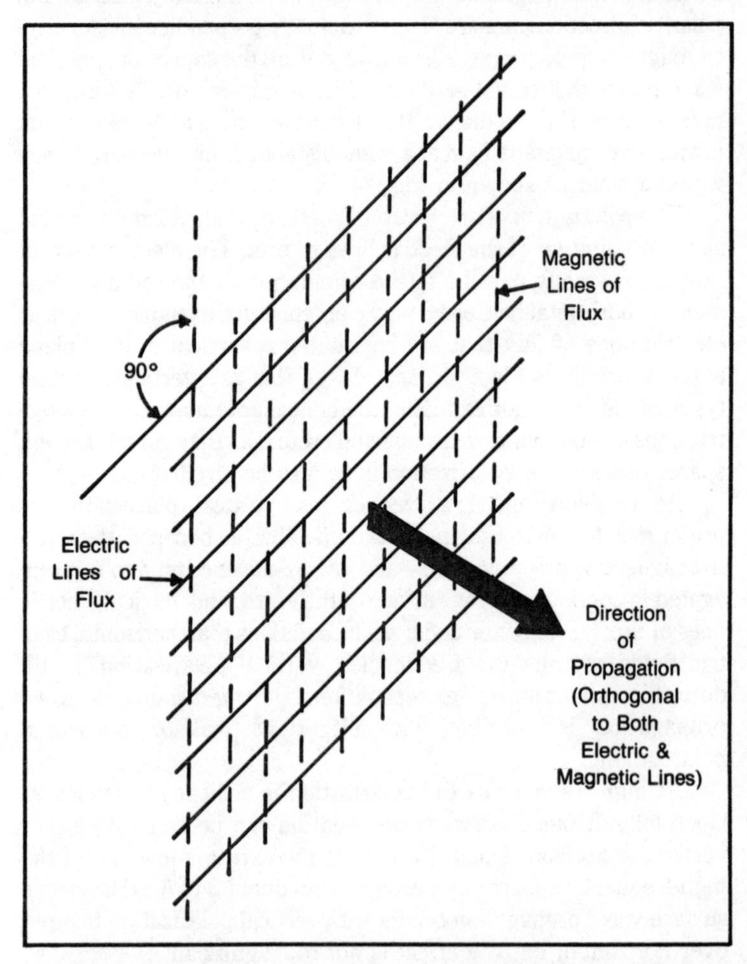

Fig. 24-18. An electromagnetic field consists of electric flux (solid lines) perpendicular to magnetic flux (dotted lines). The direction of propagation is orthogonal to both the electric and magnetic flux lines.

from the sun are randomly polarized. Polarizing filters can be used, however, to eliminate radiation that is polarized in a certain direction at these wavelengths.

Polarization can be varied in a controlled way, such as around and around as the signal is propagated through space. This is called *elliptical* polarization. If the signal strength remains constant throughout a full rotation, the polarization is called *circular*. There is an advantage to using circular polarization at radio frequencies, because, if circular polarization is used at one end of a communications circuit, it does not matter what the polarization is at the other end. This is valuable in satellite communications, for example, where the orientation of the satellite varies, and consequently the polarization of its antennas changes with time. A circularly polarized ground-station antenna eliminates the fading that would occur if a linearly polarized antenna were used.

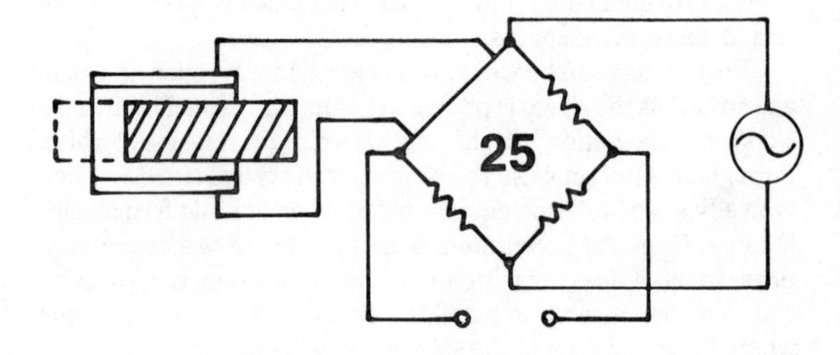

Proximity Sensors

Proximity sensors include a wide variety of transducer types described in previous chapters: magnetic and inductive transducers (Chapter 9), variable-differential transformers (Chapter 10), magnetic pickups (Chapter 11), and photoelectric devices (Chapter 12).

Other types are based on pure sensors, including sonic sensors, solid-state proximity switches, and fluid devices. At the other extreme, the probe used may be a simple contacting type.

Solid-state proximity switches are based on a sensory element and associated electronics in a single package. Because they have no moving parts, they are particularly suitable for aircraft, ship, and vehicle applications when sensory information is required. This can apply from test and research information down to everyday applications—all met by specially designed devices.

Sensors of this type are somewhat similar in principle to magnetic pickups but with considerably greater sensitivity given by more sophisticated electronic circuitry. They work on the principle of detecting the interruptions to a generated electromagnetic field produced by an oscillator when a metal object comes within the range of the field. The best known type is undoubtedly the portable metal detector, some of which can detect metal objects buried three feet or more in the ground. Other detectors of similar type can even distinguish between different metals.

EDDY-CURRENT PROXIMITY PROBE

An eddy-current proximity probe is one of the various types of noncontact pickups used for sensing displacement. It consists of a probe or core wound with a coil connected by a coaxial cable to a driver. The driver produces a high-frequency signal that is fed to the coil, which generates a magnetic field surrounding the coil tip (Fig. 25-1).

When a conductive material (of any metal) is approached by the probe tip, eddy currents are produced in the material, which has the effect of absorbing some of the field strength of the probe. The closer the tip to the object the greater the power absorbed.

Under such conditions the driver circuit measures the resulting field strength and compares it with the original to arrive at a difference signal, representing the distance between the probe tip and the surface of the object, which is converted into a standard calibrated output.

Probes of this type are capable of very sensitive and accurate measurement because they can produce an output of the order of 200 mV per thousandth of an inch. Thus, a change in relative position of probe and surface of, say, 0.0025 in. would produce a change in output voltage of 0.5 V. Response is also very rapid, so that the displacement of a vibrating surface having frequencies of vibration up to as high as 3500 Hz can readily be measured. For example, a probe of this type can readily measure vibration of a rotating shaft.

Its sensitivity can, in fact, work against it in some applications.

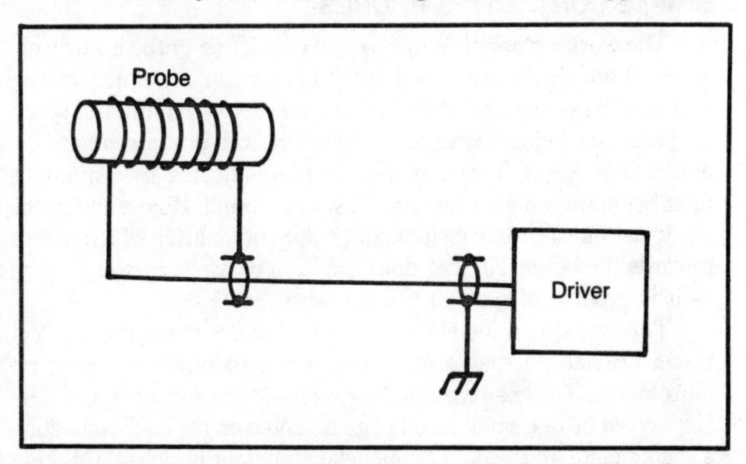

Fig. 25-1. An eddy-current proximity probe. A nearby ferrous object disturbs the magnetic field around the coil, producing eddy currents and changing the impedance of the coil.

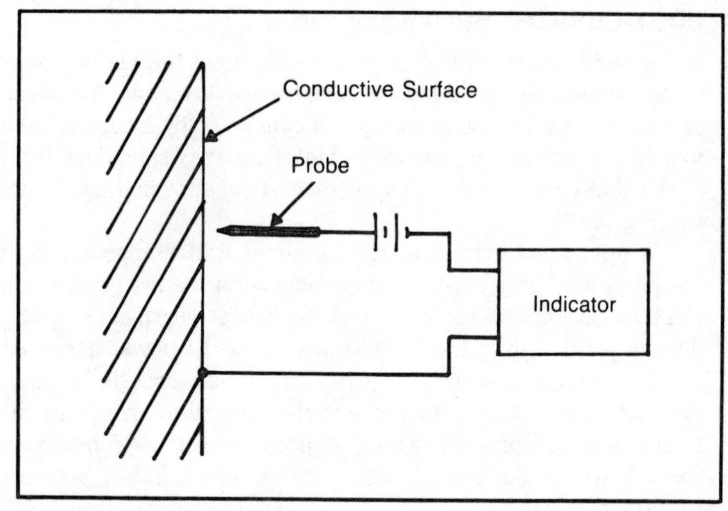

Fig. 25-2. A simple contact probe.

If it is being used to measure shaft vibration, for instance, it will also pick up and "measure" any shaft imperfections such as scratches, cracks, or dents, and even differences in plating thicknesses on a plated shaft. It is also affected by the proximity of strong electric currents or magnetic properties in the material being sensed.

SIMPLE CONTACTING PROBES

The working principle of a simple contacting probe is fairly obvious. If the object concerned is conductive, for example, connecting a metal probe and object in a simple series circuit will result in the circuit being switched on whenever the probe contacts the object (Fig. 25-2). This condition can be indicated by a motor, a light being turned on, or other ways, as required. Object and probe work just like a simple switch, and it does not matter which moves towards the other. Neither does the form of probe matter as long as it is capable of making contact with the object.

This working principle is so simple that it is often ignored. Yet it can provide an easy answer to many proximity indicating requirements. To sense when a door is closed, for example, the "object" could be one small metal plate mounted on the door contacting a spring plate on the door frame when the door is closed. (Making one plate in the form of a leaf spring can ensure possible contact by suitable bending.)

CAPACITIVE PROXIMITY PROBE

A similar principle can be employed in the noncontacting proximity probe, again where a conductive object is involved. This time the probe is in the form of a plate that, on approaching the object face to face, becomes effectively a *capacitor* with air dielectric (Fig. 25-3). However,this is not so simple.

To be able to devise a signal indicating the near-pressure of the object, we must now detect the level of *capacitance* in the near-contact position; this will be very low. For a capacitor probe area of 1 cm^2, for example, within 1 mm of the object, the effective capacitance produced will only be about 1 micromicrofarad ($1\mu\mu F$). Thus, noncontacting proximity sensors of this type do not lend themselves to simple construction. The same applies to most noncontacting proximity probes. They normally need fairly complex (but not necessarily sophisticated) electronic circuitry, either capacitor or magnetic types.

LEVEL DETECTORS

Detection of the height of liquid levels with contacting probes is, however, quite straightforward, provided the liquid is conductive (which most are, including water). The same principle as in Fig. 25-2 is used, except that two equal probes are employed, mounted in a suitable position. As the liquid level rises, it eventually touches the ends of the two probes, the resistance between

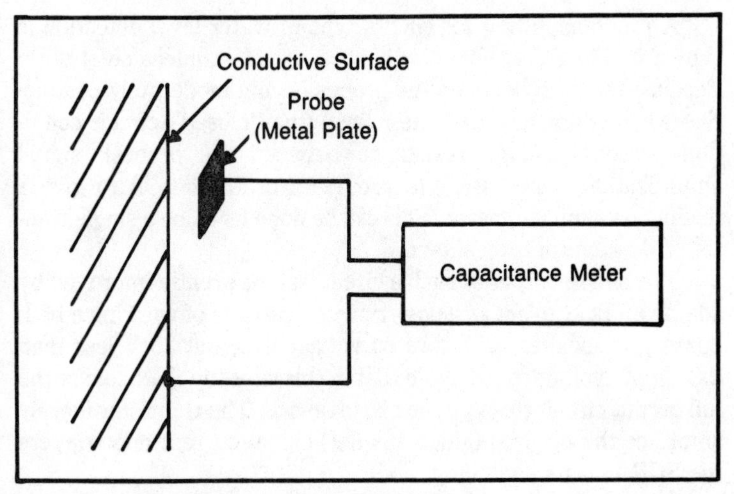

Fig. 25-3. A simple capacitive proximity probe.

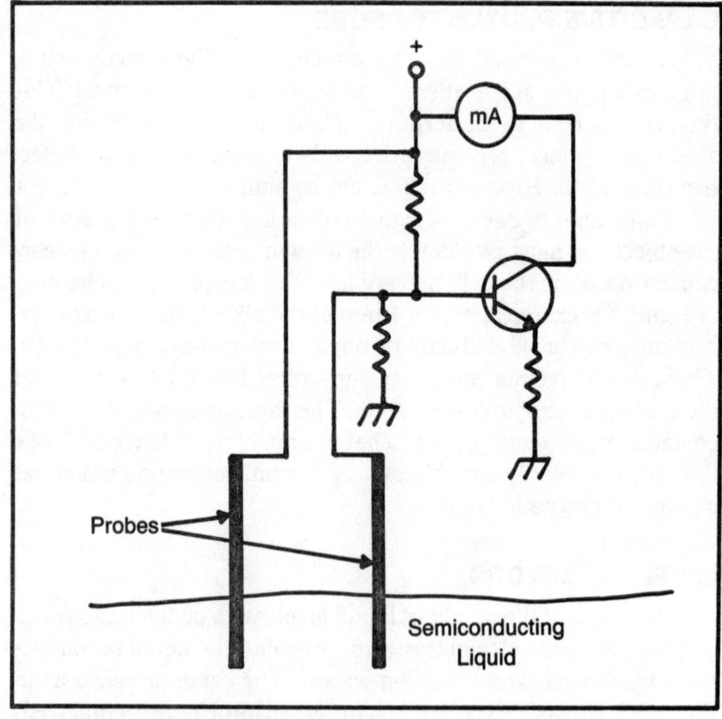

Fig. 25-4. A fluid-level-measuring device.

them changing from infinite (open-circuit condition) to some finite value.

A practical circuit for such a type of water-level detection is shown in Fig. 25-4. The two probes are of stainless steel wire. Because the path between the probes is only moderately conductive when contacting the water level (the "closed"-circuit condition presents a high resistance between the probes), signal amplification is necessary to provide a usable output to give a reading on a milliammeter. This can be done by using a single-transistor dc amplifier, as shown.

The single output of such a circuit can be greatly improved by adding an SCR to act as a power switch capable of working a bell, buzzer, or speaker, with a rated voltage of about 20 V less than the supply voltage used. Note that in this case the SCR covers the full output circuit through the alarm device. The transistor merely amplifies the original (much smaller) signal current and triggers the SCR into its switching mode.

A point to note with all immersible liquid-level probes is that if liquids containing dissolved metallic salts are involved, the input should be an ac signal to prevent plating deposits being formed in the probe(s).

PNEUMATIC SENSORS

Extremely sensitive sensors of simple form can be produced by using compressed air. They are *noncontacting* devices in the sense that the sensor itself never contacts the object it is sensing; only a compressed air jet impinges on the object.

In all cases the output signal devised is *air pressure* (or a change in air pressure), which can be used to operate a switching element

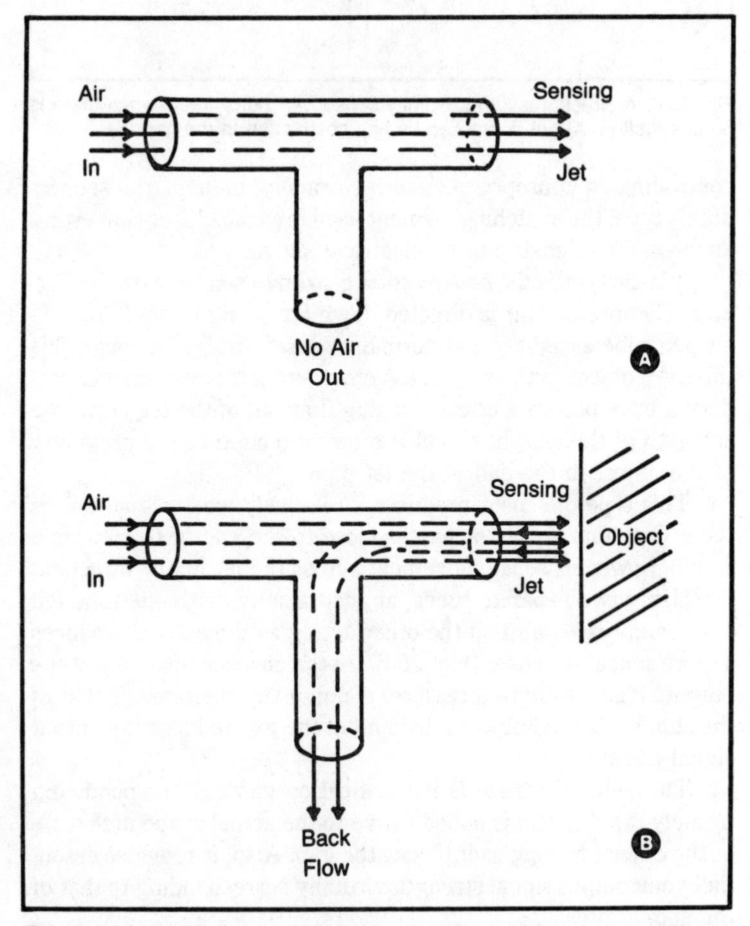

Fig. 25-5. A simple back-pressure proximity sensor. At A, no object is near the sensing jet; at B, an obstruction causes back flow.

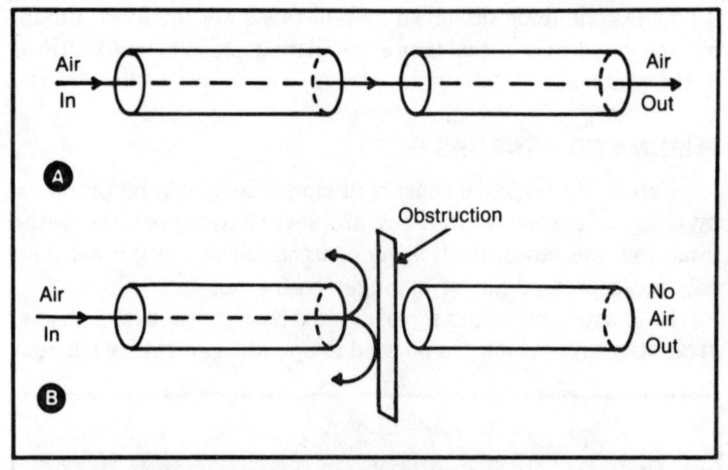

Fig. 25-6. A direct-pressure proximity sensor. Air flows freely when there is no obstruction (A) but is blocked by any obstruction in the gap (B).

controlling an appropriate circuit. Normally, in industrial sensors of this type, the switching element used is a transducer converting air pressure (signal) into an electrical signal.

A basic type is the *back-pressure* proximity sensor shown in Fig. 25-5. Compressed air is directed down the straight length of a T-shaped tube assembly and normally passes straight through this tube. An object in the way of the emergent jet, however, will produce a back-pressure effect, causing flow out of the tee arm. The strength of this output signal is directly related to the proximity of the object to the end of the jet pipe.

This type of sensor produces a relatively weak signal unless the object approaches or passes close to the end of the jet pipe. It will, however, sense movement across the jet in any direction.

Using two separate tubes, aligned axially, with one tube fed with compressed air and the other acting as a receiver, produces a more sensitive sensor (Fig. 25-6). In the absence of an object the output signal strength (pressure) is nominally the same as that of the supply. A solid object intruding in the gap reduces the output signal to zero.

This type of sensor is not critical on gap size, responds extremely rapidly, and is not sensitive to the actual shape in texture of the object, as long as it blocks the gap. Also, it provides essentially one output signal strength virtually corresponding to that of the supply pressure.

A rather more elaborate form of pneumatic proximity sensor

is shown in Fig. 25-7. Here the receiver tube is T-shaped and connected to an auxiliary supply at the far end at a pressure appreciably lower than that of the supply pressure. The object of this is to produce a back pressure that increases the output signal strength and at the same time purges the receiver of any entrained air.

This form of sensor is also used with a second supply tube at right angles to the first and directed at the jet gap (Fig. 25-8). This works on a rather different principle. In the absence of any object the jet from supply 2 impinges on the jet across the gap from supply 1, rendering this gap flow turbulent and resulting in a very low-to-nil signal output.

If now an object appears to block the jet from supply 2, the gap flow reverts to laminar. This change from turbulent to laminar flow produces a marked change in signal output level, which switches to full system pressure.

Finally, a configuration that combines the merits of the two is shown in Fig. 25-9. Here two separate supplies are used, feeding

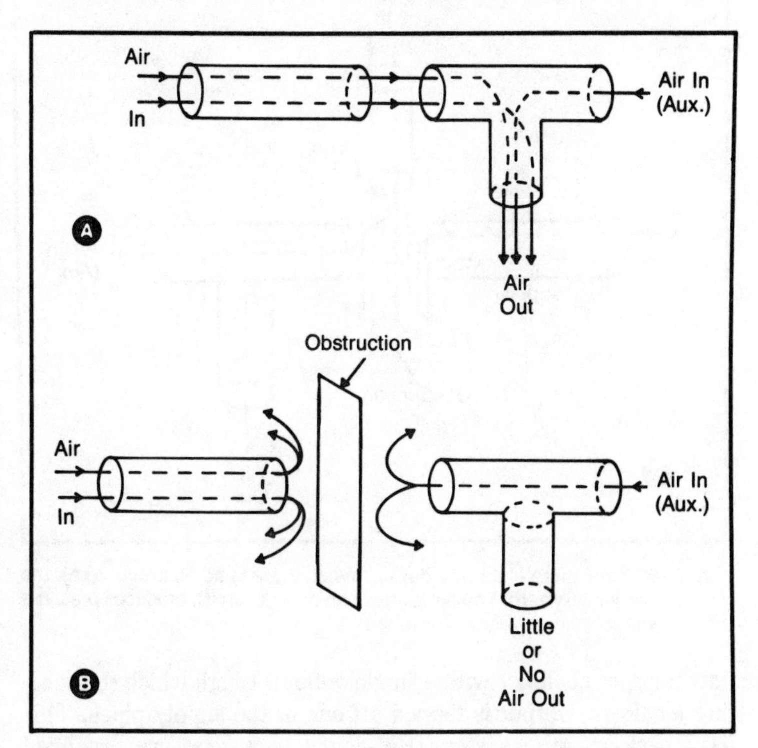

Fig. 25-7. A more elaborate form of direct-pressure sensor. At A, no obstruction; at B, obstruction in the gap.

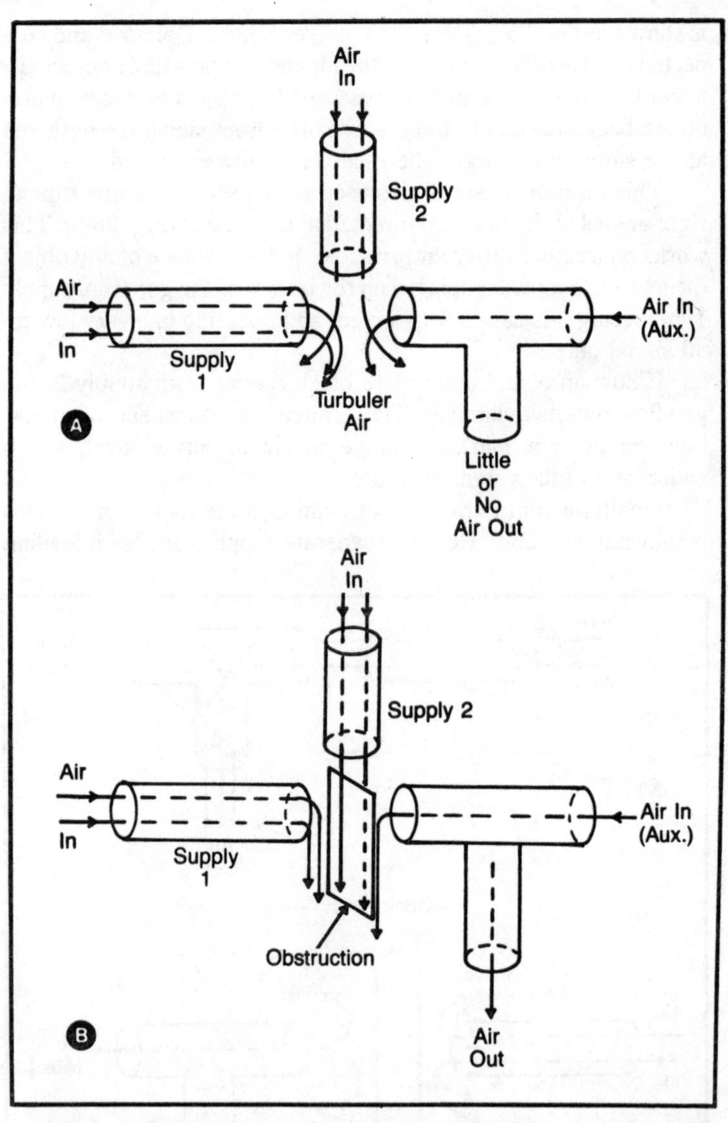

Fig. 25-8. A still more elaborate device. When there is no obstruction (A), the output is essentially zero. The presence of an obstruction (B) produces pressure at the output.

into a larger chamber with a single outlet through which the sensing jet flows. Output is tapped off one of the supply pipes. This is a back-pressure system; the actual back pressure produced depends on the two supply pressures. These can be the same, that

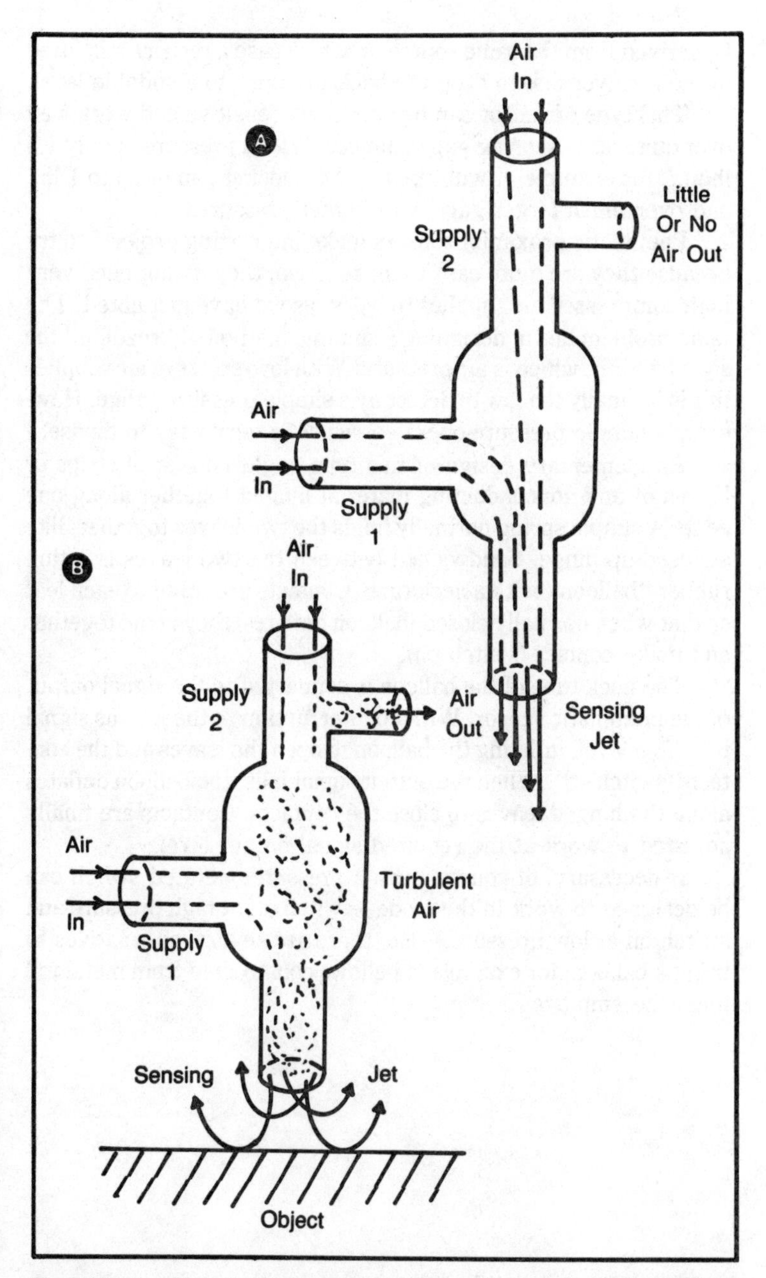

Fig. 25-9. A sophisticated back-pressure pneumatic proximity sensor. Air flows freely through the chamber and out of the sensing jet when there is no obstruction (A). The presence of an obstruction generates turbulence in the chamber and results in pressure at the output (B).

is, derived from the same source, in which case a restrictor is fitted in the receiver line to drop the back pressure to a suitable level.

This type of sensor can be extremely sensitive and work well over quite large sensing-gap distances. With a pressure of only 1.5 lb/in.^{2}, for example, it will work with a sensing gap of up to 1 in., and over much larger gaps with higher pressures.

Pneumatic proximity sensors make interesting projects to try because they are quite easy to make. Also, they do not need very high compressed air supplies to work, as we have just noted. The main problem lies in designing a suitable method of "reading" the signal output, which is air pressure. With low-pressure air supplies this is normally too low to detect by a simple pressure gauge. However, a simple pressure-operated *switch* is very easy to devise.

An elementary design of this type might consist of strips or leaves of any nonconducting material hinged together along one edge. A simple spring normally holds the two leaves together, like a closed-up hinge. Sandwiched between the two leaves is a thin rubber "balloon" with a neck tube. Contacts are fitted to each leaf so that when normally closed (balloon deflates) they come together and make contact (switch on).

The neck tube of the balloon is connected to the signal output of the pneumatic sensor. With no interruption to the jet this signal is at high level, inflating the balloon to open the leaves and the contacts (switch off). When the output signal falls, the balloon deflates along the hinged leaves to close the contacts. Contacts are finally adjusted to work at the required signal output level.

If necessary, of course, such a pressure-operated switch can be designed to work in the mode—switch off at high pressure and switch on at low pressure. Also, there are several alternatives to using a balloon; for example, a bellows constructed from metal foil might be employed.

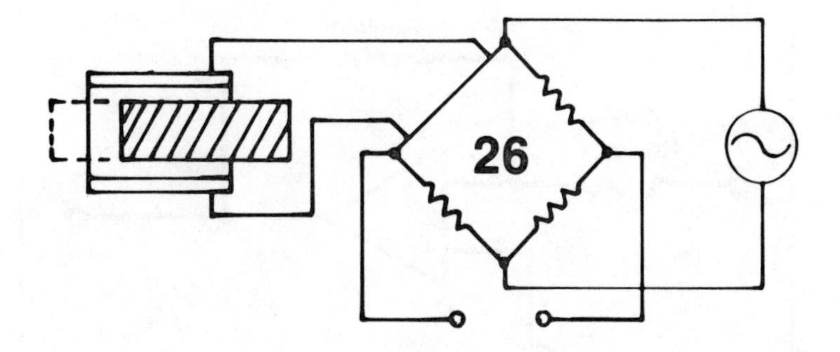

Electronic Amplifiers

There are several different kinds of amplifiers used with transducers. The particular type of amplifier depends on the nature of the signal to be amplified. We may classify amplifiers in several categories, as follows: low-level ac amplifiers, power ac amplifiers, dc amplifiers, and switching (pulse) amplifiers.

LOW-LEVEL AC AMPLIFIERS

The most important consideration in a low-level amplifier (also commonly called a preamplifier) is the noise figure. Operational amplifiers are often used as preamplifiers (Fig. 26-1). These devices can provide gain, in practice, of more than 60 dB.

The main problem with operational amplifiers in applications where extreme sensitivity is needed is the noise figure. Where noise must be kept to a minimum, it is common practice to increase the negative feedback in the operational amplifier circuit, thereby reducing the gain, and precede the operational amplifier with a low-noise field-effect-transistor (FET) circuit. An example of such an amplifier is shown in Fig. 26-2. The lowest noise figure is generally obtained with the gallium-arsenide (GaAs) FET. If broadband operation is needed, the FET circuit may be untuned, although low-pass or high-pass networks at the input and output may help reduce the overall noise figure.

Examples of FET low-level amplifiers using low-pass and high-

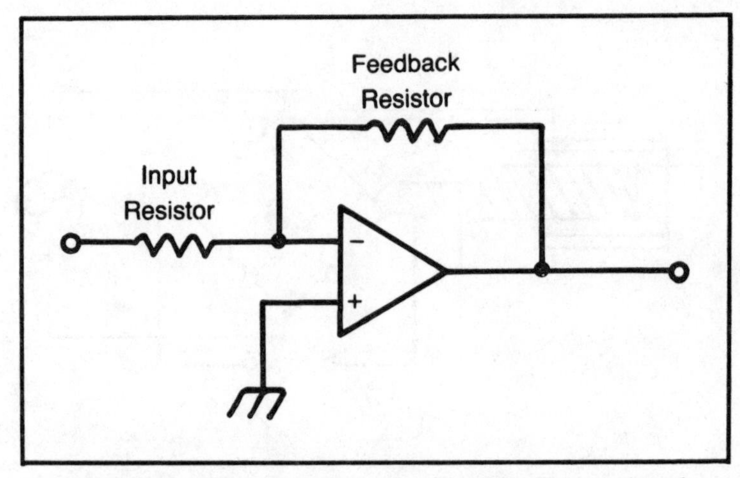

Fig. 26-1. A preamplifier using an operational amplifier. The input impedance and gain are determined by the values of the resistors.

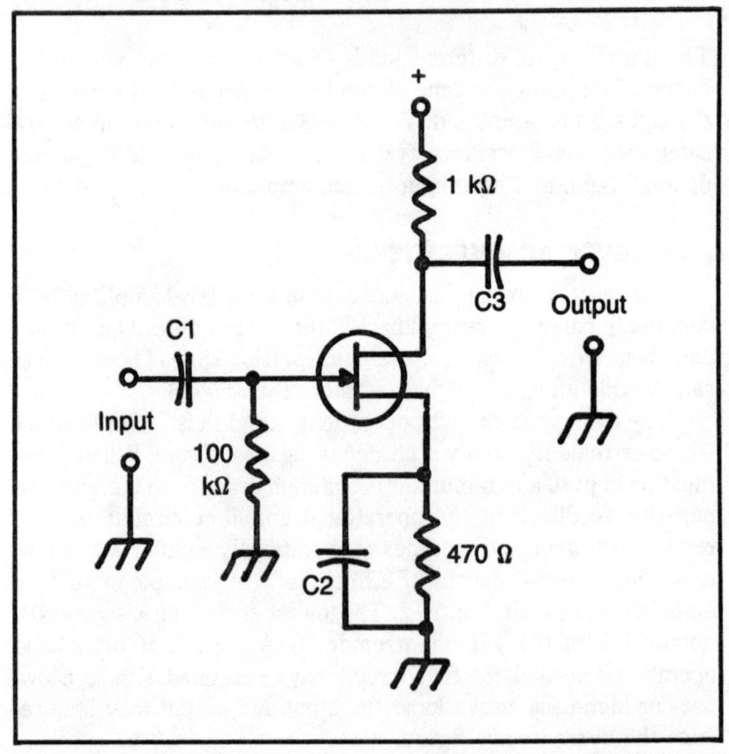

Fig. 26-2. A preamplifier using an FET. The values of C1, C2 and C3 depend on the frequency range desired.

pass networks are shown in Figs. 26-3A and B. The cutoff frequency of the network at A should be well above the highest operating frequency. A value of about 1.5 times the highest operating frequency will permit a linear response in the operating range. Similarly, the cutoff frequency of the high-pass network (B) should be well below the lowest operating frequency; an optimal value might be about 0.67 times the lowest expected operating frequency.

If the operating frequency range is quite narrow, tuned circuits may be incorporated at the input and/or output of the FET preamplifier stage. These circuits may be simple parallel-resonant LC networks (Fig. 26-3C) or they may be Chebyshev or Butterworth bandpass networks. The primary problem with parallel-resonant LC networks at both the input and output is a tendency toward oscillation of the amplifier stage. A resistor in series with the inductance of the tuned circuit will reduce this tendency. Careful attention to the physical layout of the circuit, minimizing interwiring capacitance, is important.

An example of a low-level, highly sensitive ac amplifier is shown in Fig. 26-4. This circuit is used in conjunction with an ordinary photovoltaic (solar) cell for the purpose of receiving modulated-light signals. The FET amplifier incorporates a low-pass filter having a cutoff of approximately 4.5 kHz at the input, restricting the response to the communications baseband range. A high-pass filter, with a cutoff of about 200 Hz, helps to filter out the 120-Hz ac "hum" emitted by light bulbs operating from the utility mains. The high-pass filter is at the input of the FET stage, and the low-pass filter is at the output. Pot cores are used for the coils.

POWER AC AMPLIFIERS

In a power amplifier the design depends on the amount of power needed and on whether or not linearity is important. In a radio-frequency transmitter, for example, the output power may be as low as 1 or 2 W or as high as 1 million W or so. Linearity is not important in a radio-frequency transmitter insofar as the rf waveform itself is concerned, but linearity may be essential for the modulation envelope.

Power amplifiers are classified according to the proportion of the input cycle during which current flows in the output (collector or drain). In class A amplifiers, current flows in the output circuit 100 percent of the time. Furthermore, the instantaneous output is

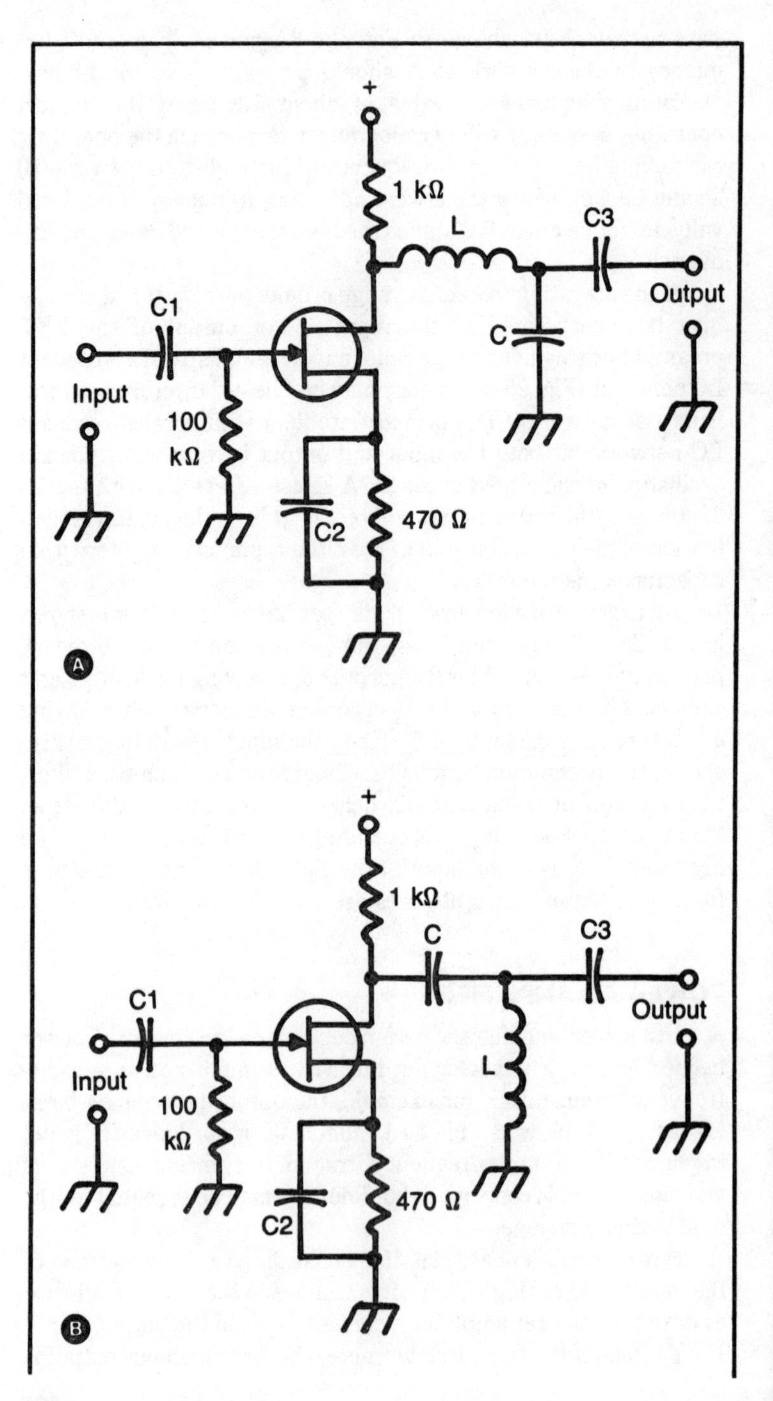

+
1 kΩ
L
C3
Output
C1
C
Input
100
kΩ
470 Ω
C2
A
+
1 kΩ
C
C3
C1
Output
Input
L
100
kΩ
470 Ω
C2
B

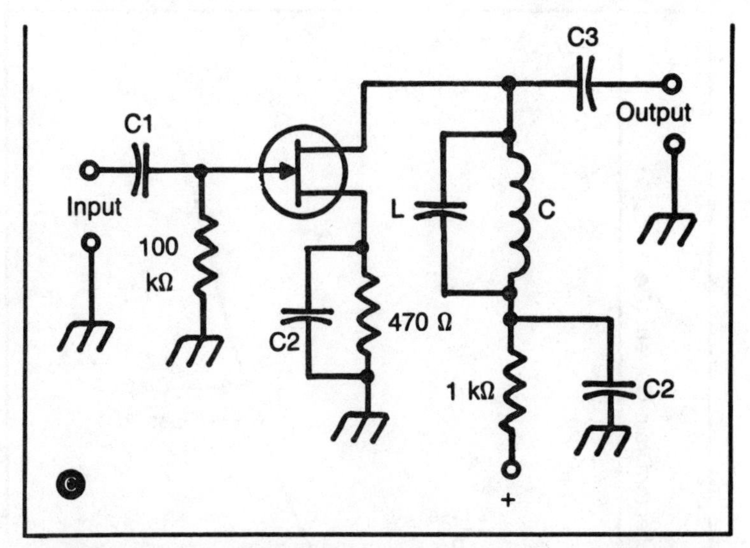

Fig. 26-3. Examples of FET amplifiers using selective networks. At A, low pass; at B, high pass; at C, bandpass.

linearly proportional to the instantaneous input throughout the cycle. A class A amplifier is biased so that, under conditions of zero input signal, the device is at the middle of the linear portion of the base/gate-voltage vs. collector/drain-current curve (Fig. 26-5).

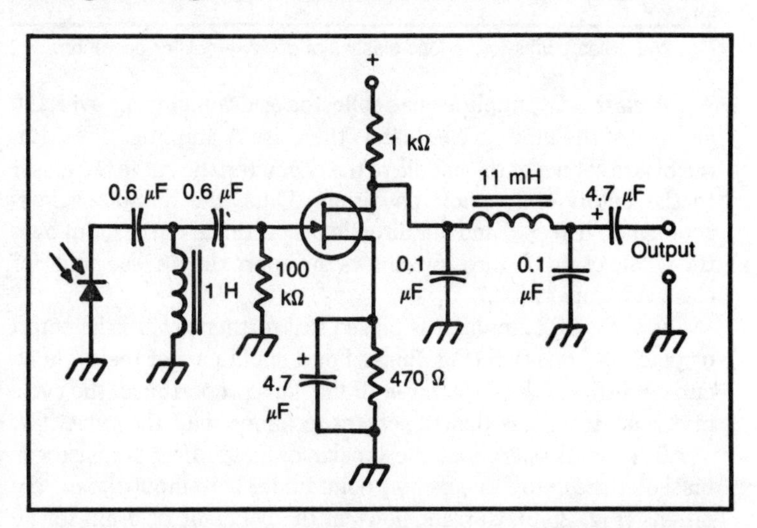

Fig. 26-4. An FET amplifier incorporating both low-pass and high-pass networks. This amplifier is used with a solar cell for reception of audio-modulated light signals.

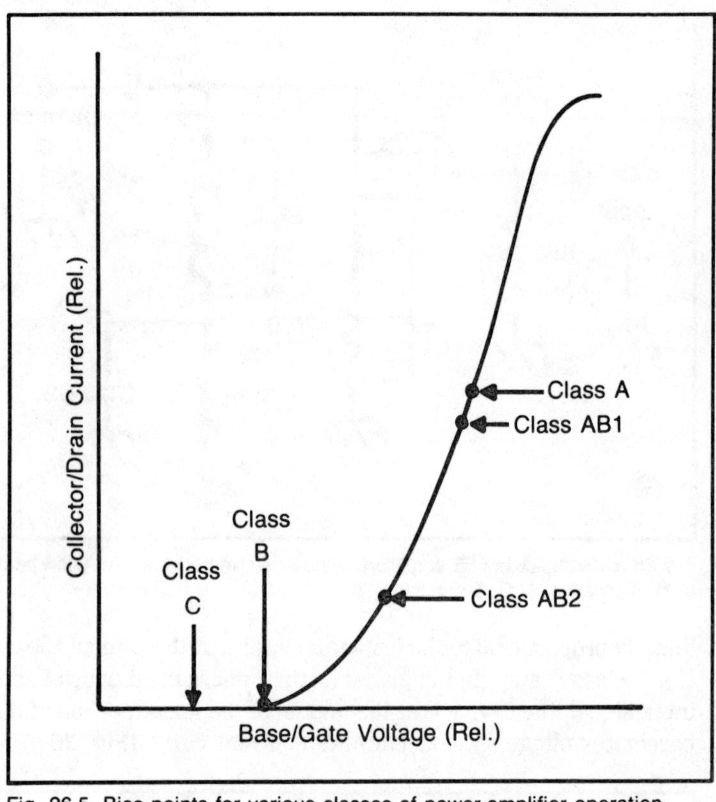

Fig. 26-5. Bias points for various classes of power-amplifier operation.

A class AB1 amplifier has collector or drain current over 100 percent of the input cycle, just as the class A amplifier does. But the bias is set not at the middle of the characteristic curve but closer to the cutoff or pinchoff condition. Thus, the output cycle is somewhat distorted and not directly proportional to the input over the whole cycle. Figure 26-5 shows the approximate bias point for class AB1 operation.

A class AB2 amplifier is biased still further toward the cutoff or pinchoff condition (Fig. 26-5). For a small part of the cycle no current flows in the collector or drain. This proportion of the cycle may vary from less than 1 percent to almost half the cycle.

In class B operation, the bipolar or field-effect transistor is biased at the cutoff or pinchoff point under zero-input-signal conditions (Fig. 26-5). Current flows in the collector or drain for 50 percent of the cycle.

In class C operation, bias is considerably beyond cutoff or

pinchoff (Fig. 26-5). Current flows during less than 50 percent of the cycle. The proportion may be as small as just a few percent.

Figure 26-6 illustrates, qualitatively, the output waveforms for these various classes of power-amplifier operation, assuming a perfect sine-wave input.

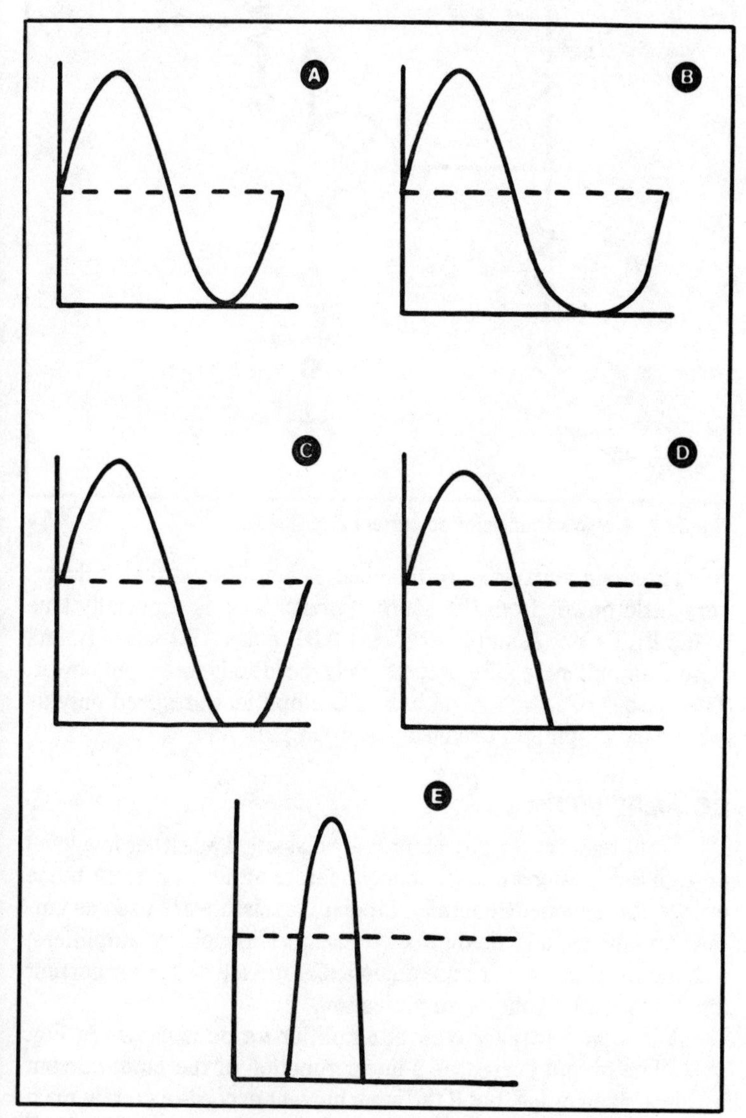

Fig. 26-6. Output waveforms for class A (A), class AB1 (B), class AB2 (C), class B (D), and class C (E) operation.

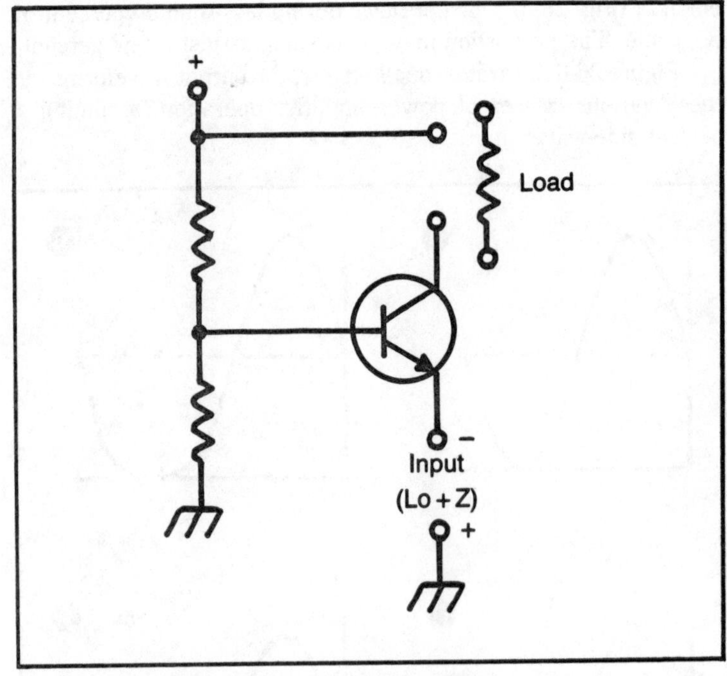

Fig. 26-7. A bipolar-transistor dc current amplifier.

The main advantage of the class A amplifier is that it draws very little power from the source (input). This is especially true of the FET class A amplifier. Class AB1, class AB2, class B, and class C amplifiers require progressively more and more input power. The class AB2, class B, and class C amplifiers are used only as secondary amplifiers driven by a preamplifier.

DC AMPLIFIERS

Amplifiers for dc may be broadly classified as either low level or high level. Such amplifiers may operate either in current mode or in voltage mode. Generally, bipolar transistors are used as current amplifiers, and field-effect transistors as voltage amplifiers, although it is possible to use a particular device either for current amplification or voltage amplification.

A bipolar-transistor current amplifier for dc is shown in Fig. 26-7. The output current is a linear function of the input current within a given range, but if the input current exceeds a certain maximum, the output current will increase less and less rapidly as the condition of saturation is approached (Fig. 26-8). Under zero-input-current conditions the transistor is biased exactly at cutoff.

An FET dc voltage amplifier is shown in Fig. 26-9. As with the current amplifier, the output is directly and linearly proportional to the input. Under conditions of zero input voltage the FET is pinched off. This is accomplished by means of a large-value resistor connected from the gate of the FET to a source of negative voltage. You may recognize the circuit at Fig. 26-9 as a FET voltmeter (without the meter!).

As the input voltage increases, the output voltage increases in linear proportion up to a certain point. Then the output voltage increases less and less rapidly, and finally it levels off (Fig. 26-8).

There is essentially no difference in the circuit configuration for high-level current or voltage amplifiers for dc as compared with low-level amplifiers. The only real difference is in the size of the bipolar transistor or FET.

Direct-current amplifiers may be cascaded to obtain enhanced gain.

Operational amplifiers are often used for amplification of dc, especially in conjunction with such devices as servomotors. The general configuration for a dc amplifier using an operational amplifier is shown in Fig. 26-10.

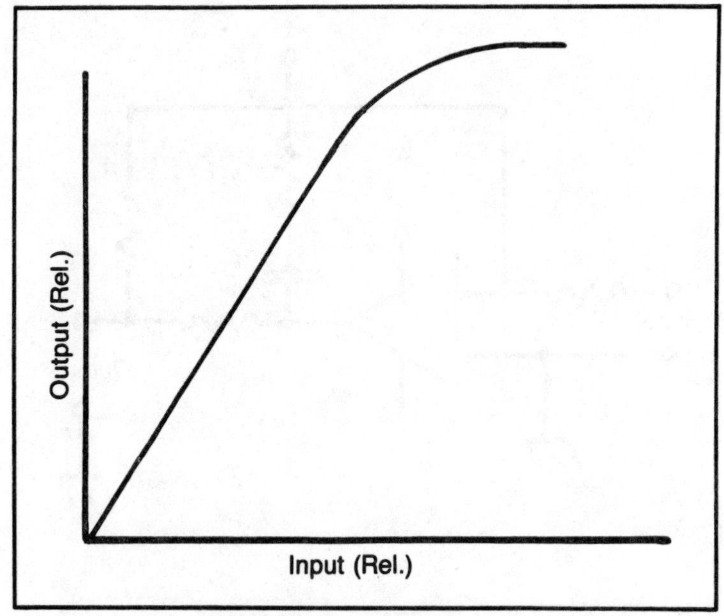

Fig. 26-8. Output-vs.-input curve for dc current or voltage amplifiers.

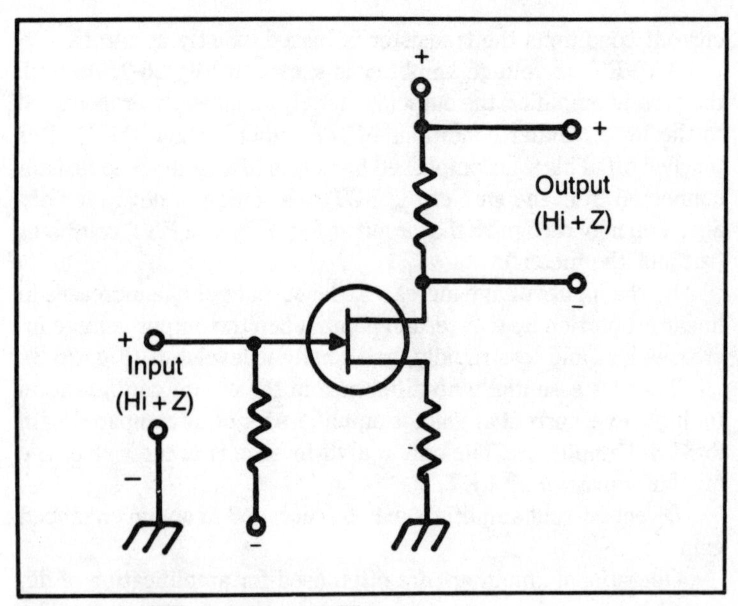

Fig. 26-9. An FET dc voltage amplifier.

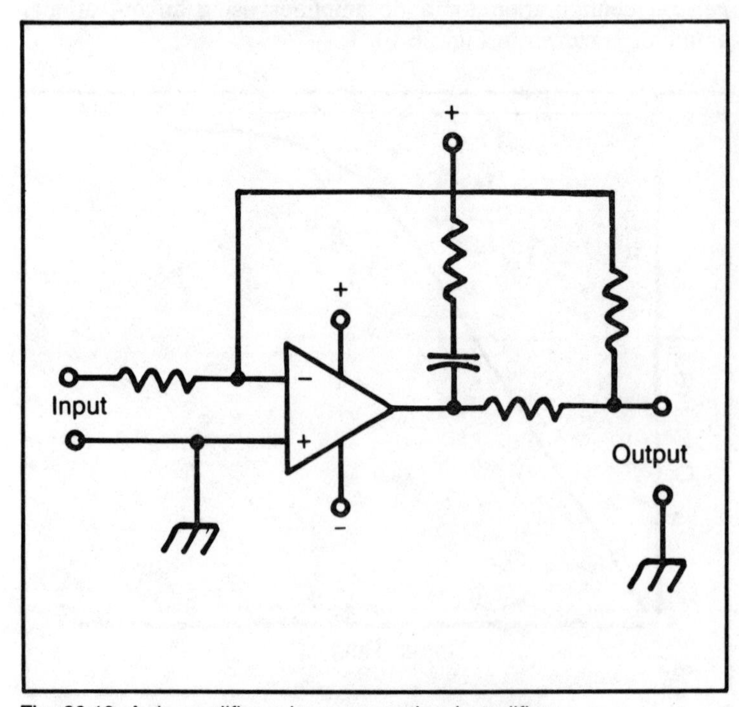

Fig. 26-10. A dc amplifier using an operational amplifier.

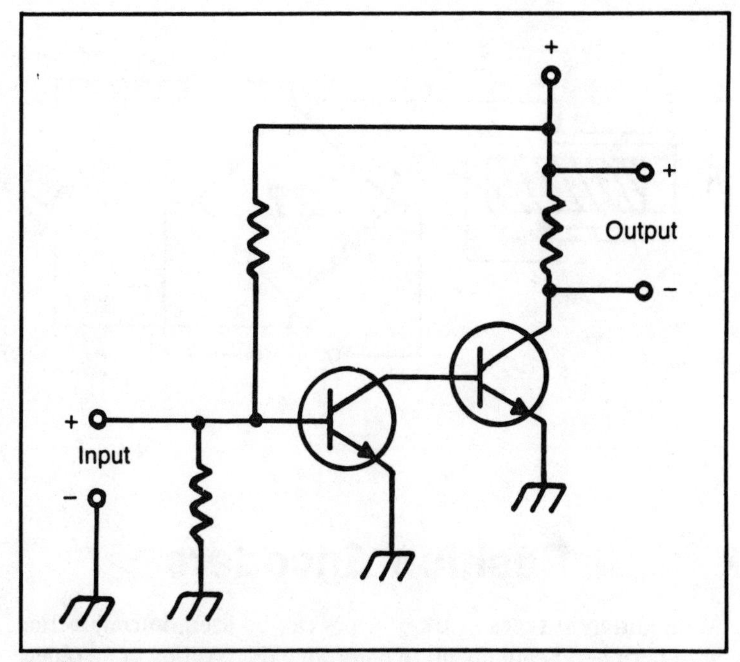

Fig. 26-11. A simple switching amplifier. This circuit is used to amplify voltage pulses at moderate to high speeds.

SWITCHING (PULSE) AMPLIFIERS

Switching amplifiers are similar to low-level ac amplifiers, except that the switching amplifier must have broad bandwidth. This is necessary because of the rapid rise and decay of square pulses. The switching amplifier may or may not be driven into saturation (full conduction) during the conducting part of the cycle. Switching amplifiers are sometimes referred to as class D amplifiers if they are saturated in the "on" condition and cut off or pinched off during the "off" condition.

A simple switching amplifier is shown in Fig. 26-11. It consists of two bipolar transistors biased slightly beyond cutoff. The first transistor , Q1, is biased so that a small positive-going pulse (on the order of 0.2 V or more) will cause it to conduct. The output voltage is somewhat larger, about 2 V, and this pulse drives Q2 into saturation. An operational amplifier may be substituted for Q1 if high sensitivity is needed.

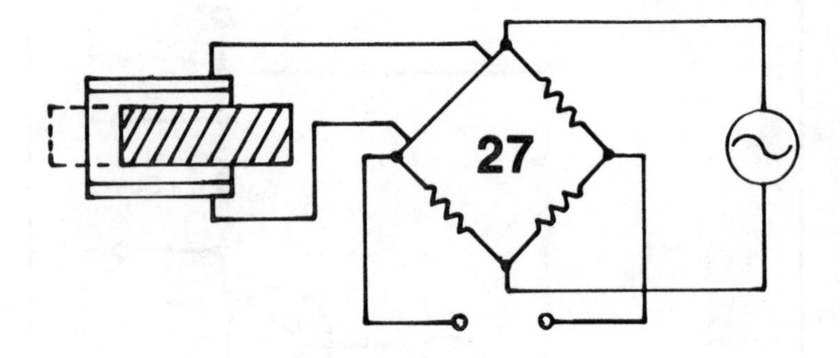

Position Encoders

Many different types of transducers can be used, in conjunction with peripheral equipment, to determine the position of an object in one, two, or three dimensions. The number of transducers increases as the number of dimensions increases; for example, we can expect that it will require more devices to locate a satellite in space as opposed to determining the linear separation between two objects.

In this chapter we will examine several types of position encoders. The examples given here are just a small cross section of all the different methods of position encoding.

ACOUSTIC MEASUREMENT OF DISTANCE

Electroacoustic transducers (speakers, microphones, and variations) can be used to determine the distance between two points, the location of an object on land or at sea, or the position of an object in the air or under water. Generally, these types of position encoder make use of the fact that acoustic waves in various media travel at a fairly constant, known speed.

Perhaps you have stood at the top of a cliff in a so-called "echo canyon" and listened to your shouts come back to you. Sound waves travel at about 1100 ft/sec in the atmosphere of our planet. Thus, if the sound returned to you in, say, 3 sec., you would know that the echo barrier was 1.5 "sound seconds," or about 1650 feet, away.

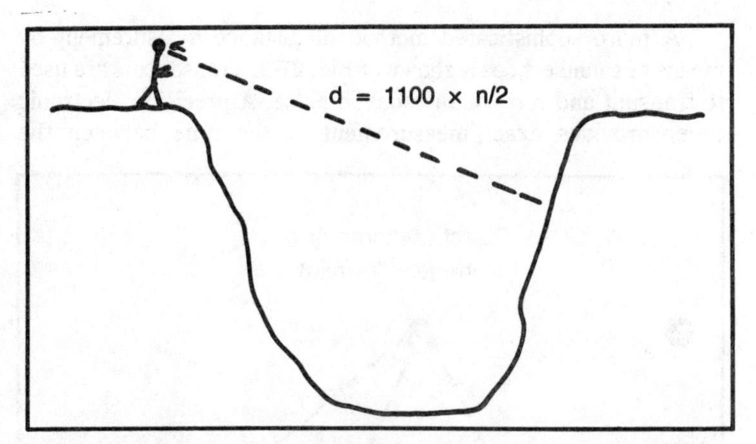

Fig. 27-1. A primitive method of determining distance by acoustic means. The distance d is proportional to the delay n for the echo to return.

In general, an echo delay of n seconds would mean a distance d, in feet, of

$$d = 1100 \times n/2$$

This is shown in Fig. 27-1.

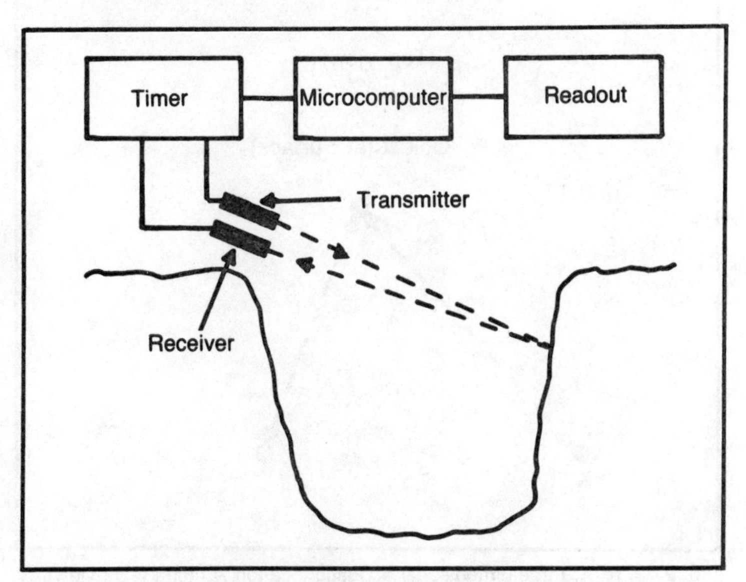

Fig. 27-2. A more sophisticated method of distance determination by means of sound.

A more sophisticated method of distance measurement by means of sound echoes is shown at Fig. 27-2. Transducers are used to transmit and receive an acoustic pulse. A precision electronic timer provides exact measurement of the time between the

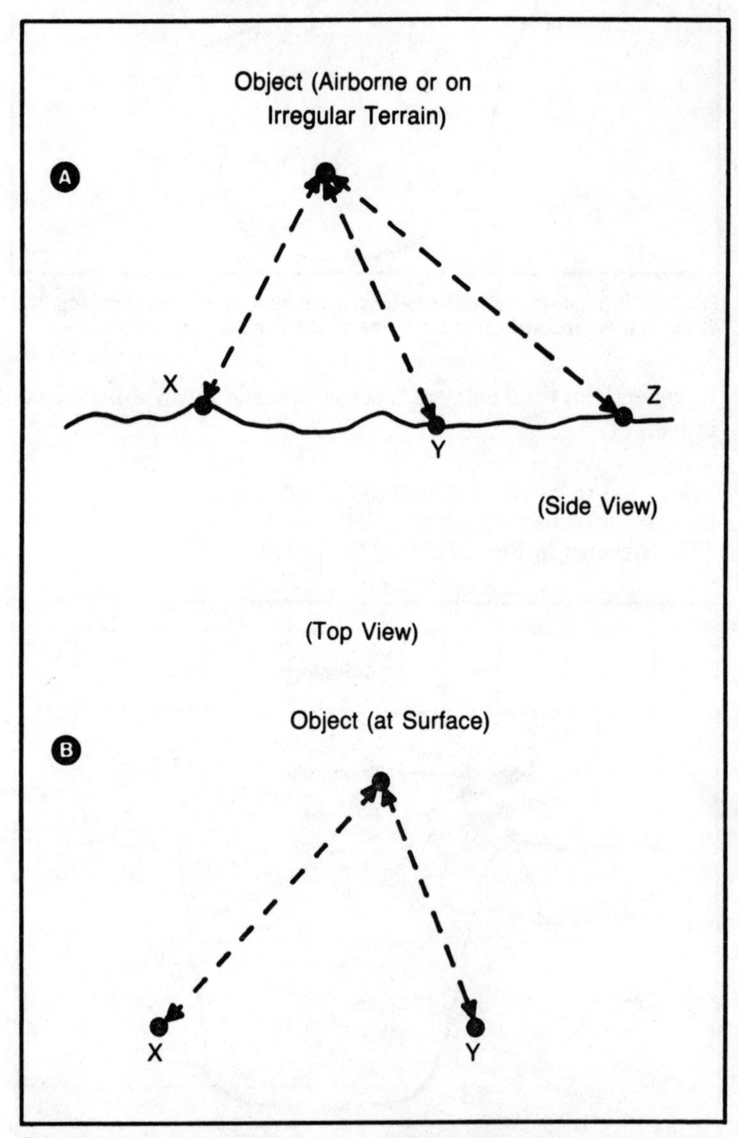

Fig. 27-3. At A, three-dimensional acoustic location scheme (side view). At B, two-dimensional acoustic location scheme (top view, level terrain). Transmitting/receiving stations are represented by X, Y, and Z.

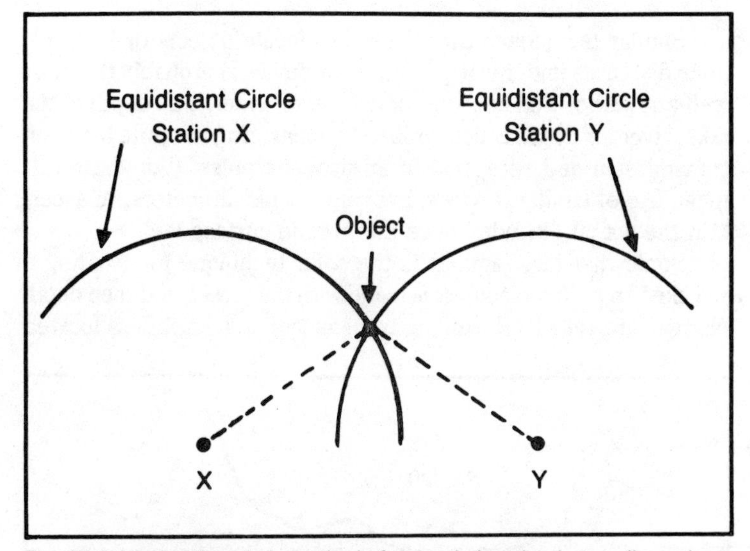

Fig. 27-4. Intersecting-circle method of acoustic location in two dimensions. (This method will work in some cases but not, in general, if stations X and Y are an opposite sides of the object.)

transmission and the reception of the pulse. A microcomputer device interprets the information and sends the appropriate signal to the readout. If the speed of sound is known (for the particular altitude and weather conditions) to a high degree of accuracy, the distance to a far-off object can be found to within better than 1 percent.

The apparatus of Fig. 27-2 tells us nothing about the location of the acoustic reflector; it only provides one-dimensional information. In order to precisely locate the point of reflection in space, it is necessary to use three different transmitting/receiving stations located some distance apart (Fig. 27-3A). If the terrain is flat, then only two transducers are needed (Fig. 27-3B).

In the three-dimensional case (Fig. 27-3A) the three transmitting/receiving stations must be located sufficiently far apart from each other that a clear intersection point can be found and calculated. This will be the intersection among three spheres in space. In the two-dimensional case the point of the reflecting object is found by the intersection of two or three circles. The spheres or circles have radii that are determined according to the preceding distance formula. The three-dimensional case is somewhat difficult to illustrate. Figure 27-4 shows the principle for the two-dimensional case.

Similar techniques can be used to locate objects or to determine distances under water. The sonar device is probably the most well-known position encoder for use under water. The depth of the lake, river, or ocean is determined by measuring the time between transmission and reception of an acoustic pulse. (Some aircraft make use of similar devices, known as sonic altimeters, to ascertain the actual altitude above the ground surface.)

Somewhat less familiar is the sofar technique for locating a wrecked ship. Radio contact is made with the vessel, and then depth charges are set off. Receiving transducers and amplifiers located

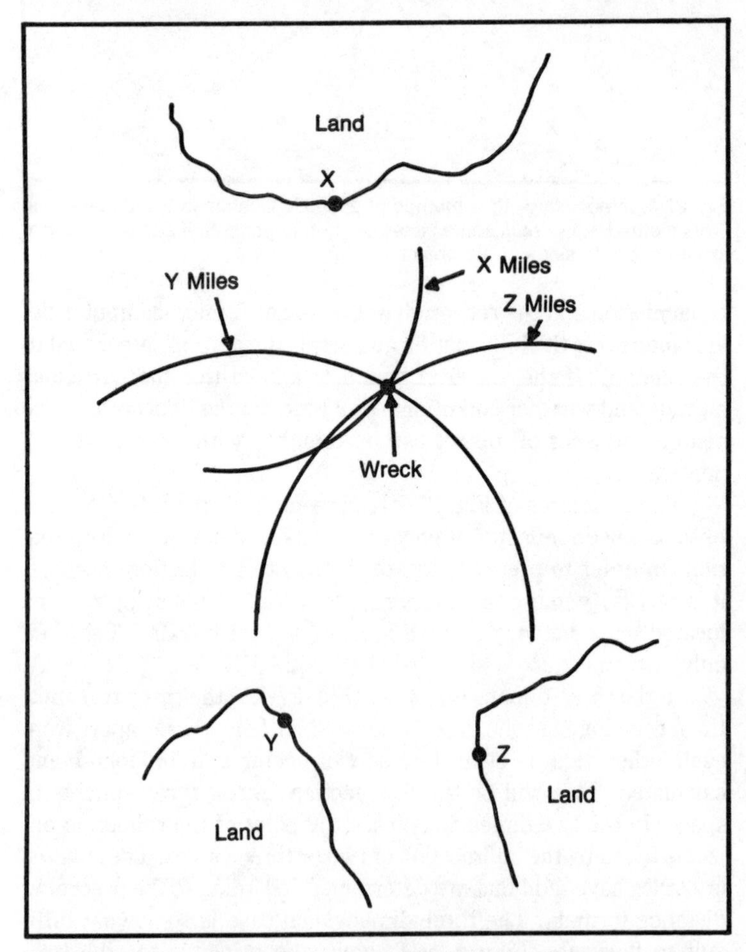

Fig. 27-5. Location of a wrecked ship from land-based stations. Three stations allow unambiguous determination of the ship's position.

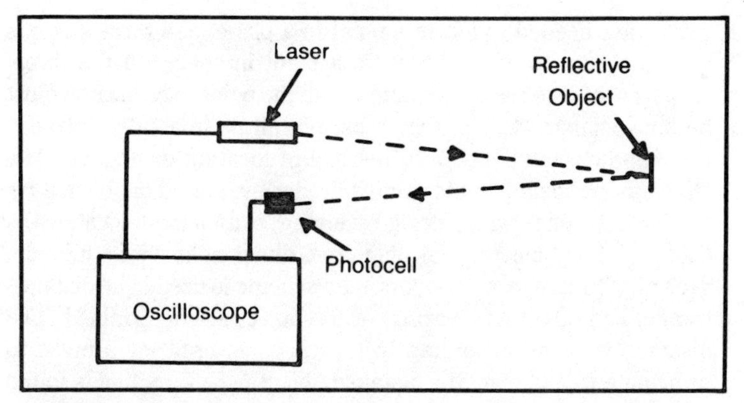

Fig. 27-6. A laser and photocell can be used to measure the distance to a reflecting object.

at three widely separated shore points are used to determine the distance to the ship (Fig. 27-5). The intersection point of the three circles is the location of the wreck.

ELECTROMAGNETIC POSITION ENCODING

Distances can be measured by using electromagnetic waves in the same way as acoustic waves. Radio signals, infrared radiation, visible light, and even ultraviolet rays can be employed. Lasers are commonly used for this purpose because of their long-distance, narrow-beam propagation qualities. Electromagnetic radiation travels through space at a precisely known speed of 186,282 mi/sec (299,792 km/sec). This is true in the atmosphere and in a vacuum such as outer space.

A visible-light (laser) apparatus for distance measurement is shown in Fig. 27-6. A pulse is sent out by the laser, and the instant of this pulse marked by an oscilloscope or high-speed timer. A photoelectric or photovoltaic cell detects the return (reflected) pulse. The time difference is determined; if the delay is n seconds, the distance d to the object is given by

$$d = 186{,}282 \times n/2$$

in miles. If n is given in microseconds, the distance d in miles is

$$d = 0.186282 \times n/2$$

As is the case with acoustic waves, two transmitting/receiving setups are needed to locate a point in a plane, and three stations are necessary to uniquely locate a point in space. In the three-dimensional case the three stations and the point to be located must be noncoplanar; that is, they must not all lie in a single plane.

Another commonly used method of locating objects by electromagnetic means is the familiar radar system. This device requires only one transmitting/receiving station and locates the position of an object or objects in two dimensions with a high degree of accuracy. A polar-coordinate scheme is used. The distance (range) and direction (bearing) of the object are determined. The distance is found according to the time required for a pulse to propagate to and from the distance object. The direction is found by means of a rotating antenna emitting frequent pulses in a narrow beam at a short wavelength.

Two radar sets can be used—one in the horizontal plane, and the other in the vertical plane—to determine the location of an object in three-dimensional space. The range and bearing are found by using horizontal radar. Then the vertical radar is operated in a plane passing through the line representing the bearing as indicated by the horizontal radar (Fig. 27-7). Such a system can be useful, for example, for locating satellites in space. The celestial coordinate system might be used; the "horizontal" radar would be

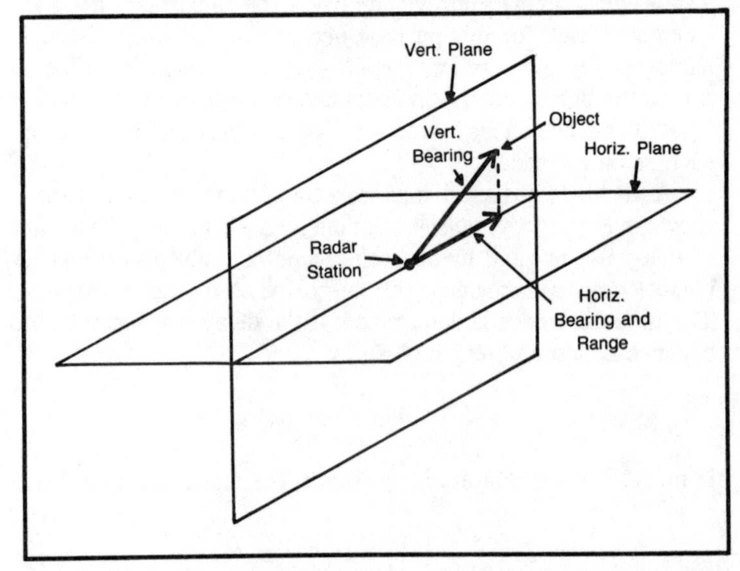

Fig. 27-7. Determination of exact position in three dimensions, using two radar sets.

operated in the plane of the celestial equator, and the "vertical" radar in a perpendicular plane, corresponding to the right ascension of the object as found by the "horizontal" radar.

An object equipped with tricorner reflectors can be located with extreme accuracy if lasers are used. The distance can be found to within a tiny fraction of 1 percent from three different transmitting/receiving stations. (Using lasers, scientists have ascertained the distance to the moon to within a few inches.)

OTHER POSITION SENSORS

Potentiometers, capacitors, and inductors can be used as position sensors. The potentiometer is commonly employed for determining angular position (see Chapter 3). Capacitors are sometimes employed for determining position on a small scale (Chapter 5). Inductors can be employed in a similar way on a small scale with ferrous materials (Chapter 9, Fig. 9-3).

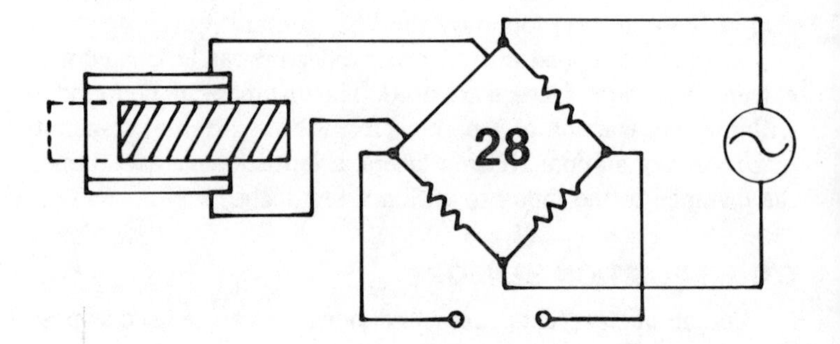

Vibration Measurement and Monitoring

Three different types of transducers are used for measuring vibration. These are

- noncontacting proximity probes (normally eddy-current type)
- velocity pickups
- accelerometers (piezoelectric, capacitance, or strain gauge, with the piezoelectric type normally being preferred).

Each type has its advantages and disadvantages. Choice to a large extent thus depends on the particular application and the most significant parameter to be measured. The three parameters involved in vibration are *amplitude* or displacement, *frequency* or repetition rate, and *phase* in the case of multiple vibrations, which gives the relationship of one vibration to the other(s).

Directly measurable quantities are displacement, velocity, and acceleration, which are all interrelated and also related to frequency. Equal displacements at two different frequencies, for example, do not result in equal vibration velocities or vibration acceleration.

VIBRATION MEASUREMENT

Probably the simplest way of measuring vibration is by displacement (and also determining the frequency). But this is not necessarily the best way. Measuring vibration velocity or vibration acceleration may be better in some cases, and if displacement (amplitude) is also needed to be known, this can be derived from velocity and acceleration. Similarly, if acceleration is measured, then integrating this signal once will give vibration velocity. With suitable circuitry an accelerometer can thus provide both acceleration and velocity readout.

Without going too deeply into details, the advantages and disadvantages of measuring vibration displacement, velocity, or acceleration can be summarized as follows:

□ Measuring *displacement* gives a signal that is proportional to displacement only. The advantages are

- easy to understand
- easy to specify
- low-impedance signal easy to handle and display

The disadvantages are

- difficult to measure directly
- not related directly to vibration energy
- low sensitivity
- difficult to install. Linear variable-differential transformers or potentiometers require coupling to base reference.

□ Measuring *velocity* gives a signal that is proportional to displacement and frequency of vibration. The advantages are

- directly related to vibration energy
- sensor easy to install
- low-impedance signal easy to handle
- sensor normally supplied critically damped

The disadvantages are

- limited frequency response (8 Hz to 1 kHz),
- relatively large sensor required (not necessarily a disadvantage in industrial applications)

□ Measuring *acceleration* gives a signal that is proportional to displacement energy over $(\text{frequency})^2$. The advantages are

- wide frequency response
- can use small lightweight sensors of robust construction

The disadvantages are

- not directly related to vibration energy
- more complex instrumentation signalling equipment required
- low output at very-low frequencies
- susceptible to thermal effects

Measurement of vibration velocity thus enables vibration energy to be determined, independently of the frequency of vibration, for any given system of constant mass. Velocity is proportional to the frequency of vibration for any given system of constant mass and constant amplitude. This is also illustrated by the frequency relationship diagrams of Fig. 28-1.

This relationship is readily demonstrated by electromagnetic vibrations, such as used for calibration purposes. Here, when frequency is increased at a constant power level (constant energy for unit time), vibration displacement decreases, vibration velocity remains constant, and vibration acceleration increases.

Vibration velocity measurement, therefore, is particularly applicable where the mechanical system concerned may vibrate at an unknown or variable frequency, or where the fundamental frequency of the system cannot be predicted.

A further advantage offered by vibration velocity measurement is that the sensor is normally critically damped so that the chance of resonance affecting the readings is eliminated. With an undamped sensor and the presence of a broad spectrum of vibration, false readings may be obtained. However, vibration acceleration measurement is commonly preferred because of its wide frequency response and the fact that the sensor(s) used can be quite small, lightweight, and robust.

With the above in mind we can then look at the three different types of transducers involved and their advantages and disadvantages.

□ Proximity Probe (Displacement Pickup). The advantages are

- noncontacting
- measures motion directly
- measures in engineering terms

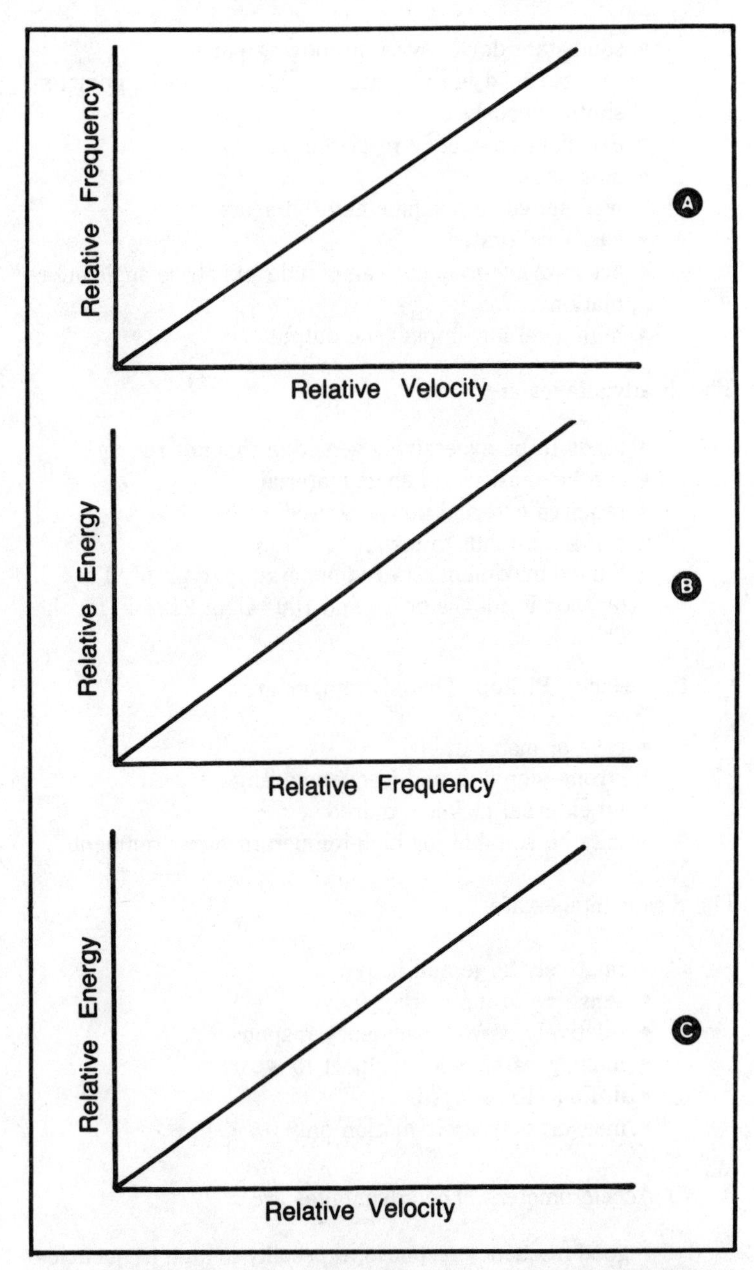

Fig. 28-1. For a system having constant mass and constant amplitude of vibration, measurement of vibration velocity will yield a direct indication of vibration energy. This is indicated by the linear relationships at A and B; the result (C) is that vibration energy is linearly proportional to vibration velocity.

- solid-state device with no moving parts
- measures dynamic motion and average position simultaneously
- excellent frequency response
- small size
- well suited to machine environments
- easily calibrated
- accurate low-frequency amplitude and phase-angle information
- high-level low-impedance output

The disadvantages are

- tends to be excessively sensitive to shaft runout
- can be sensitive to shaft material
- requires external power source
- can be difficult to install
- limited maximum service temperature (typically 175° C. or 350° F. for the probe and 100° C. or 212° F. for the driver)

☐ Velocity Pickup. The advantages are

- ease of installation
- strong signal in mid-frequency range
- no external power required
- may be suitable for high-temperature environment

The disadvantages are

- relatively large and heavy
- sensitive to input frequency
- relatively narrow frequency response
- moving part device (subject to wear)
- difficult to calibrate
- measures dynamic motion only

☐ Accelerometers. The advantages are

- good frequency response, especially to high frequencies
- small and lightweight
- strong signal in high-frequency range
- may be suitable for high-temperature environment

The disadvantages are

- sensitive to input frequency
- relatively expensive
- difficult to calibrate
- requires external power source
- sensitive to spurious vibrations
- impedance matching necessary; also some filtering for monitoring applications

From the above we can draw up a selection guide for transducer selection for vibration measurement; see Table 28-1. The first deciding factor is the mode of measurement required. An accelerometer will measure vibration acceleration, for example, but it can also give vibration velocity (by integrating the signal once) and vibration displacement (by integrating the signal twice). Similarly, a velocity pickup will give acceleration velocity and also velocity displacement (by integrating the signal once).

Other parameters that can affect transducer selection are

Speed of Machine, or Dominant Frequencies to Be Measured. This parameter basically serves to indicate the limited range of the velocity pickup and its insensitivity to low-frequency vibrations. Various problems within a machine can generate vibration frequencies from one-half to 50 times rotating speed. This must also be taken into account when determining the frequency to be measured.

Temperature at Mounting Point. The typical maximum operating temperature limitation is listed for each type of transducer. In general, if there is a range listed for the maximum operating temperature, the higher temperature units are more expensive. The 20 ° C. (160 ° F.) listed under accelerometers reflects an accelerometer with an integral charge amplifier.

Cable Length, Strength of Signal. All three types of pickups require a good grade of twisted, shielded, transducer cable. The accelerometer is listed as 30-300 mm (100-1000 ft). The 30-m (100-ft) limit is for a transducer with a charge output; the 300-m (1000-ft) limit refers to an accelerometer with an integral charge amplifier. It should be noted here that all three types of transducers generate a relatively low-level ac signal. Proper transducer cable installation is critical to the overall operation of the system. Transducer cable runs near or parallel to high-voltage or high-current cable can induce false signals into the system.

Table 28-1. Selection Parameters for Transducers.

Transducer selection parameters		Preferred pickup		
		Non contact pickup	Velocity pickup	Accelero-meter
Mode of measurement	(i) acceleration			x
	(ii) velocity		x	x
	(iii) displacement	x	x	x
Type of bearings	(i) sleeve	x		
	(ii) antifriction(-ball)		x	x
	(iii) antifriction (-roller)		x	x
Speed of machine, or dominant frequencies to be measured	(i) 1-15 Hz	x		x
	(ii) 15-1500 Hz	x	x	x
	(iii) 1500 Hz up	x		x

Temperature at mounting point	Note temperature limitations of each type of probe	175° C. (350° F.) Max	200-250° C. (400-500° F.) Max	70-290° C. (160-550° F.) Max
Cable length, strength of signal	Note typical pickup-to-monitor cable length limitations	Up to 450 m (1500 ft)	Up to 300 m (1000 ft)	30-300 m (100-1000 ft)
Installation requirements		Easy	Hard	Hard
Relative mass, rotor to case	Rotor mass case mass		x	x
	Rotor mass case mass	x		

Installation Requirements. Accelerometers and velocity pickups are normally installed by 1/4 - 28 stud on the machine and thereby rate "easy" for ease of installation. The noncontact probe with its necessary probe tip clearances rates "hard" for hard to install. The noncontact probe is difficult to install on a retrofit program or a machine that is in service. In addition to complexity of installation, physical space limitations for mounting the transducer must also be considered.

Relative Mass. When the mass of the case of the machine is much greater than the mass of the rotor, such as in a boiler feed pump, the shaft forces may not be sufficient to cause significant case vibration, and a noncontact probe would be preferable. With light case machines, the case tends to follow the shaft vibration, and a velocity pickup or accelerometer is adequate.

Machine Limitations/Past Machine Problems. Past machine problems or specific machine problems that are being protected against should also be kept in mind. If a particular machine has been destroying motor mounts or connecting ductwork, a pickup that measures overall case motion (velocity pickup or accelerometer) would be preferred. If you are trying to monitor a rotor's position within a housing very accurately to protect against possible mechanical interference, a noncontact probe would be preferred.

Unusual Installation Problems. Various environmental factors can affect transducer selections. Is the machine located in an unusual place such as on a ship or very flexible platform? Is the machine subject to many starts and stops? Will the transducer be subjected to salt air, corrosive chemicals, or other unusual substances? Do the pickups need to be protected against physical damage? These are all important questions when applied to transducer selection.

User Experience. An often-ignored factor is past experience with a certain type of pickup. If a plant and its crew are experienced in the installation and characteristics of a velocity-type pickup, there will be fewer problems with the installation than if a different transducer were selected.

VIBRATION MONITORING

Vibration monitoring is now widely applied to machines so that any deterioration of machine condition can be detected at an early stage, and necessary remedial action taken in advance of impending failure. The more important the machine the more effective vibration monitoring can prove to be in avoiding costly shutdowns. It has also been used in aircraft for some 20 years; aircraft types employing vibration monitoring as a standard feature include the Lockheed Tri Star, Boeing 747, Boeing 757, Boeing 767, Douglas DC-10, European Airbus A300, Concorde, and Tornado.

As far as machines are concerned, the most widespread application of vibration monitoring is to cover all phases of shaft motion relative to bearings and free space, using proximity probes, velocity pickups, and instruments. Specific points to be monitored include (in order of importance)

- bearings
- bearing housings
- casings
- foundations or mountings
- connected auxiliaries.

Here it should be noted that vibration monitoring is only part (although usually the most important part) of what is known as *machinery health monitoring* covering the whole spectrum of machine working. This can involve both conventional instrumentation and further types of transducers to monitor machine temperature, rotational speed, shaft phase angle, and position, as well as process variables. Such a system may also be fully computerized.

BASIC VIBRATION MONITORING SYSTEM

The two basic forms of vibration monitoring systems are

☐ Built-in instrumentation providing a measure of vibration in terms of overall level, coupled to alarm and shutdown devices in the event of the vibration level rising to a certain level. Such a system of monitoring provides continuous protection but no specific information other than that vibration has increased, indicating wear or a potential fault.

☐ Vibration measurement by analytical instruments to provide a vibration signature. This would normally be done periodically but, in the case of extremely critical machines, could be continuous, although continuous monitoring with built-in instrumentation plus periodic checks with analysis machines would be more realistic. Signature analysis provides a much more sophisticated preventive maintenance capability, because the deterioration of specific machine components can be isolated while the machine is running.

CONTINUOUS OR PERIODIC MEASUREMENTS?

Parameters also fall into two distinct categories: those that should be continually monitored, and those where only periodic measurement is necessary. A parameter not considered important enough for continuous monitoring may be considered important enough to require a very reliable means of measurement on a periodic basis. An example could be machine-housing vibration. A machine could be continuously monitored by shaft-observing proximity probes but may require an accurate analysis of shaft-vs.-housing vibrations during certain running conditions, startup for example. Because a permanently installed transducer usually provides a more reliable measurement than any hand-held transducer, the housing measurement transducers can be permanently installed, without continuous monitors, for machine analysis.

The selection of parameters to be monitored depends on the level of sophistication desired for the monitoring system and the various mechanical considerations particular to a specific machine design. It is equally important to note that a transducer chosen for monitoring one parameter can sometimes be employed to provide the measurement for monitoring a second parameter. Examples of this are an axial position sensor, which can be used to measure axial vibration as well, and a shaft-observing radial-vibration proximity probe, which can also be used to measure shaft radial position, an indicator of alignment conditions.

Specific Machine Requirements

In general, it is important to recognize that in order to determine the optimum protection system for machinery, each piece of machinery must be evaluated individually. Often, insufficient data is available for a detailed analysis of a particular machine's expected behavior under normal and malfunction conditions. It then becomes necessary to use your best engineering judgment and experience in determining what should be monitored. Often the user company has a machinery specialist group to provide the function of monitoring system specification. However, the user can also rely on the machinery manufacturer, the engineering consultant/contractor, and/or the machinery protection system manufacturer to accomplish this function.

Table 28-2. Recommended Monitoring for Machines.

Machine	Monitors	Parameters monitored
Electric motors	X-Y proximity probes Keyphasor probe Temperature indicators	(i) axial vibration (ii) position measurements (periodicially) (iii) casing vibration (iv) speed, phase angle, and timing (v) bearing and oil temperatures vi) rotor and stator winding temperatures
Pumps	X-Y proximity probes Keyphasor probe	(i) axial vibration (ii) shaft motion relative to bearings (iii) shaft phase angle (unless directly coupled) (iv) bearing and oil temperatures (v) casing vibration (vi) casing temperature
Fans	X-Y proximity probes	(i) shaft vibration (ii) bearing housing vibration (iii) casing vibration
Gears	X-Y proximity probe at each bearing	(i) axial vibration (ii) input shaft (iii) outut shaft (iv) thrust loads (axial probes) (v) gear teeth interaction (vi) casing vibration (vii) bearing ad oil temperatures

As a general guide, some specific recommendations for common machines are given in Table 28-2.

Continous Monitoring Equipment

This type of equipment is usually modular, so that protection systems can be matched to specific machine requirements. Typical modules with meter indication of levels are as follows:

- power supply/control
- displacement monitor (say, for thrust or quasi-static displacement monitoring) used with eddy-current proximity probes
- vibration displacement monitor, single or dual channel, used with eddy-current proximity probe
- vibration velocity monitor, used with moving-coil pickup
- vibration acceleration monitor, used with piezoelectric accelerometers
- Rpm monitor, used with either eddy probe or fiber-optic tachometer, indicates rpm and trips at preset overspeed

Continous-vibration monitors are usually based on the measurement of overall vibration level. The vibration signals are rectified and smoothed, resulting in a dc level. Typically both quasi-static and vibration monitors contain facilities for adjustment of preset levels for alarm and shutdown and meter indication of these trip levels when required. Examples of uses are

- continuous comparison between set levels and monitored level and consequent contact closure and indication of alarm trip status by lamps.
- voting logic where necessary for auto showdown.
- check and control circuits for bypass (for calibration checks), reset, scale multipliers (for run-through criticals), indicator lamp tests, first failure alarm.

Modules of this type can be powered by a common power module, or they can be individually powered. Signal conditioning, such as low-pass, high-pass, and bandpass filtering, is often included to improve the signal-to-noise ratio or to pinpoint particular frequency bands. Sometimes tuned bandpass tracking filters are incorporated to allow monitoring of vibrations related to particular shaft frequencies.

Periodic Monitoring Equipment

Periodic monitoring consists of logging measurements at predetermined intervals from transducers identical in type and location to those used for permanent monitoring systems.

Typical monitoring equipment ranges from relatively simple overall reading meters to relatively complex vibration analyzers, usually frequency analyzers. The data is collected at periods appropriate to the machine and its previous history, such as monthly, weekly, daily, hourly, or in critical situations, even continuously. The measurements can be logged manually, plotted in analogue form, or processed digitally in a computer system.

Most equipment in this category is transportable from machine to machine and site to site. Some typical methods, in order of sophistication, are

☐ Hand-held overall vibration-level meter with low-pass, high-pass, or fixed bandpass filtering to suit the signal-to-noise requirements and/or to pinpoint particular frequencies.

☐ Manually tunable bandpass filter with level meter to pinpoint specific frequencies.

☐ Tracking filter-based equipment that can provide both frequency analysis at a given machine running speed (rpm) and the vibration level of a given machine order versus speed. This type of equipment is based on the principle of automatically tuning a narrow bandpass filter (fixed bandwidth usually) in such a way that the center frequency of the filter is locked to an external tuning signal originating either from an oscillator (for frequency analysis) or from a tachometer signal generated at a given multiple of the rotational speed of a shaft (for order analysis).

☐ Time compression real-time analyzer (RTA). Typically, this type of analyzer can produce one amplitude-vs.-frequency spectrum in 50 ms that can be plotted in analog form on a standard X-Y plotter in a few seconds or fed to a computer for further processing, such as scaling, comparison, storage, or readout. This method of measurement is therefore very fast and, over the last few years, has become widely used as a preventive maintenance tool. An extensive methodology has developed around the real-time analyzer for this purpose.

Accessories that can extend the analysis capability of the basic analyzer include

- ensemble averager for signal enhancement and averaging time-varying data.
- signal ratio adapter, for order analysis and order tracking.
- frequency translator, for high-resolution "zoom-in."

The RTA provides a relatively simple means of gathering a sequence of standardized plots for a given machine/transducer location so that the trend of levels of specific vibration frequencies and/or orders can be simply determined.

It plays another important role in the care of rotating machines. It is often a major tool in the research and development and early installation stages of a machine, leading ultimately to the establishment of criteria of acceptable machine vibration levels for subsequent monitoring purposes.

□ Digital signal processor (DSP). A "hard-wired" FFT-based dual-input analyzer can be used in the same way as a time compression real-time analyzer. Its analysis capability is more comprehensive, including, for instance, a time-averaging capability, which is a powerful form of machine vibration analysis for such things as gear boxes or similar devices.

□ Computer-aided vibration monitoring system. Both the time compression RTA and DSP can be used in conjunction with digital computers so that automatic scanning of large numbers of transducers on a machine complex can be handled effectively in this way. The computer system can be programmed to manipulate and store data from the RTA or DSP. The high speed and economy of dedicated analyzers are thus combined with the flexibility of a digital computer and related storage peripherals to very efficiently handle large-scale machine-monitoring installations.

□ Lastly, the analysis manipulation and storage of machine monitoring data can be performed on a purely software-based computer system. The analyzing functions are slower on this type of system as compared with dedicated analyzers (such as the RTA or DSP) on a cost-comparative basis, and, of course, the software system requires the backup of experienced computer personnel.

Although the first five methods lend themselves to portable operation, the last two are essentially static systems and, hence, more applicable to large machinery installations.

It will be seen that for periodic monitoring, there is an emphasis on frequency analysis in most of the methods listed previously. The ability to monitor vibration levels at particular frequencies for a given machine condition is very powerful, because often one transducer—for example, a velocity pickup or accelerometer on a casing—can be used to indicate the status of several parts of the machine (for example, rotor unbalance, bearing, and gearbox).

The same information can be used for malfunction diagnosis because it offers a very positive method of identifying a potential failure mechanism within the frequency range of the transducer used in the measurement.

Thus, a spectrum analyzer may be used to relate the frequency components of the noise spectrum to some specific mechanical event or pattern in the machine as it operates. Single-channel real-time spectrum analyzers are used to obtain this amplitude-vs.-frequency, or amplitude-vs.-order, information.

If several noise sources exist within common surroundings, the analysis problem becomes more difficult. To identify which noise source is contributing the most to the overall noise measurement requires two-channel analysis capability for mutual-property investigation of the data signals. Cross-correlation techniques have been used to separate noise sources in a composite noise signal. Recently, frequency-domain mutual-property characteristics, coherent output power, and the coherence function have provided the capability to identify noise sources and their respective contribution to a total power measurement.

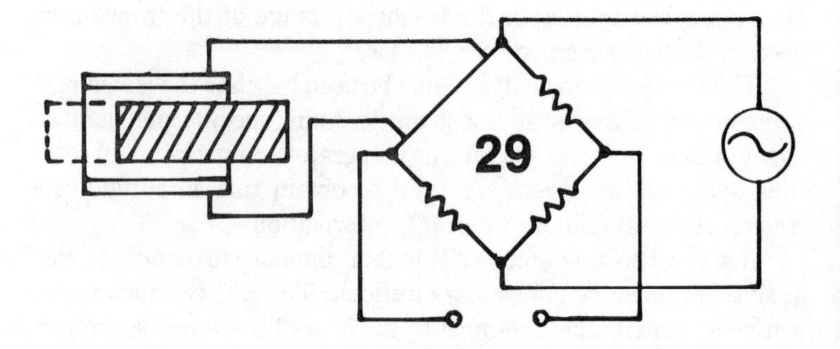

Processors and Displays

Circuitry related to different types and applications of transducers has already been described in various chapters. This chapter summarizes the basic systems required and alternative methods of readout, recording, or display.

Our starting point is the strength and type of signal generated by the transducer itself. Normally it will need amplification (although not necessarily so) and possibly filtering to eliminate phasing errors. The resulting dc signal output can then be used to give a readout by an analog instrument, or, alternatively, it can be subjected to signal processing as required (Fig. 29-1).

The most common requirement in signal processing is converting a dc analog signal into a digital signal. Three basic methods that may be employed are

- ☐ successive approximation
- ☐ voltage-to-frequency conversion
- ☐ voltage-to-time conversion

SUCCESSIVE APPROXIMATION DIGITIZING

This form of analog-to-digital converter employs a digital divider to feed back to the amplifier known increments of reference voltage in fixed sequence. At each step the amplifier then decides whether that increment is switched on, or ignored,so that a volt-

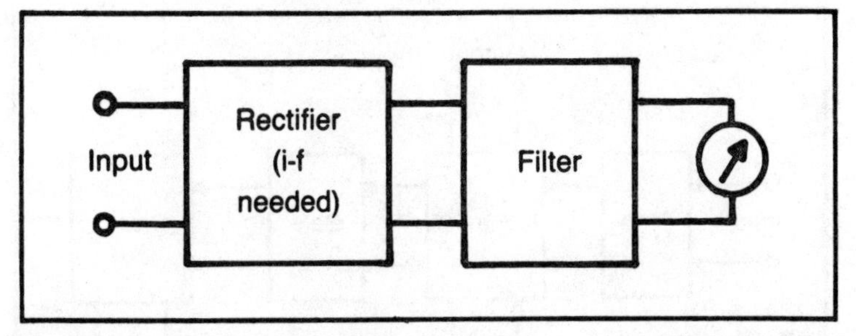

Fig. 29-1. A simple circuit for direct measurement of an analog transducer output.

age is built up approximately to the value of the input.

A block diagram of this form of working is shown in Fig. 29-2. The actual circuitry involved can be quite complex, incorporating automatic sequencing and capable of very high conversion speeds. Accuracy of conversion depends on the accuracy and stability of the reference obtained from the digital divider.

VOLTAGE-TO-FREQUENCY CONVERSION

With voltage-to-frequency conversion the converter produces a train of pulses with a frequency proportional to the voltage input. Pulses are accumulated in a counter for a fixed period, and at the end of that period the total is readout.

Figure 29-3 shows a block diagram of this type of converter.

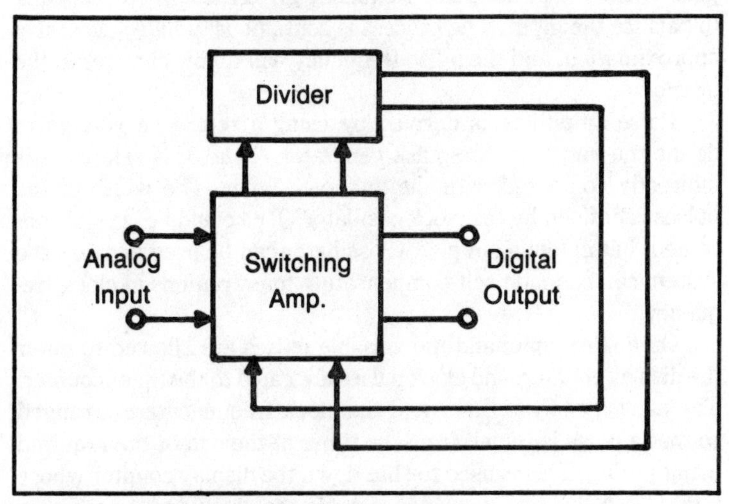

Fig. 29-2. Successive approximation digitizing.

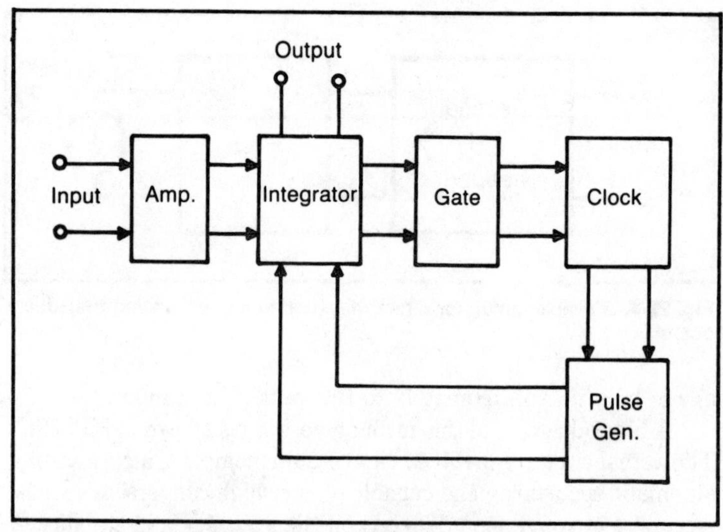

Fig. 29-3. Voltage-to-frequency conversion.

An integrator produces a voltage ramp whose rate of rise varies with the output from the amplifier. As the ramp reaches a defined level, the gate opens and allows the clock to trigger the pulse generator. The pulse generator then resets the integrator and a new ramp commences. Thus, a train of pulses is produced that are filtered to give a dc level, and this is fed back to the input amplifier where it is compared with the unknown. The loop therefore adjusts itself so that the pulse frequency gives the correct feedback to balance the input. The process is continuous, unlike successive approximation, and the pulse frequency will follow changes in the unknown.

Pulse amplitude is derived by using a reference voltage to define the output of the pulse generator. Thus, this reference is indirectly compared with the unknown input. The width of the pulses is defined by the clock oscillator. This could be crystal controlled, but in fact a simple LC oscillator can be used because the system can be made self-compensating for variations in clock frequency.

On a start command the variable pulses are allowed to enter the display counter, and clock pulses are gated to the timer counter. The capacity of the timer and the clock frequency are arranged so that a pulse is issued from the timer at the end of the required count period. This is used to shut down the display counter whose total is decoded and displayed. Should the clock drift to a higher frequency, the filtered dc fed back to the input will tend to fall; this will cause more pulses to be generated by the converter loop. But, since the clock frequency has increased, the court period will be shorter. Thus, the display counter will accumulate pulses faster, but for less time. Hence, the displayed reading is unaltered.

This technique is particularly suitable for measuring small voltages and is readily adapted to provide a range of sensitivities.

VOLTAGE-TO-TIME CONVERSION

For voltage-to-time conversion the input is allowed to charge a capacitor for a fixed time. With the input removed, a reference voltage is then used to discharge the capacitor, the time taken for discharge being a measure of the input.

In order to achieve linear characteristics the capacitor is associated with an integrator, the working of which is as follows. When the integrator is connected to the input its output "ramps up" at a rate that is directly proportional to the value of the input. After a fixed time the switch changes over and connects the reference in place of the input. It is arranged that the reference voltage is of opposite polarity to that of the input, so that the integrator output now "ramps down" at a fixed rate that is determined by the value of the reference. The time taken to complete ramp-down is a direct measure of the unknown. The ramp-down time can be measured, and the ramp-up time defined. A block design of this form of circuit is shown in Fig. 29-4.

The sequence starts switching the integrator to the input by the control. As soon as the integrator leaves zero, the gate is opened and clock pulses pass into the counter. After a time that is dependent upon the clock frequency and the counter "full-house," the counter fills and sends a pulse back to control. This causes the input to be exchanged for the reference, and ramp-down commences. Meanwhile, clock pulses are still fed to the counter, which has started again from zero. As the end of ramp-down is reached, the gate closes, and the counter now holds a total that is a measure of the unknown input. This is decoded and passed to the display.

Because the unknown input is applied to an integrator, the voltage stored on the integrator capacitor at the end of ramp-up is directly proportional to the integral of the input. Thus, if the input contains a periodic component such as 50 Hz, the integral of this component is zero when the ramp-up time is equal to its period (20 cms for 50 Hz). The result is similar to that obtained with the V

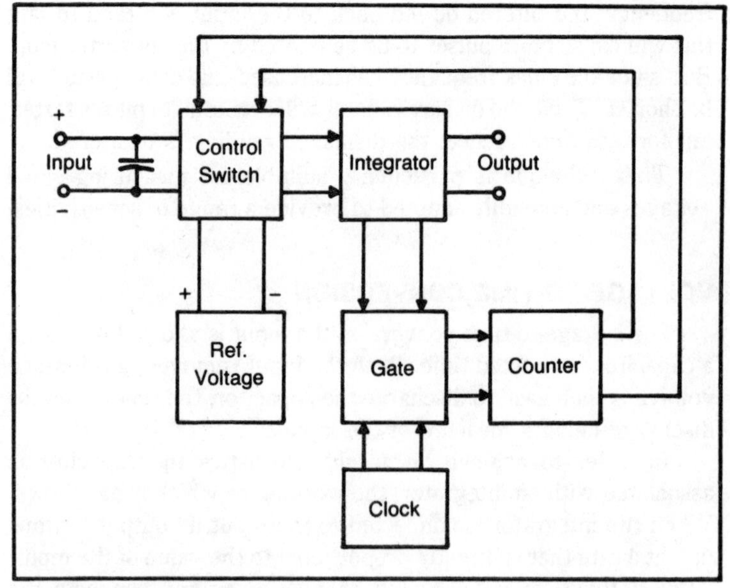

Fig. 29-4. Voltage-to-time conversion.

to f arrangement, in that there is inherent rejection of series-mode interference. With the ramp-up time fixed at 20 ms, if the mains' supply varies about 50 Hz, as commonly occurs, the rejection will be less than ideal. But as the ramp-up time is controlled by the clock, the clock frequency can be adjusted, so that ramp-up always lasts for one mains period. One DVM achieves this by comparing the ramp-up time with the mains and putting appropriate corrections into the clock by means of a digital servo.

This technique permits very thorough isolation of the input circuits, while the rest of the voltmeter is grounded. The isolation and series-mode rejection can be so effective that with some products interference can virtually be ignored.

DATA LOGGING

Recording measurement results digitally on magnetic tape is also the basis of *data logging*. Basically this technique samples the parameters at regular intervals and records such data for subsequent offline input to a microprocessor, computer, or other data processing equipment. Such a system is shown in block form in Fig. 29-5. It is composed of a *transducer* to provide the signal input and a *signal conditioner* or *processor* to convert the input signal to a form acceptable to the *data logger*, the output from which is passed to the data processor or analyzer.

Normally if more than one transducer is involved, a separate conditioner is required for each transducer. Outputs from the conditioners are then scanned by the data logger at preset intervals (that is, sampled sequentially), and the results are recorded in digital form on magnetic tape. For fast systems, inputs may have to be sampled and held at the beginning of the scan so that recorded information for a scan all corresponds to the same point in time. Thus, for optimum flexibility the signal-conditioning system needs to be programmable.

Another major requirement of the signal-conditioning system is intelligence. Many logging applications require some inputs to be logged only as background information, whereas others are logged more frequently, or to be logged only when input signals exceed preset levels. With the availability of today's large-scale integrated circuits, it is fairly easy to design this level of intelligence into a conditioning unit accommodated on a single printed circuit board.

The final step in the data-signal-handling procedure is that of transferring the data from the recording medium to the appropriate analysis system, which normally takes the form of either a mainframe computer, which also performs many other tasks, or a dedicated minicomputer or microprocessor-controlled system. In the former case a suitable media-conversion system is usually required, capable of reading the logger's cassette or cartridge and outputting the data in suitable format for input to the mainframe.

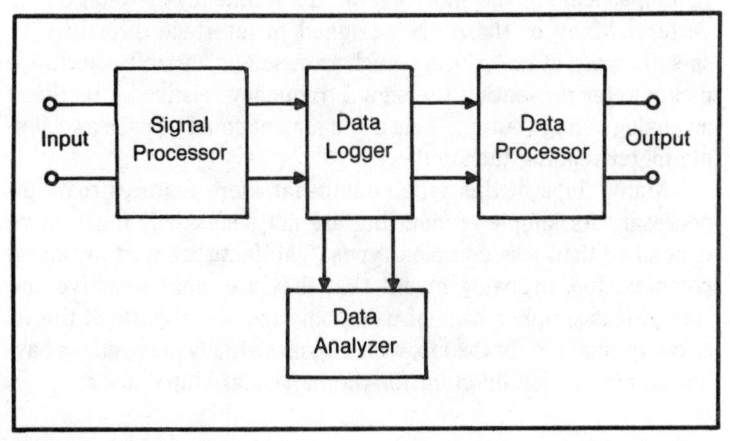

Fig. 29-5. Data logging.

The hardware involved in such a system could well be comparable with that required for the complete dedicated analyzer but with much simpler software programming.

COMPUTERS AND ANALYZERS

Computers are readily capable of calculating data directly from transducer signal readout and can present almost any type of information required. They can also employ many different techniques of filtering, windowing, zooming, and averaging, or they can extract individual frequency components such as modal parameters or graphical modeling all at very high speeds. Modern real-time analyzers can, in fact, sample at rates of greater than 250,000 times per second. They represent a highly specialized subject, however, that is outside the scope of this book.

Simple Techniques

Analog-to-digital converters are available in the form of inexpensive IC chips, requiring only a minimum of additional components to produce a working circuit. There are numerous ICs of this type produced (A/D converters), many of which are designed to interface directly with a microcomputer data lens. There are also IC A/D converters that incorporate a display driver for powering an LED readout. Because these incorporate specific circuits and pin numbering, general circuit designs can be given. To use them, purchase a type for which an application diagram is available.

Various other useful forms of converters are also available as IC chips. Among the most useful are frequency-to-voltage converters. Many of these are designed to interface directly with specific types of transducers, such as magnetic variable-reluctance pickups, for presenting the signal frequency received directly as an analog voltage output. This arrangement could be used as a simple meter-reading tachometer.

Many chips of this type contain far more features than are necessary for simple readout but are not necessarily much more expensive than less complex types. The fact that they are more complex, too, probably means that they are more sensitive and stable. It is simply a case of using only the actual parts of the IC circuitry required for the job, which again virtually necessitates having an application diagram for the particular chip chosen.

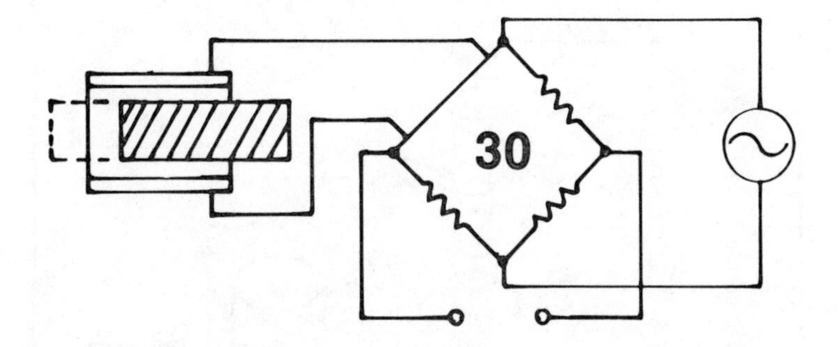

Testing the Behavior of Structures

If we summarize the behavior of transducers and sensors, they both have a common purpose—generating signals related to a particular parameter under study. The traditional way of observing signals is to view them in the *time domain*. That is to say, observation in the time domain is a record of what is happening to a particular parameter over the period of time.

As an elementary example, Fig. 30-1 shows a weight suspended by a spring that, if set in motion, will combine to oscillate up and down. The up-and-down motion gradually diminishes with time. A pen attached to the weight bears against a strip of paper mounted on a cylinder that is rotated at a constant rate. A trace will then be produced on the paper, representing the displacement of the weight plotted against time, or a *time-domain* view of displacement.

Although this form of time-domain recording is commonly used, the more common method is to measure the parameter (in this case, displacement) in such a way that it can be turned into electrical signals. These are much easier to record or display, and they usually present a more practical method of measurement. The direct recording system of Fig. 30-1 would not work in practice, for example. It needs to be associated with a transducer to render mechanical displacement in terms of proportional electrical signals.

In terms of electrical signals and parameters that vary in value with time, virtually any type of record is then essentially a *waveform*, although it may not look like one. The picture presented by the

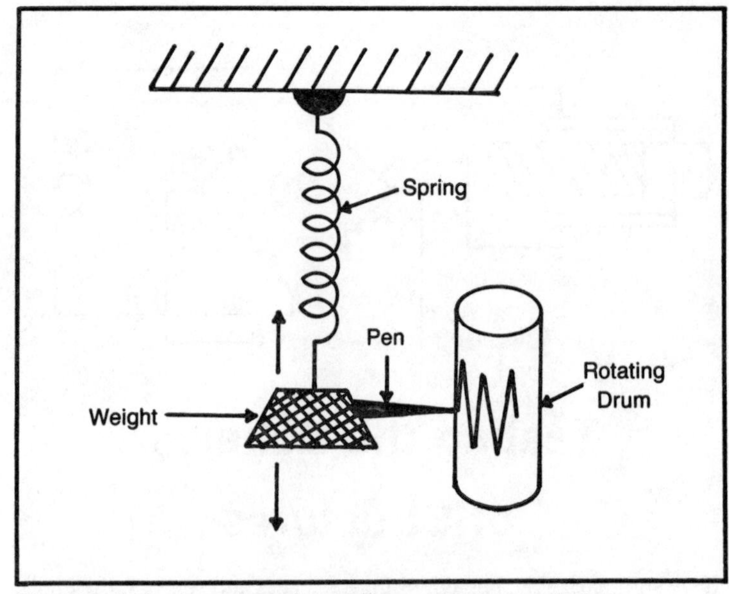

Fig. 30-1. A simple method of recording mechanical oscillation.

time-domain record may look quite irregular or ragged, as in Fig. 30-2A. However, any waveform that exists in the real world can be broken into a series of component parts, each of which is a simple sine wave differing in amplitude (displacement), frequency, and position relative to each other in terms of time. Thus, a ragged appearance of Fig. 30-2A is simply a (complex) waveform that could be duplicated by adding a large number of individual sine waves, the exact number depending on how many individual component waves are necessary to generate the final complex form.

If this seems far fetched, it is not. This sort of thing can be worked out mathematically by a method, developed more than 100 years ago by Jean Baptiste Fourier, known as *Fourier analysis* or *Fourier transform*, although the mathematics involved is highly complex (and nowadays normally done by a computer). And the idea of analysis in terms of component simple waveforms is further supported by the quantum theory, which, basically, holds that *everything* (even solid, stationary objects) can be represented in terms of waveforms.

Let's make things simple by accepting that Fourier analysis does logically work, and let's take as an example the quite simple waveform of Fig. 30-2B, which can be reproduced by adding just two waves together. The same principle then holds no matter how many individual sine waves are involved in the makeup of the original waveform being considered.

The original waveform, as it might be measured, is shown in terms of amplitude plotted against time. If the two component sine waves are extracted, they would appear as shown in the Fig. 30-3.

Now suppose, the record is redrawn as a *three-dimensional* design with a frequency axis added at right angles to amplitude and time (Fig. 30-4A). In the conventional two-dimensional design this axis was already there but invisible, because we were looking directly along it. But if we now change one direction of view through 90° to look directly at the frequency axis, the time axis disappears, and we see a record of amplitude against *frequency* (Fig. 30-4B). The picture in this case is quite different: just two straight lines showing the amplitude of the *component* sine waves at their specific frequencies. This is the *frequency-domain view*, where each sine wave component is separated and individually displayed.

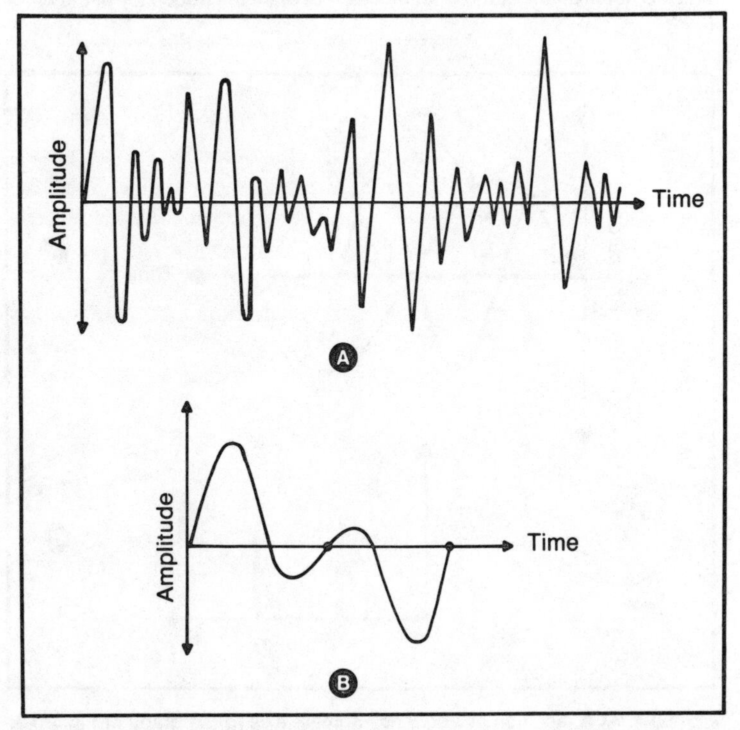

Fig. 30-2. At A, a complex waveform. Despite its complexity it can be reduced to a combination of sine waves. At B, a less complicated waveform. It consists of just two sine waves.

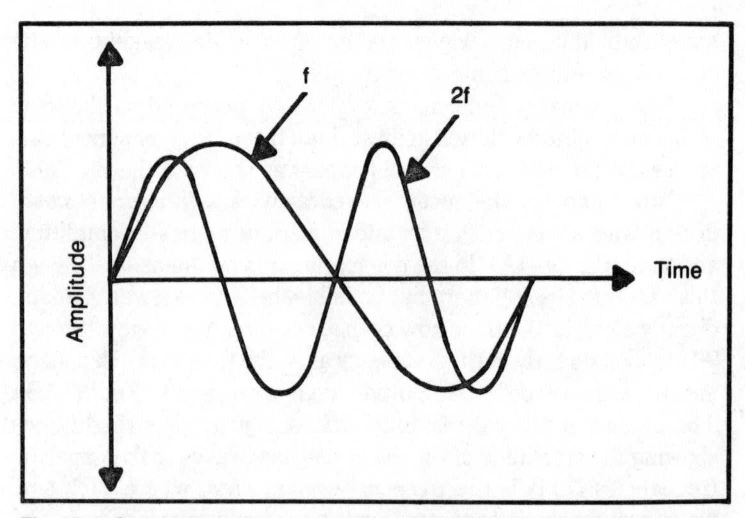

Fig. 30-3. Composition of the waveform at Fig. 30-2B. It consists of two waves that are equal in amplitude, with one having twice the frequency of the other (f, 2f).

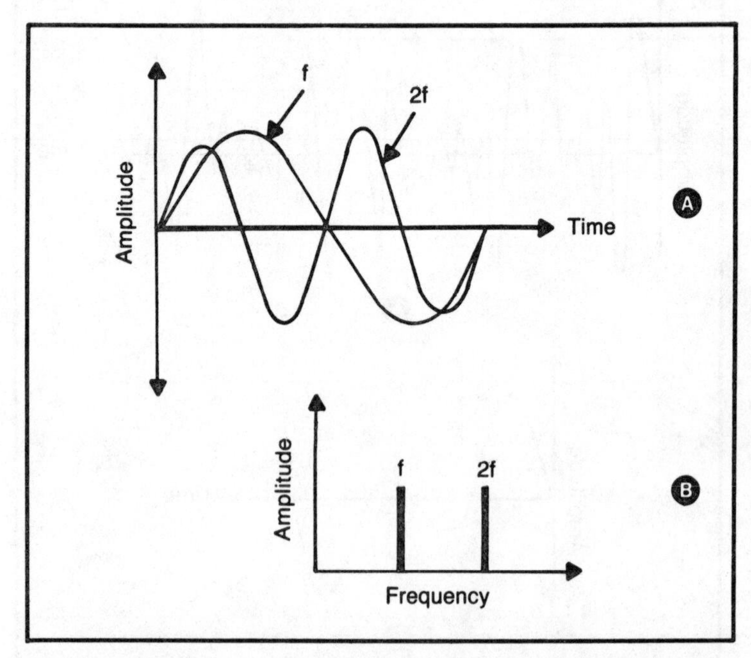

Fig. 30-4. At A, addition of frequency-domain axis to the graph of Fig. 30-3. (The frequency-domain axis is at right angles to both the time and amplitude axes.) At B, graph of amplitude as a function of frequency.

In changing from one domain to another we have neither gained nor lost information (except phase information), but we have only expressed this information differently. In other words, it gives another look at the subject, which can be very useful. In particular, the frequency-domain view can often show the presence of small components not visible in the time-domain view because they are masked by larger ones. Also, if necessary, phase information can be recovered.

THE MODAL DOMAIN

There is a third domain, the modal domain, that is of particular interest for analyzing the behavior of mechanical structures (seeing "logically" how they behave). To describe this simply, we will start with the behavior of a tuning fork, which generates a simple form of vibration and, as a result, an (apparently) single tone note. In a time-domain view the sound will appear as a sine wave lightly damped (Fig. 30-5A). However, a view in the frequency domain (Fig. 30-5B) will show that in addition to a predominant single frequency, the tuning fork also generates a number of smaller peaks at equally spaced frequency intervals; these are called *harmonics*. The mean tone, in fact, is generated by the first mode of vibration of the tuning fork, the first harmonic peak by the second mode of vibration of the tuning fork, the first harmonic peak by the second mode of vibration, and so on.

To determine the total vibration of the tuning fork, you would have to simultaneously measure the vibration at several points along its structure. For simplicity, we will use just three. Figure 30-6 shows a frequency-domain picture of the individual frequencies and their appearance at points 1, 2, and 3. Sharp peaks all occur at the same frequencies, independent of the measuring points, the only difference being the relative size or amplitude of their progress.

Another interesting point to emerge from the simple tuning form model is the effect of adding weights to the two tines to provide damping (more rapid decay of vibration). Adding weights to the ends of the tines would provide damping of the first mode of vibration (the main tone) without affecting the amplitude of the harmonics. Adding weight to the middle of the tines would provide minimal damping of the main tone but marked damping of the harmonics. This is a principle of vital importance in providing damping in structures that vibrate.

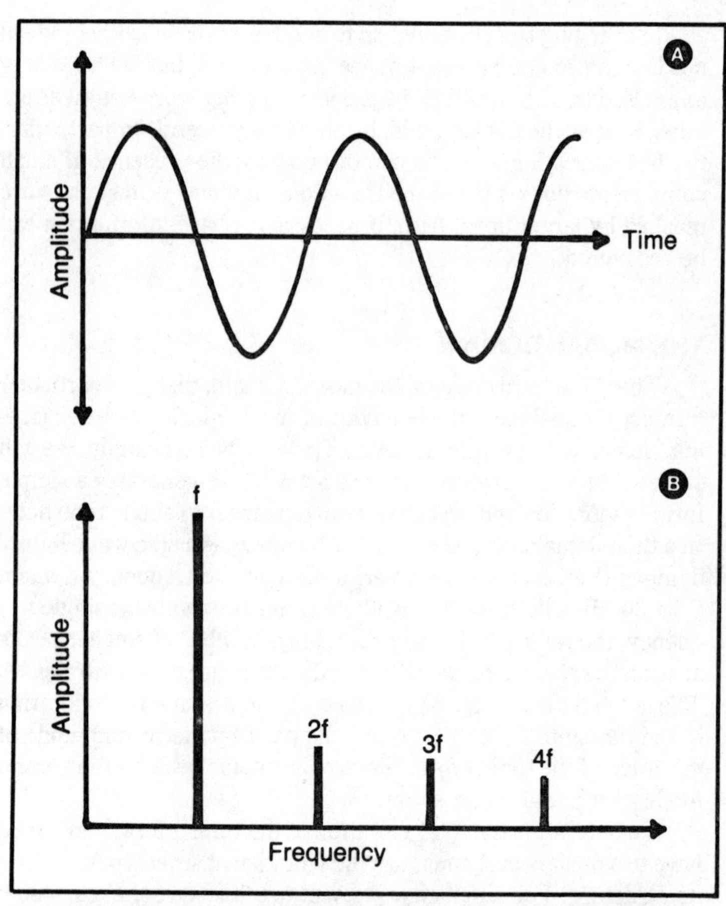

Fig. 30-5. At A, waveform from a tuning fork as it would appear on an oscilloscope. At B, amplitude-vs.-frequency rendition shows that harmonic energy is present.

MODAL ANALYSIS

Because any real waveform can be represented by the sum of much simpler waveforms, any real vibration in a structure or body can be represented by the sum of much simpler vibrations. The method of delivering the shape and magnitude of the structural deformities in each vibration mode is known as *modal analysis*. There are two basic techniques used:

☐ Exciting only one mode at a time.
☐ Computing the individual modes of vibration from the total vibration.

Let's take a tuning fork as a simple example. To excite just the first mode, we need two shakers driven by a sine wave and attached to the ends of the tines. Varying the frequency of the generator near the first-mode resonance frequency would then give its frequency, damping, and mode shape.

In the second mode the ends of the tines do not move, so to excite the second mode, we must have the shakers to the centers of the tines. If the ends of the tines are anchored, vibrations will then be constrained to the second mode alone.

In more realistic, three-dimensional problems, it is necessary to add many more shakers to ensure that only one mode is excited. The difficulties and expense of testing with many shakers has limited the application of this traditional modal analysis technique.

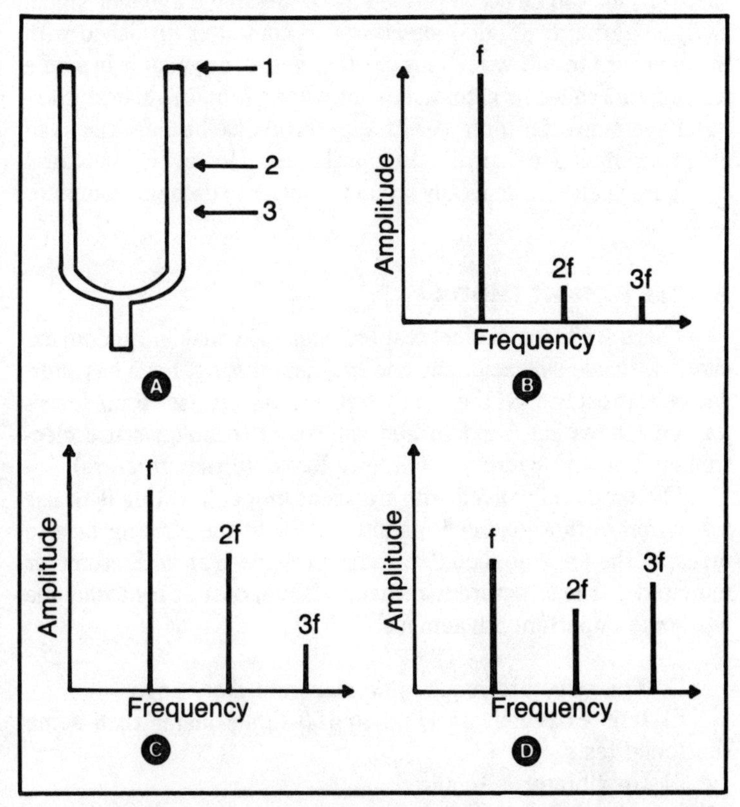

Fig. 30-6. Amplitude-vs.-frequency measurements may be taken at various points along the length of a tuning fork, as shown at A. At B, rendition for point 1; at C, for point 2, halfway up the tuning fork; at D, for point 3, one-third of the way up the tuning fork.

To determine the modes of vibration from the total vibration of the structure, you need to determine the frequency response of the structure at several points and compute at each resonance the frequency, damping, and what is called the residue (which represents the height of the resonance). This is done by a curve-fitting routine to smooth out any noise or small experimental errors. From these measurements and the geometry of the structure, the mode shapes are computed and drawn on a CRT display or a plotter. If drawn on a CRT, these displays may be animated to help the user understand the vibration mode.

It now becomes obvious that a modal analyzer requires some type of network analyzer to measure the frequency response of the structure and a computer to convert the frequency response to mode shapes. This can be accomplished by connecting a dynamic signal analyzer through a digital interface to a computer furnished with the appropriate software. This capability is also available in single instruments called structural dynamics analyzers. In general, computer systems offer more versatile performance because they can be programmed to solve other problems. However, structural dynamics analyzers generally are much easier to use than computer systems.

EXCITATION METHODS

There are three distinct test techniques available: random excitation, sinusoidal excitation, and transient testing. Each has alternative approaches. Of these only *transient testing* uses transducers for impact testing (random and sinusoidal techniques use electromagnetic and hydraulic vibrators to excite the structure).

The method involved with transient (impact) testing is to use a hammer with a load cell mounted close to the striking face to measure the force applied to the structure, with an accelerometer mounted on the structure to measure its response. This technique has some important advantages:

- The structure requires no elaborate mountings.
- It is extremely fast, up to 100 times faster than some sinusoidal tests.
- No vibrator is required.

The main drawback with this method is that the input-force power spectrum is not as easily controllable as when a vibrator is used, and there can be significant variations between successive blows, which can cause nonlinearities to be excited. To some extent the bandwidth of the input-force spectrum can be selected by changing the material of the hammer head; a softer head will give a longer impulse and, hence, more energy at the lower frequencies. If a hard head is used, the total energy will be spread over a wide frequency range, and the excitation energy density will be low, leading to measurements with poor signal-to-noise ratios, particularly for massive, heavily damped structures.

HAMMER TECHNOLOGY

The basic technology involved in hammer testing involves measurement of the force and response, yielding signals that can be processed in an FET analyzer to display the amplitude of the vibratory random force ration versus frequency. Force is applied to the structure by the hammer through a load cell, with response measured by a suitable response transducer (such as an accelerometer), signal information being fed to suitable signal conditioning equipment. Briefly, the conditioning equipment amplifies and digitizes the signals, which are then Fourier transformed and sampled, and the cross spectrum and two power spectrums are computed and averaged. Frequency response and coherency functions are then computed from the averaged power and cross spectrum.

Hammers

The two most important characteristics of the hammer are its weight and tip hardness. The frequency content of the force produced by a hammer blow is inversely proportional to its weight and directly proportional to the hardness of the tip. Hammer weight also determines the magnitude of the force power, so weight is the primary parameter governing choice of hammer. Tip hardness can then be chosen to achieve the required pulse-time duration.

Actual sizes of hammers used may range from a fraction of an once to many pounds, depending on the size of the object or structure to be tested. Force and response transducers are incorporated in the hammer head. These need to be of special rigid type, normally piezoelectric. Resonances in the hammer structure are a source of spurious spikes in the force signal spectrum that may need correction in the computer circuitry, or that can be eliminated by using a *modaling tuned* hammer, a recent development.

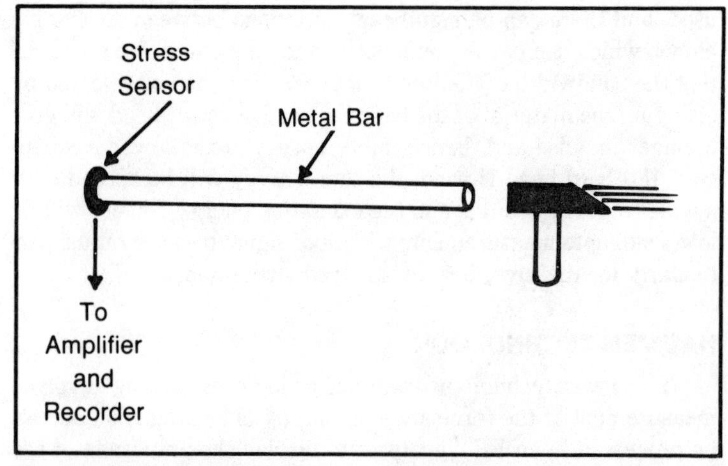

Fig. 30-7. A Hopkinson-bar apparatus for measuring mechanical stress resulting from acoustic propagation.

Hopkinson Bars

A related device is the so-called *Hopkinson bar*, also used for impulse testing as well as calibrating sensors, measuring proportions of materials, and researching fast strain effects. This consists of a log, straight, circular bar of metal with a stress sensor (transducer) at one end (Fig. 30-7).

If the sensor end is rested on a rigid surface and the other end of the bar is tapped with a (plain) hammer, a compression stress-strain wear is produced in the bar traveling down it at the speed

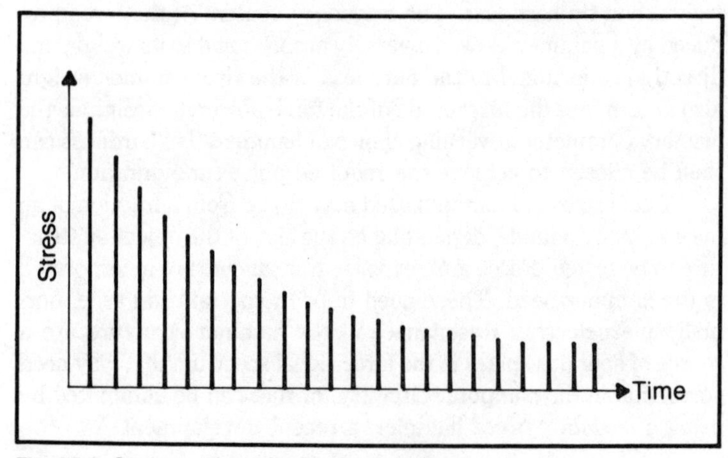

Fig. 30-8. Stress-vs.-time plot for Hopkinson bar.

of sound through the bar material. The original compression wear is then reflected from the fixed end,then as a (united) tension wave from the free end, putting on the stress sensor on the second pass. This process is repeated until all the energy is finally described as friction. Figure 30-8 shows typical signal data generated by the stress sensor (transducer).

Index

Index

U

V

W

X

Z